AF677728

Fatty Acids from Marine Organisms, 2nd Edition

Fatty Acids from Marine Organisms, 2nd Edition

Guest Editors

Giuseppina Tommonaro
Annabella Tramice

MDPI

Basel • Beijing • Wuhan • Barcelona • Belgrade • Novi Sad • Cluj • Manchester

Guest Editors

Giuseppina Tommonaro
National Research
Council-Institute of
Biomolecular Chemistry
Pozzuoli
Italy

Annabella Tramice
National Research
Council-Institute of
Biomolecular Chemistry
Pozzuoli
Italy

Editorial Office
MDPI AG
Grosspeteranlage 5
4052 Basel, Switzerland

This is a reprint of the Special Issue, published open access by the journal *Marine Drugs* (ISSN 1660-3397), freely accessible at: https://www.mdpi.com/journal/marinedrugs/special_issues/KW29SIG7QZ.

For citation purposes, cite each article independently as indicated on the article page online and as indicated below:

Lastname, A.A.; Lastname, B.B. Article Title. *Journal Name* **Year**, *Volume Number*, Page Range.

ISBN 978-3-7258-5507-0 (Hbk)
ISBN 978-3-7258-5508-7 (PDF)
https://doi.org/10.3390/books978-3-7258-5508-7

© 2025 by the authors. Articles in this book are Open Access and distributed under the Creative Commons Attribution (CC BY) license. The book as a whole is distributed by MDPI under the terms and conditions of the Creative Commons Attribution-NonCommercial-NoDerivs (CC BY-NC-ND) license (https://creativecommons.org/licenses/by-nc-nd/4.0/).

Contents

marine drugs

MDPI

Article

A Comparative Study of the Fatty Acid Profile of Non-Edible and Edible Tissues of Raw and Processed Common Octopus (*Octopus vulgaris*)

Luis Freiría-Martínez [1], Marcos Trigo [1], Ricardo Prego [2] and Santiago P. Aubourg [1,*]

[1] Department of Food Technology, Marine Research Institute (CSIC), c/E. Cabello 6, 36208 Vigo, Spain; lfreiria@iim.csic.es (L.F.-M.); mtrigo@iim.csic.es (M.T.)

[2] Department of Oceanography, Marine Research Institute (CSIC), c/E. Cabello 6, 36208 Vigo, Spain; prego@iim.csic.es

* Correspondence: saubourg@iim.csic.es

Abstract: A comparative study of the fatty acid (FA) composition of non-edible (viscera) and edible (mantle and arm) tissues of octopus (*Octopus vulgaris*) was carried out. According to the specimen size, three different groups (1–2 kg, 2–3 kg, and 3–4 kg, respectively) were taken into account. The effect of the cooking process (40 min at 90 °C) and frozen storage (4 months at −18 °C) was analyzed. In all kinds of samples, the polyunsaturated FA (PUFA) group was the most abundant ($p < 0.05$) and monounsaturated FAs were the least abundant ($p < 0.05$). Lower ($p < 0.05$) ω3-PUFA, ω3/ω6 ratio and docosahexaenoic acid values were detected in viscera (35.4–41.9%, 3.0–4.5%, and 12.7–17.5%, respectively) than in edible tissues (44.4–52.5%, 4.1–6.1%, and 24.3–30.1%, respectively). Conversely, higher ($p < 0.05$) eicosapentaenoic acid content was detected in viscera (19.6–21.9%) than in the edible tissues (17.2–19.3%). In most cases, the cooking process and frozen storage led to an average decrease in the PUFA and ω3-PUFA content and to an increase in the saturated FA presence. In agreement with current nutritional recommendations, all tissues showed great levels of highly valuable indices regarding the lipid fraction. The study proves that viscera, a waste substrate, can be considered a relevant source for food and pharmaceutical industrial requirements.

Keywords: octopus; viscera; arm; mantle; cooking; frozen storage; ω3 fatty acids; ω3/ω6 ratio; EPA; DHA

1. Introduction

Seafood is known to provide high contents of important constituents for the human diet. The great biological and chemical diversity in marine fish and invertebrate makes them a relevant source of highly valuable constituents susceptible to be used in a wide range of applications [1,2]. In agreement with a great number of studies focused on the employment of marine-enriched diets, the marine fatty acid (FA) profile has been found to be responsible for different health benefits (i.e., decreases in cardiovascular, neurological and inflammatory disorders) [3–5]. Notably, the consumption of eicosapentaenoic acid (EPA) and docosahexaenoic (DHA) acid consumption has been associated with low prevalence of different diseases related to neurodegenerative and cardiovascular problems [6,7].

The chemical composition of marine species has proved to encompass wide variations resulting from endogenous (genetic, anatomical, and physiological) and exogenous (water temperature, season, and feeding availability) factors [8]. Endogenous factors influence the distribution of biochemical constituents within different body tissues of marine species [9–11]. Among these constituents, lipid content and composition exhibit the greatest variability, differing significantly among edible tissues in wild invertebrates [12,13], fatty [14] and lean [15,16] fish species, and farmed fish [17,18].

The processing of marine species leads to a wide quantity of undesired by-products. In general, heads, blood, viscera, skin, or tails are obtained at different steps of seafood processing and constitute an important drawback for environmental contamination [19–21]. However, seafood by-products have been reported to be an important source of major components like proteins, minerals, lipids, and vitamins, in addition to minor constituents such as chitin, enzymes, pigments, and collagen [22–24]. Notably, the highest levels of high-added-value constituents is often present in marine tissues or parts that are often discarded [25,26].

Cephalopods are considered to be a highly interesting biological group because of their great nutritional value for human health and for their great commercial value [27–29]. Among cephalopods, the different species of octopus are considered as a highly nutritional seafood that can be commercialized in a wide range of products [30,31]. In the present research, a comparative study of FA composition was carried out between non-edible (viscera as a whole) and edible (mantle and arm) tissues of common octopus (*Octopus vulgaris*). The aim of the study was to prove the valuable FA composition of all tissues with a special stress on the viscera tissue, a waste product resulting from commercial processing. On the basis of the specimen sizes, three different groups were considered separately. Additionally, the effects on the FA profile of the cooking process and frozen storage were analysed.

2. Results and Discussion

2.1. FA Composition of Raw Samples: Effect of Tissue

The FA profiles of the different tissues were analyzed in the raw samples corresponding to the three sizes, i.e., Group I (1–2 kg per specimen; small-sized specimens), Group II (2–3 kg per specimen; medium-sized specimens), and Group III (3–4 kg per specimen; large-sized specimens). Throughout the whole study, each group was considered separately in order to carry out the comparison among tissues.

For the viscera tissue, the most abundant FA ($g\cdot100\ g^{-1}$ total FAs) in all groups was EPA (19.5–22.0 range), followed by DHA (12.7–17.5 range), C16:0 (13.9–15.2 range), C18:0 (9.3–11.2 range), and C20:4ω6 (6.1–7.6 range). A previous seasonal study on several non-edible tissues of octopus (*O. vulgaris*) detected the same main FAs [32]; thus, the following decreasing sequences were detected for the digestive gland and the ovary: DHA > EPA > 16:0 > C18:0 > C20:4ω6 and DHA > 16:0 > EPA > C20:4ω6 > C18:0, respectively. The same main FAs (DHA, EPA, C16:0, C20:4ω6, and C18:0) than in the present study were also detected as the most abundant in octopus (*O. vulgaris*) by-products considered as a whole after lipid extraction by using low-toxicity solvents (acetone, ethanol, and ethyl acetate) [33].

Regarding the composition of non-edible tissues obtained from other cephalopod species, the same most abundant FAs (DHA, C16:0, and EPA) as in the current study were detected in total by-products obtained from Patagonian squid (*Doryteuthis gahi*) during a seasonal study [34]. However, C18:1ω9, DHA, and C20:1ω11 FAs were found to be the most abundant in cuttlefish (*Sepiella maindroni de Rochebrum*) viscera [35], and C16:0, EPA, and C18:0 FAs were observed as the most abundant in *Sepia officinalis* viscera (i.e., stomach, intestines, and pyloric caeca) [36]. A different distribution of the main FAs than in the present work was also detected by Singh et al. [28] in squid (*Loligo formasana*) ovary; in their study, DHA, EPA, and C20:4ω6 were found as the major FAs. Total viscera obtained from squid (*Illex argentinus*) showed the following decreasing sequence for the most abundant FAs: DHA > C16:0 > C18:1ω9 > EPA [37].

Regarding mantle and arm tissues in the present research, the most abundant FA (g·100 g^{-1} total FAs) in all groups was DHA (22.0–26.1 and 23.9–28.5 ranges, respectively), followed by C16:0 (16.8–17.0 and 14.3–19.1 ranges, respectively), EPA (16.7–18.6 and 17.2–19.0 ranges, respectively), C20:4ω6 (6.3–9.4 and 6.2–9.0 ranges, respectively), and C20:1ω9 (3.3–3.5 and 3.4–3.5 ranges, respectively). A similar FA distribution was detected in previous research related to the edible tissues of the current species of octopus. This distribution was observed in the mantle [32,38] and arm [32,39] tissues and in the edible tissues considered as a whole [40–43].

With the aim of better focusing on possible composition changes, discussion of FA values will be addressed in the present study to FA groups (saturated FAs, STFAs; monounsaturated FAs, MUFAs; PUFAs; ω3-PUFAs) and FA ratios (total ω3/total ω6; polyene index, PI; flesh-lipid quality, FLQ). Additionally, and on the basis of the importance of ω3-PUFAs, analysis of the content of single ω3-PUFAs (EPA; DHA; docosapentaenoic acid, DPA) will also be discussed.

2.1.1. FA Groups

Values obtained for the STFA, MUFA, and PUFA groups in raw samples are shown in Tables 1–3, respectively. In all specimen sizes, the PUFA group showed to be the most abundant (ca. 47–62 g·100 g^{-1} total FAs range) and the MUFA group provided the lowest values ($p < 0.05$) (ca. 10–24 g·100 g^{-1} total FAs range).

Table 1. Saturated fatty acid (STFA) content (g·100 g^{-1} total FAs) * in different kinds of raw and processed tissues corresponding to different specimen sizes **.

Specimen Size	Raw or Processed Tissue	Tissue		
		Viscera	Mantle	Arm
Group I	Raw	31.82 ± 3.33 Aa	31.93 ± 2.82 Aa	34.48 ± 0.09 Aa
	Frozen	32.56 ± 0.41 Aa	36.27 ± 0.61 Bb	35.92 ± 0.11 Bc
	Cooked	NA ***	33.74 ± 1.81 Aab	35.05 ± 0.34 Ab
	Cooked-Frozen	NA	37.70 ± 0.35 Ac	37.54 ± 0.18 Ad
Group II	Raw	31.83 ± 0.23 Ba	35.14 ± 0.43 Ca	28.82 ± 0.94 Aa
	Frozen	32.47 ± 0.35 Aa	36.63 ± 0.12 Bb	37.90 ± 0.60 Cb
	Cooked	NA	36.49 ± 0.38 Bb	31.30 ± 2.84 Aa
	Cooked-Frozen	NA	37.06 ± 0.05 Ac	37.94 ± 0.13 Ab
Group III	Raw	29.16 ± 1.49 Aa	35.51 ± 0.24 Ba	34.71 ± 1.15 Ba
	Frozen	32.42 ± 0.19 Ab	36.17 ± 0.16 Bb	36.80 ± 0.02 Bb
	Cooked	NA	35.72 ± 0.37 Aab	35.79 ± 0.14 Aa
	Cooked-Frozen	NA	37.53 ± 0.29 Ac	37.06 ± 0.18 Ab

* Average values± standard deviations of three replicates (n = 3). Within each group and for each row, different capital letters (A–C) indicate significant differences ($p < 0.05$) among tissues. Within each group and for each column, different lowercase letters (a–d) indicate significant differences ($p < 0.05$) as a result of processing. ** Specimen sizes: Group I (1–2 kg per specimen), Group II (2–3 kg per specimen), and Group III (3–4 kg per specimen). *** NA: not analyzed.

Table 2. Monounsaturated fatty acid (MUFA) content (g·100 g^{-1} total FAs) * in different kinds of raw and processed tissues corresponding to different specimen sizes **.

Specimen Size	Raw or Processed Tissue	Tissue		
		Viscera	Mantle	Arm
Group I	Raw	18.01 ± 0.92 Ba	10.05 ± 0.08 Aa	10.29 ± 0.03 Ab
	Frozen	20.40 ± 0.88 Bb	10.14 ± 0.04 Aa	10.26 ± 0.28 Ab
	Cooked	NA ***	10.17 ± 0.33 Aa	10.24 ± 0.44 Ab
	Cooked-Frozen	NA	10.32 ± 0.10 Ba	9.14 ± 0.17 Aa
Group II	Raw	17.03 ± 0.04 Ba	10.90 ± 0.27 Aa	10.0 ± 0.46 Aa
	Frozen	17.03 ± 44 Ba	10.15 ± 0.20 Aa	10.51 ± 0.29 Aa
	Cooked	NA	10.86 ± 0.33 Ba	9.97 ± 0.29 Aa
	Cooked-Frozen	NA	10.64 ± 0.08 Ba	10.02 ± 0.08 Aa
Group III	Raw	23.71 ± 0.65 Bb	10.43 ± 0.25 Ab	9.99 ± 0.31 Ab
	Frozen	18.42 ± 0.25 Ba	10.11 ± 0.07 Ab	9.82 ± 0.09 Ab
	Cooked	NA	10.30 ± 0.07 Ab	10.03 ± 0.14 Ab
	Cooked-Frozen	NA	9.22 ± 0.07 Ba	8.45 ± 0.13 Aa

* Average values± standard deviations of three replicates (n = 3). Within each group and for each row, different capital letters (A,B) indicate significant differences (p < 0.05) among tissues. Within each group and for each column, different lowercase letters (a,b) indicate significant differences (p < 0.05) as a result of processing. ** Specimen sizes as expressed in Table 1. *** NA: not analyzed.

Regarding the STFA group, no differences (p > 0.05) were detected among tissues for the small-sized specimens (Group I). Conversely, higher average values were detected in the mantle tissue than in the other tissues in the medium- and large-sized samples (Groups II and III, respectively); differences were found to be significant (p < 0.05) in both cases in comparison to the viscera tissue. A marked higher MUFA value (p < 0.05) was obtained in the viscera tissue when compared to the two other tissues; meanwhile, no differences (p > 0.05) were observed between mantle and arm tissues. In the case of the PUFA group, the opposite distribution as for the MUFA group was detected. Thus, the viscera tissue depicted the lowest values (p < 0.05) in all sizes (Groups I, II, and III). Comparisons of arm and mantle tissues revealed higher PUFA values (p < 0.05) in the mantle tissue for small-sized specimens; conversely, higher levels (p < 0.05) were detected in the arm tissue in the case of medium-sized octopus.

Regarding non-edible tissues, the same FA group distribution as in the current study was previously detected in octopus (*O. vulgaris*) ovary during a seasonal study carried out in specimens obtained in the Atlantic coast [32]; conversely, the digestive gland showed the following decreasing sequence: PUFAs > MUFAs = STFAs. During a recent study focused on lipid extraction with low-toxicity solvents [33], PUFAs were found to be the most abundant group, and MUFAs depicted the lowest values in total non-edible tissues from octopus (*O. vulgaris*). Also in agreement with the present study, the PUFAs > STFAs > MUFAs decreasing sequence was detected for FA groups in non-edible tissues obtained from Patagonian squid (*D. gahi*) [34] and viscera obtained from squid (*I. argentinus*) [37], cuttlefish (*S. officinalis*) [36], and Giant squid (*Dosidicus gigas*) [44].

A similar distribution of the three FA groups than in the present case was already detected in different edible tissues of octopus (*O. vulgaris*). This result has been observed in specimens also captured in the European Atlantic coast for mantle and arm tissues [32], and for edible parts considered as a whole [40,43]. Additionally, a similar FA distribution was also observed in specimens captured in the Mediterranean Sea for the mantle [38], arm [39], and edible parts [41], as well as in specimens obtained in the Brazilian coast [42].

Table 3. Polyunsaturated fatty acid (PUFA) content (g·100 g^{-1} total FAs) * in different kinds of raw and processed tissues corresponding to different specimen sizes **.

Specimen Size	Raw or Processed Tissue	Tissue		
		Viscera	Mantle	Arm
Group I	Raw	50.17 ± 2.24 Ab	58.03 ± 1.75 Cc	55.26 ± 0.12 Bc
	Frozen	47.04 ± 0.87 Aa	53.59 ± 0.64 Bb	53.79 ± 0.34 Ba
	Cooked	NA ***	56.10 ± 1.90 Abc	54.72 ± 0.17 Ab
	Cooked-Frozen	NA	51.98 ± 0.42 Aa	53.33 ± 0.31 Ba
Group II	Raw	51.14 ± 0.19 Ab	53.96 ± 0.20 Bc	61.10 ± 1.22 Cb
	Frozen	50.51 ± 0.13 Aa	53.22 ± 0.22 Bbc	51.59 ± 0.88 Aa
	Cooked	NA	52.65 ± 0.28 Aab	58.73 ± 3.09 Bb
	Cooked-Frozen	NA	52.30 ± 0.05 Ba	52.05 ± 0.04 Aa
Group III	Raw	47.12 ± 2.14 Aa	54.06 ± 0.46 Ba	55.30 ± 1.86 Bab
	Frozen	49.15 ± 0.18 Aa	53.72 ± 0.12 Ba	53.38 ± 0.11 Ba
	Cooked	NA	53.98 ± 0.32 Aa	54.48 ± 0.24 Ab
	Cooked-Frozen	NA	53.25 ± 0.34 Aa	54.18 ± 0.08 Bb

* Average values± standard deviations of three replicates (n = 3). Within each group and for each row, different capital letters (A,B,C) indicate significant differences ($p < 0.05$) among tissues. Within each group and for each column, different lowercase letters (a,b,c) indicate significant differences ($p < 0.05$) as a result of processing. ** Specimen sizes as expressed in Table 1. *** NA: not analyzed.

Marine lipids are reported to include many beneficial constituents for the human health, especially related to their high values on ω3-PUFAs [3,4]. On the basis of the great significance of ω3-PUFAs, its content was studied in the current research. Thus, a similar distribution was detected for the ω3-PUFAs (Figure 1) as for the PUFA group (Table 3). Values obtained in raw samples of the different size groups were included in the 35.4–41.9, 44.4–46.6, and 44.30–52.5 g·100 g^{-1} total FAs ranges for viscera, mantle and arm tissues, respectively (Figure 1). The lowest levels ($p < 0.05$) were detected in the viscera tissue of all sizes. Additionally, the mantle tissue showed higher values ($p < 0.05$) than the arm tissue for specimens of small sizes; conversely, higher average values were observed in the arm tissue for samples corresponding to medium- and large-sized specimens. In spite of such differences among the tissues considered, present ω3-PUFA levels in specimens of all size groups can be considered to be highly valuable regarding nutritional and healthy requirements [2,4,8].

Previous research has also shown high values of ω3-PUFAs in non-edible zones of the present species of octopus. Expressed as g·100 g^{-1} total FAs, the ovary and digestive gland depicted the 40.4–47.0 and 35.8–43.5 ranges, respectively, during a seasonal study of Atlantic octopus (*O. vulgaris*) [32]. A recent lipid extraction with low-toxicity solvents revealed values included in the 36.8–38.3 range for octopus (*O. vulgaris*) by-products [33]. Concerning non-edible tissues of other related cephalopod species, Kacem et al. [36] showed a 21.4–26.1 range in cuttlefish (*S. officinalis*) viscera from two different catching times and a ca. 37.4 value was reported for total viscera of squid (*I. argentinus*) [37]. A seasonal study on total squid (*D. gahi*) by-products indicated a 46.1–48.6 range [34].

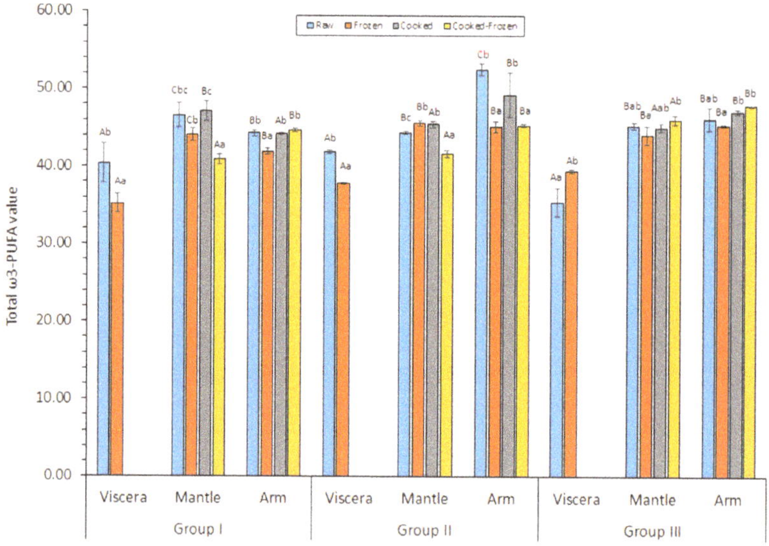

Figure 1. Determination of the total ω3 polyunsaturated fatty acid (ω3-PUFA) content (g·100 g^{-1} total FAs) in different kinds of raw and processed tissues corresponding to different specimen sizes. Mean values of three replicates (n = 3); standard deviations are indicated by bars. Within each group and for each raw/processed substrate, different capital letters (A,B,C) indicate significant differences ($p < 0.05$) among tissues. Within each group and for each tissue, different lowercase letters (a,b,c) indicate significant differences ($p < 0.05$) as a result of processing. Specimen groups are as expressed in Table 1. The viscera tissue was not subjected to the cooking process.

Previous studies have also reported a high ω3-PUFA presence in edible tissues corresponding to the current species of octopus. Thus, similar values (expressed as g·100 g^{-1} total FAs) were detected in the edible tissue considered as a whole by Bonafe et al. [42] (39.4), Özoğul et al. [39] (41–47 range), and Zlatanos et al. [41] (37.7). Regarding single tissues, arms corresponding to octopus (*O. vulgaris*) captured in the Mediterranean Sea revealed a 46–49 range value [38]. The seasonal study of the ω3-PUFA presence showed ranges of 49.8–54.2 and 50.3–55.3 for arm and mantle, respectively, corresponding to specimens obtained in the European Atlantic coast [32]. Conversely, a low ω3-PUFA value (17.3) was detected by Biandolino et al. [45] in the edible tissue of farmed specimens from the Ionian Sea.

2.1.2. Total ω3/Total ω6 FA Ratio

It is well-known that Western countries, in general, do not consume necessary levels of ω3-PUFAs, so a great attention has been accorded to the ω3/ω6 ratio of foods included in the human diet [3,4]. With the aim of preventing several health disorders (cardiovascular, neurological, and inflammatory), the World Health Organization (WHO) recommends nowadays a higher ratio than 1:10 in the human diet [46]. Furthermore, the European Nutritional Society indicated that notable health benefits could be achieved if an ω3/ω6 ratio of 1:5 or higher was provided in the human diet [47].

Values for ω3/ω6 ratios obtained in raw samples in the current study were included in the 3.0–4.5, 4.1–5.2, and 4.1–6.1 ranges for viscera, mantle, and arm tissues, respectively

(Figure 2). The analysis of the ω3/ω6 ratio values revealed no differences ($p > 0.05$) among tissues in specimens corresponding to the small-sized group. Conversely, the arm tissue depicted higher levels ($p < 0.05$) than its counterpart viscera tissue for medium- and large-sized specimens. In the case of large-sized samples, the mantle tissue provided a higher ω3/ω6 ratio value ($p < 0.05$) than the viscera one. In spite of differences mentioned, and in agreement with current recommendations provided by nutritional organizations, ω3/ω6 ratio values obtained in all cases can be considered as highly valuable for the human diet.

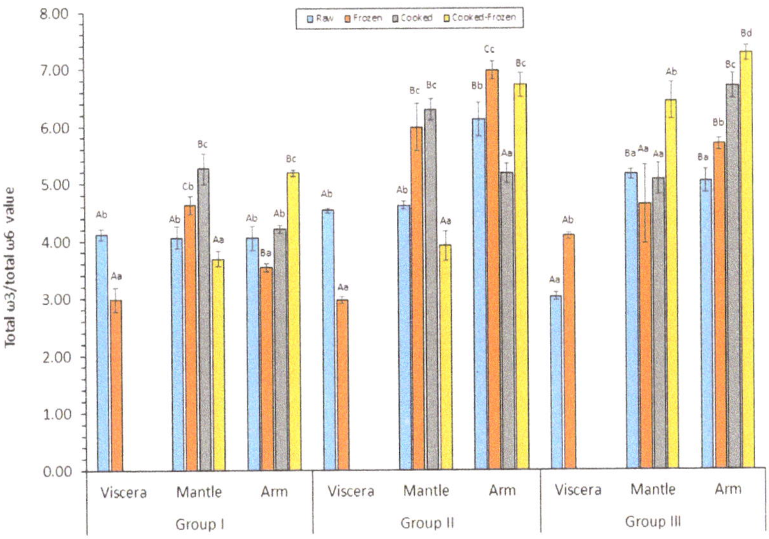

Figure 2. Determination of the total ω3 polyunsaturated fatty acid (ω3-PUFA) content (g·100 g^{-1} total FAs) in different kinds of raw and processed tissues corresponding to different specimen sizes. Mean values of three replicates ($n = 3$); standard deviations are indicated by bars. Within each group and for each raw/processed substrate, different capital letters (A,B,C) indicate significant differences ($p < 0.05$) among tissues. Within each group and for each tissue, different lowercase letters (a,b,c,d) indicate significant differences ($p < 0.05$) as a result of processing. Specimen groups are as expressed in Table 1. The viscera tissue was not subjected to the cooking process.

Compared to the present values obtained for viscera, Sieiro et al. [32] observed higher ω3/ω6 ratio values for the ovary and the digestive gland (4.2–8.3 and 5.0–9.1 ranges, respectively) during a seasonal study of common octopus (*O. vulgaris*). Conversely, total by-products from the same marine species showed values included in the 3.2–3.7 range when extracted with low-toxicity solvents (ethanol, acetone, and methyl acetate) [33]. In the case of other cephalopod species, higher ω3/ω6 ratio values were detected for squid (*I. argentinus*) viscera (7.6–8.0 range) [37] and squid (*D. gahi*) by-products (12.1–13.3 range) [34] than in the present viscera tissue. Conversely, lower values (1.5–2.4 range) were detected by Kacem et al. [36] for cuttlefish (*S. officinalis*) viscera from specimens corresponding to two different catching times.

Higher ω3/ω6 ratios than found in the current work were obtained during a seasonal study of different edible tissues obtained from Atlantic European octopus (*O. vulgaris*) [32], in which arm and mantle tissues provided values included in the 9.5–12.8 range. Similar values to those in the present research were reported for the same species in the muscle tissue by Oliveira et al. [43] (5.3), Özoğul et al. [38] (3.8–5.6), and Zlatanos et al. [41] (3.6).

Conversely, Biandolino et al. [45] showed a notably lower ω3/ω6 ratio value (1.7) in the arm tissue of farmed octopus (*O. vulgaris*).

2.1.3. Single ω3-PUFAs

Among ω3-PUFAs, EPA and DHA have received a great attention, in agreement with their beneficial health properties. Thus, EPA consumption has been associated with low prevalence of circulatory, coronary, and inflammatory diseases [48]. On the other side, DHA has been reported to be responsible for the prevention of neurodegenerative diseases and correct fetal development, and functioning of the nervous system and visual organs in the fetus [49]. In spite of its lower content in seafood than EPA and DHA, DPA is attracting an increasing attention because of its presence in human brain and its high levels in human milk, which implies a potential impact during pregnancy and early development [50,51]. Moreover, DPA has been reported to be related to the improvement of cardiovascular and metabolic diseases [52,53]. Consequently, and on the basis of their great significance to human health, the current research on the FA profile will now be focused on the presence of these three ω3-PUFAs, i.e., EPA, DHA, and DPA.

In the present study, the EPA value was included in the 19.6–21.9, 18.4–18.8, and 17.2–19.3 g·100 g^{-1} total FAs ranges for viscera, mantle, and arm tissues, respectively, in the different size groups (Table 4). The following decreasing sequence ($p < 0.05$) was observed for the EPA content of tissues corresponding to small-sized specimens: viscera > mantle > arm. In the case of medium- and large-sized groups, the lowest values ($p < 0.05$) were detected in the mantle, while the highest average levels were obtained in the viscera tissue.

Table 4. Determination of the eicosapentaenoic acid (EPA) content (g·100 g^{-1} total FAs) * in different kinds of raw and processed tissues corresponding to different specimen sizes **.

Specimen Size	Raw or Processed Tissue	Tissue		
		Viscera	Mantle	Arm
Group I	Raw	20.83 ± 0.83 Cb	18.57 ± 0.24 Bc	17.22 ± 0.24 Ab
	Frozen	17.19 ± 1.00 Ba	17.33 ± 0.19 Bb	16.30 ± 0.25 Aa
	Cooked	NA ***	18.18 ± 0.25 Ac	18.49 ± 0.15 Ac
	Cooked-Frozen	NA	16.03 ± 0.59 Aa	18.29 ± 0.72 Bbc
Group II	Raw	21.93 ± 0.03 Cb	18.41 ± 0.25 Ab	19.73 ± 0.79 Bab
	Frozen	17.47 ± 0.19 Aa	20.20 ± 0.56 Bc	21.56 ± 1.13 Bb
	Cooked	NA	19.31 ± 0.36 Ac	19.06 ± 0.51 Aa
	Cooked-Frozen	NA	17.06 ± 0.76 Aa	19.39 ± 0.70 Ba
Group III	Raw	19.58 ± 0.62 Ba	18.75 ± 0.18 Abc	19.33 ± 0.14 Ba
	Frozen	19.15 ± 0.10 Ba	17.52 ± 1.03 Aab	19.19 ± 0.23 Ba
	Cooked	NA	17.91 ± 0.38 Aa	20.21 ± 0.17 Bb
	Cooked-Frozen	NA	19.10 ± 0.32 Ac	20.01 ± 0.03 Bb

* Average values± standard deviations of three replicates ($n = 3$). Within each Group and for each row, different capital letters (A,B,C) indicate significant differences ($p < 0.05$) among tissues. Within each Group and for each column, different lowercase letters (a,b,c) indicate significant differences ($p < 0.05$) as a result of processing. ** Specimen sizes as expressed in Table 1. *** NA: not analyzed.

Results concerning DHA are depicted in Table 5. A notably lower value ($p < 0.05$) (expressed as g·100 g^{-1} total FAs) was detected for all sizes (Groups I, II, and III) in the viscera tissue (12.7–17.5 range) than in the edible tissues. Differences between arm (24.3–30.1 range) and mantle (24.3–26.1 range) values were only observed in medium-sized samples, with arm tissue showing a higher ($p < 0.05$) content.

Table 5. Determination of the docosahexaenoic acid (DHA) content (g·100 g^{-1} total FAs) * in different kinds of raw and processed tissues corresponding to different specimen sizes **.

Specimen Size	Raw or Processed Tissue	Tissue		
		Viscera	Mantle	Arm
Group I	Raw	16.01 ± 2.22 Aa	26.14 ± 2.21 Bbc	25.15 ± 0.15 Ba
	Frozen	15.53 ± 0.34 Aa	25.13 ± 0.63 Cb	23.77 ± 0.22 Bb
	Cooked	NA ***	27.29 ± 1.16 Bc	23.49 ± 1.08 Aab
	Cooked-Frozen	NA	23.43 ± 0.67 Aa	24.94 ± 1.12 Aab
Group II	Raw	17.46 ± 0.14 Aa	24.25 ± 0.35 Bb	30.06 ± 1.28 Cc
	Frozen	18.05 ± 0.09 Ab	24.46 ± 0.41 Bb	23.60 ± 1.38 Ba
	Cooked	NA	24.57 ± 0.15 Ab	27.44 ± 2.12 Abc
	Cooked-Frozen	NA	22.99 ± 0.57 Aa	25.69 ± 0.94 Bab
Group III	Raw	12.70 ± 1.05 Aa	24.50 ± 0.23 Ba	24.29 ± 1.34 Bab
	Frozen	16.8 ± 0.125 Ab	24.40 ± 1.03 Bab	25.74 ± 0.37 Bb
	Cooked	NA	25.66 ± 0.30 Ab	24.94 ± 0.15 Aa
	Cooked-Frozen	NA	25.20 ± 0.24 Ab	25.67 ± 0.08 Ab

* Average values ± standard deviations of three replicates ($n = 3$). Within each group and for each row, different capital letters (A,B,C) indicate significant differences ($p < 0.05$) among tissues. Within each group and for each column, different lowercase letters (a,b,c) indicate significant differences ($p < 0.05$) as a result of processing. ** Specimen sizes are as expressed in Table 1. *** NA: not analyzed.

Regarding DPA, values were notably lower than in the case of the two other ω3-PUFAs (Table 6). Thus, ranges of 2.6–3.5, 1.5–1.9, and 1.9–2.5 g·100 g^{-1} total FAs were obtained for viscera, mantle, and arm tissues, respectively, in the different size groups. The lowest average values were detected in the mantle tissue for all sizes of specimens; differences with other tissues were found significant ($p < 0.05$) for Groups II and III. The highest average values were obtained in viscera; remarkably, differences with the arm tissue were significant ($p < 0.05$) in small- and large-sized samples.

Previous research regarding non-edible tissues of the current species of octopus showed a different decreasing tendency (i.e., DHA > EPA > DPA) than in the current study for the presence of the three ω3-PUFAs. Thus, the following ranges were detected during a seasonal study [32]: 24.8–31.6, 13.4–18.1, and 0.3–1.1 g·100 g^{-1} total FAs, respectively, in ovary tissue; in the same study [32], the digestive gland provided the following value ranges: 16.7–28.7, 14.4–18.1, and 0.1–1.0 g·100 g^{-1} total FAs, respectively. In agreement with the present research, the EPA value (21.0–22.4 g·100 g^{-1} total FAs range) showed to be higher than that of DHA (14.0–14.5 g·100 g^{-1} total FAs range) in lipid extracts obtained from total by-products by employing low-toxicity solvents (ethanol, acetone, and ethyl acetate) [33]. In the case of other cephalopod species, total viscera obtained from squid (*I. argentinus*) revealed values ca. 16.4, 9.3, and 0.5 g·100 g^{-1} total FAs for DHA, EPA, and

DPA, respectively [37]. During a seasonal study carried out on Patagonian squid (*D. gahi*) by-products [34], 29.5–30.8, 15.9–17.2, and 0.5–0.6 g·100 g^{-1} total FAs ranges were detected for DHA, EPA, and DPA, respectively. A similar distribution (35.0–39.0, 13.1–14.3, and 0.7–0.8 g·100 g^{-1} total FAs ranges, respectively) was observed in the arms and tentacles of European squid (*Loligo vulgaris*) [54]. Conversely, a seasonal study carried out on cuttlefish (*S. officinalis*) viscera led to ranges of 6.3–9.1, 7.1–11.6, and 1.66 g·100 g^{-1} total FAs for DHA, EPA and DPA.

Table 6. Determination of the docosapentaenoic acid (DPA) content (g·100 g^{-1} total FAs) * in different kinds of raw and processed tissues corresponding to different specimen sizes **.

Specimen Size	Raw or Processed Tissue	Tissue		
		Viscera	Mantle	Arm
Group I	Raw	3.54 ± 0.48 Bb	1.85 ± 0.11 Ab	1.93 ± 0.03 Aa
	Frozen	2.45 ± 0.10 Ca	1.61 ± 0.06 Aa	1.86 ± 0.05 Ba
	Cooked	NA ***	1.64 ± 0.01 Aa	2.26 ± 0.11 Bb
	Cooked-Frozen	NA	1.94 ± 0.16 Ab	2.21 ± 0.11 Ab
Group II	Raw	2.59 ± 0.05 Bb	1.54 ± 0.03 Aa	2.50 ± 0.45 Bab
	Frozen	2.25 ± 0.05 Ba	1.99 ± 0.02 Ab	1.91 ± 0.09 Aa
	Cooked	NA	1.59 ± 0.15 Aa	2.76 ± 0.20 Bb
	Cooked-Frozen	NA	2.02 ± 0.02 Ab	1.95 ± 0.07 Aa
Group III	Raw	3.12 ± 0.15 Ca	1.84 ± 0.07 Abc	2.45 ± 0.28 Bb
	Frozen	3.47 ± 0.04 Cb	1.93 ± 0.07 Ac	2.29 ± 0.04 Bb
	Cooked	NA	1.50 ± 0.03 Aa	1.98 ± 0.12 Ba
	Cooked-Frozen	NA	1.71 ± 0.03 Ab	2.21 ± 0.02 Bb

* Average values ± standard deviations of three replicates (n = 3). Within each group and for each row, different capital letters (A,B,C) indicate significant differences ($p < 0.05$) among tissues. Within each group and for each column, different lowercase letters (a,b,c) indicate significant differences ($p < 0.05$) as a result of processing. ** Specimen sizes are as expressed in Table 1. *** NA: not analyzed.

Previous research has already shown the current decreasing tendency, i.e., DHA > EPA > DPA, in the edible tissues of the present species of octopus. Thus, Zlatanos et al. [41] detected values of 20.1, 13.6, and 2.0 g·100 g^{-1} total FAs, respectively, in muscle tissue. In a seasonal study [32], ranges of 28.3–32.9, 19.1–21.4, and 0.2–0.5 g·100 g^{-1} total FAs, respectively, were detected in arm and mantle tissues. Oliveira et al. [43] showed values of 100.4, 77.2, and 7.0 mg·100 g^{-1} edible tissue, respectively. In a different species of octopus, *Eledone moschata*, values of 24.7, 16.7, and 1.8 g·100 g^{-1} total FAs, respectively, were detected in the mantle tissue [55]. During a seasonal study of musky octopus (*E. moschata*) [39], the edible tissue showed values included in the 21.0–28.2 and 7.9–12.2 g·100 g^{-1} total FAs ranges for DHA and EPA, respectively.

2.2. Effect of Cooking on the FA Composition

The cooking treatment was applied to raw mantle and arm tissues but not to viscera. As for samples corresponding to the raw stage (Section 3.1), changes produced in the FA profile will be discussed on the basis of the cooking effect on FA groups and ratios as well as on single ω3-PUFAs.

2.2.1. FA Groups

As a result of the thermal treatment, an increased average value of the STFA presence was observed in all tissues and for specimens corresponding to all sizes (Table 1). Differences were found to be significant ($p < 0.05$) in the mantle of medium-sized samples and in the arm of small-sized samples. In the case of MUFAs (Table 2), the evaluation of the average values did not provide any definite tendency and no effect ($p > 0.05$) of the cooking process could be implied on any of the tissues for any of the sizes considered. A general decrease in the average PUFA level was found after the cooking process (Table 3). This decrease was found to be significant ($p < 0.05$) in the mantle of medium-sized specimens and in the arm of small-sized specimens. Regarding the ω3-PUFA value (Figure 1), the only significant change ($p < 0.05$) detected was a content increase in the mantle tissue of medium-sized samples.

2.2.2. FA Ratios

The ω3/ω6 ratio showed an average increase with the cooking process in small-sized specimens, with differences being significant ($p < 0.05$) in the mantle tissue (Figure 2). A definite trend could not be concluded in the two other tissues, although increases in this ratio were obtained in the mantle of medium-sized specimens and in the arm of large-sized samples.

The PUFA presence in the human diet has been found to be strongly related to nutritional value, digestibility, and preserving properties [56,57]. In order to evaluate the PUFA content variation during seafood processing and storage, the polyene index (PI), defined as the DHA+EPA/C16:0 ratio of FA concentrations, has widely been employed [37, 58]. Technologists focused on seafood consider this ratio to be a valuable and practical tool for the assessment of the lipid fraction damage and therefore, to the quality loss of seafood during different steps of processing.

PI values obtained in the present research are expressed in Figure 3. Raw samples showed values included in the 2.3–2.6, 2.2–2.7, and 2.2–3.3 ranges for viscera, mantle and arm tissues, respectively. No differences ($p > 0.05$) among tissues could be detected in small- and large-sized specimens (Figure 3). Conversely, the following decreasing sequence was observed in medium-sized specimens: arm > viscera > mantle. Regarding the effect of the cooking process, the average PI revealed a general decrease (Figure 3). This decrease was significant ($p < 0.05$) in the arms of small-sized samples and in the large-sized samples from the mantle tissue.

The FLQ index has also been employed in order to assess the quality of the dietary lipid source [59–61]. FLQ values obtained in the current study are presented in Table 7. Raw samples depicted values included in the 47.7–65.0, 74.4–80.9 and 73.5–99.2 ranges for viscera, mantle, and arm tissues, respectively. In all sizes of specimens, a lower value ($p < 0.05$) was depicted in viscera than in both edible tissues. The cooking process did not lead to a general trend in the mantle tissues. Thus, decreased values ($p < 0.05$) were detected in frozen small- and medium-sized samples and an increase ($p < 0.05$) in frozen big-sized specimens was obtained. Regarding the arm tissue, an increase ($p < 0.05$) in frozen small-sized samples and a decrease ($p < 0.05$) in raw medium-sized specimens were observed.

Current values obtained for mantle and arm tissues have shown to be higher than those reported for muscle corresponding to freshwater and farmed fish species. Thus, FLQ value of ca. 34.8, 37.2, and 41.3 were detected for roach (*Rutilus rutilus*), perch (*Perca fluviatilis*), and pike (*Esox lucius*) muscle [61]. During a seasonal study, Senso et al. [59] reported a 19.4–31.3 value range for farmed gilthead sea bream (*Sparus aurata*). A ca.

24.5–36.4 value range was obtained in several freshwater fish species (roach, *R. rutilus*; bream, *Abramis brama*; pike, *E. lucius*; Eurasian perch, *P. fluviatilis*) [60].

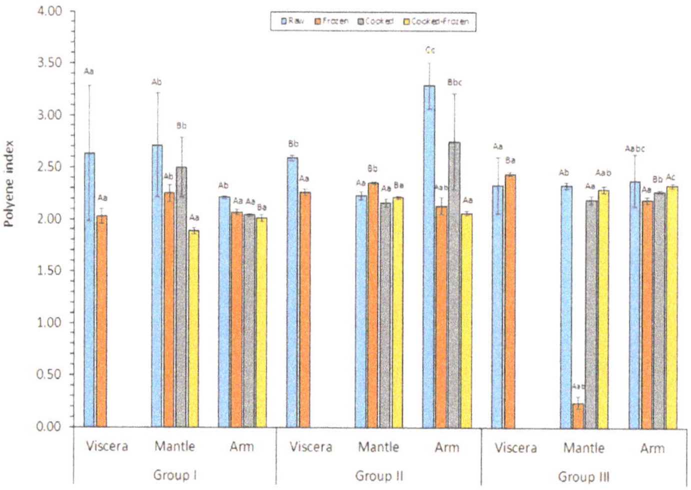

Figure 3. Determination of the total ω3 polyunsaturated fatty acid (ω3-PUFA) content ($g \cdot 100\ g^{-1}$ total FAs) in different kinds of raw and processed tissues corresponding to different specimen sizes. Mean values of three replicates ($n = 3$); standard deviations are indicated by bars. Within each group and for each raw/processed substrate, different capital letters (A,B,C) indicate significant differences ($p < 0.05$) among tissues. Within each group and for each tissue, different lowercase letters (a,b,c) indicate significant differences ($p < 0.05$) as a result of processing. Specimen groups are as expressed in Table 1. The viscera tissue was not subjected to the cooking process.

Table 7. Determination of the flesh-lipid quality (FLQ) index * in different kinds of raw and processed tissues corresponding to different specimen sizes **.

Specimen Size	Raw or Processed Tissue	Tissue		
		Viscera	Mantle	Arm
Group I	Raw	58.32 ± 5.15 Ab	80.86 ± 6.25 Bbc	73.52 ± 1.44 Bb
	Frozen	48.63 ± 2.34 Aa	73.79 ± 2.44 Cb	66.86 ± 2.22 Ba
	Cooked	NA ***	83.39 ± 2.17 Bc	72.35 ± 3.18 Aab
	Cooked-Frozen	NA	65.18 ± 3.27 Ba	76.15 ± 3.16 Ab
Group II	Raw	64.99 ± 1.12 Ab	74.40 ± 2.09 Bb	99.16 ± 3.26 Cb
	Frozen	55.09 ± 2.35 Aa	80.70 ± 3.20 Bc	82.27 ± 4.08 Ba
	Cooked	NA	78.19 ± 2.10 Abc	86.92 ± 4.38 Ba
	Cooked-Frozen	NA	66.81 ± 3.43 Aa	82.08 ± 3.39 Ba
Group III	Raw	47.67 ± 3.51 Ab	76.21 ± 3.42 Bab	77.37 ± 3.21 Ba
	Frozen	56.25 ± 1.41 Aa	72.18 ± 4.41 Ba	81.59 ± 1.40 Cab
	Cooked	NA	77.09 ± 2.37 Aab	82.32 ± 1.17 Bab
	Cooked-Frozen	NA	79.53 ± 2.17 Ab	84.09 ± 1.12 Ab

* Average values± standard deviations of three replicates ($n = 3$). Within each group and for each row, different capital letters (A,B,C) indicate significant differences ($p < 0.05$) among tissues. Within each group and for each column, different lowercase letters (a,b,c) indicate significant differences ($p < 0.05$) as a result of processing. ** Specimen sizes are as expressed in Table 1. *** NA: not analyzed.

2.2.3. Single ω3-PUFAs

Increased EPA values ($p < 0.05$) were obtained with the cooking process for the arm of small-sized octopus and for the mantle of medium-sized samples (Table 4); conversely, large-sized specimens revealed a decrease ($p < 0.05$) in the mantle tissue. Concerning the DHA value (Table 5), a general increase in the average value with cooking was obtained in the mantle of specimens corresponding to all sizes; this increase was found to be significant ($p < 0.05$) in the mantle of large-sized samples. For the arm tissue, a definite trend could not be concluded for the DHA level. Regarding the DPA presence, a decreased value ($p < 0.05$) was obtained in the mantle tissue of small- and large-sized samples and in the arm tissue of large-sized samples; conversely, small-sized specimens showed a DPA level decrease in the arm tissue.

2.2.4. Previous Related Studies Regarding the Effect of Cooking on Cephalopod Tissues

Different damage pathways have been pointed out in previous studies, resulting from the cooking treatment of seafood in general. Among them, heat degradation of nutrients, oxidation of vitamins and lipids, and protein toughening can be mentioned [62,63]. As a result, nutritional and sensory quality losses have been indicated, especially when over-processing is carried out. In such studies, great attention has been given to the evolution of PUFA compounds, as being especially prone to lipid oxidation development [58] and consequently, lead to a wide range of negative effects on nutritional and healthy values [64,65].

Previous work related to the effect of cooking on non-edible tissues corresponding to cephalopod species can be considered scarce. Toyes-Vargas et al. [44] detected the loss of FA values after the cooking process of Giant squid (*D. gigas*) viscera; thus, decreased values ($g \cdot kg^{-1}$ dry tissue weight) were detected for EPA (from 16.9 to 14.0), DHA (from 19.4 to 18.9), C20:4ω6 (from 3.6 to 3.4), and total STFAs (from 38.2 to 31.2). A decreasing tendency for the DHA, EPA and PI values was detected in squid (*I. argentinus*) viscera when increasing the heating time and the temperature of processing [37]; conversely, the authors observed no effect on the ω3/ω6 ratio.

Previous research regarding the effect of the cooking process and thermal treatment, in general, on the FA profile of edible tissues of cephalopod species can also be considered scarce. A marked effect on the FA profile of octopus (*O. vulgaris*) muscle was observed by Czech et al. [66] after subjecting it to a frying process with sunflower oil. The authors observed a remarkable content decrease ($g \cdot 100\ g^{-1}$ total FAs) in the fried product in EPA (from 14.0 to 0.5), DHA (from 29.6 to 2.5) and DPA (from 0.9 to 0.1) values and in the ω3/ω6 ratio (from 4.0 to 0.1). No differential losses in FA groups were detected by Oliveira et al. [43] by subjecting gutted octopus (*O. vulgaris*) to the boiling process. Thus, losses of 89.5%, 92.1%, 89.0% and 88.9% for STFA, MUFA, PUFA, and ω3-PUFA groups, respectively, were detected; additionally, a decrease from 5.25 to 5.17 was produced for the ω3/ω6 ratio and from 2.44 to 2.34 for the PI as a result of the thermal treatment.

2.3. Effect of Frozen Storage on the FA Composition

The effect of the frozen storage was studied on the raw samples of the viscera, mantle, and arm tissues and on the cooked samples of the mantle and arm tissues. As for samples corresponding to the raw and cooked conditions, changes produced in the FA profile will be discussed on the basis of the effect of the frozen storage on FA groups and ratios, as well as on single ω3-PUFAs.

2.3.1. FA Groups

Frozen storage led to a significant increase in STFA content in viscera (Table 1); this difference was significant ($p < 0.05$) in specimens corresponding to the large-sized group. Regarding mantle and arm tissues, a notable increase in the average STFA content was detected in all size groups as a result of the frozen storage. This tendency was observed both for raw as for cooked samples.

The effect of frozen storage on MUFA content varied by sample size: it increased significantly ($p < 0.05$) in small-sized viscera but decreased ($p < 0.05$) in large-sized ones. Meantime, the mantle tissue showed no effect ($p > 0.05$) for small- and medium-sized specimens; however, an average decrease was found in large-sized specimens, which was significant ($p < 0.05$) for cooked samples. A different behavior was obtained for the arm tissue according to the size of specimens. Thus, small- and large-sized samples showed a decrease in the average value, while medium-sized samples experienced an increase; a significant effect ($p < 0.05$) of the frozen period was detected in cooked small- and large-sized groups.

A decrease ($p < 0.05$) of the PUFA value was detected in viscera corresponding to the small- and medium-sized samples as a result of frozen storage; conversely, an average value increase was observed for large-sized specimens. Regarding the mantle tissue, a general decrease in the average PUFA level was observed; this decrease was found to be significant ($p < 0.05$) in medium-sized samples both for raw and cooked samples. A general decrease in the average PUFA value was observed in the arm tissue; this decrease was significant ($p < 0.05$) for small- and medium-sized specimens corresponding to the cooked condition.

A decrease ($p < 0.05$) in the total ω3-PUFA value was observed in viscera from small- and medium-sized samples; conversely, large-sized samples saw an increase ($p < 0.05$) as a result of the frozen period. Mantel and arm tissues showed a decrease ($p < 0.05$) of the ω3-PUFA value in raw and cooked samples of Group II. However, a definite trend for both tissues could not be concluded for this FA group regarding specimens corresponding to Groups I and III.

2.3.2. FA Ratios

A notable decrease ($p < 0.05$) in the ω3/ω6 ratio was observed as a result of the storage period in the viscera tissue corresponding to small- and medium-sized samples. Conversely, viscera corresponding to large-sized octopus showed an increased value ($p < 0.05$). Regarding the mantle tissue, a decreased value ($p < 0.05$) was obtained for cooked small- and medium-sized samples, but an increased value ($p < 0.05$) was detected in raw medium-sized samples and cooked large-sized samples. For the arm tissue, an increased value ($p < 0.05$) was obtained in most cases; the exception was the raw small-sized batch that exhibited a decrease ($p < 0.05$) after the storage period.

The PI of the viscera tissue indicated an average decrease for specimens corresponding to the small- and medium-sized batches; differences were significant ($p < 0.05$) for the medium-sized ones. Small-sized specimens showed an average PI value decrease in the mantle and arm tissues as a result of the frozen storage; differences were significant ($p < 0.05$) for cooked mantle and raw arm. Concerning medium- and large-sized samples, an increased PI was observed after the frozen period in the mantle tissue in most cases; differences were significant ($p < 0.05$) for raw medium-sized samples. For the arm tissue, a general decrease in the average PI was detected in medium- and large-sized samples; conversely, cooked specimens corresponding to Group III showed a content increase ($p < 0.05$) with frozen storage.

The FLQ value of the viscera tissue indicated a significant decrease ($p < 0.05$) in specimens of all sizes (Table 7). The effect of the frozen storage on the mantle tissue did not lead to a general trend. Thus, a value decrease ($p < 0.05$) was detected in cooked small- and medium-sized samples and a value increase in raw medium-sized specimens. For the arm tissue, decreased values were obtained in raw small- and medium-sized specimens.

2.3.3. Single ω3-PUFAs

Regarding the EPA value, viscera samples showed an average decrease with the frozen storage; differences were significant ($p < 0.05$) in small- and medium-sized samples. For mantle samples, a decreased value was detected after the frozen storage in most cases; differences were significant ($p < 0.05$) in raw and cooked small-sized samples and in cooked medium-sized ones. Conversely, the mantle tissue corresponding to raw medium-sized and cooked large-sized samples showed an increased ($p < 0.05$) EPA level as a result of the frozen storage. In the case of the arm tissue, the effect of frozen storage varied by specimen size. Thus, small- and large-sized samples depicted an average decrease, while medium-sized ones indicated an increase.

The DHA value in viscera showed an increase ($p < 0.05$) after the frozen storage in medium- and large-sized samples. Regarding the mantle tissue, an average value decrease was detected for this PUFA group in most cases; differences were found to be significant ($p < 0.05$) in cooked small- and medium-sized batches. A decreased DHA value was generally observed in the arm tissue of small- and medium-sized samples after the frozen period; this decrease was found to be significant ($p < 0.05$) in raw samples. Regarding large-sized octopus, an increasing average DHA content was observed in the arm tissue as a result of the frozen storage; differences were significant ($p < 0.05$) in cooked samples.

For the DPA level, a different trend was detected according to the size of specimens; that is, the viscera tissue showed a decrease ($p < 0.05$) in small- and medium-sized samples and an increase ($p < 0.05$) in large-sized ones. Regarding the mantle tissue, an average increase was seen in most cases. Conversely, an average decrease in the DPA value was detected in the arm tissue of all groups as a result of frozen storage; differences were only significant ($p < 0.05$) in cooked medium-sized samples.

2.3.4. Previous Related Studies Regarding the Effect of Frozen Storage on Cephalopod and Invertebrate Tissues

Frozen storage mostly inhibits microbial development, but fish constituents may undergo different kinds of deteriorative mechanisms such as the formation of aggregates, protein insolubility, mechanical damage, and development of lipid damage [67,68]. Among them, lipid hydrolysis and oxidation, and therefore FA damage, have been recognized as important factors that influence quality changes during the frozen storage of seafood in general [69,70]. Notably, the development of the lipid damage mechanisms has been revealed to be more important when increasing the time and temperature of the storage period [15,71].

Regarding non-edible tissues of cephalopods and invertebrate species in general, previous research focused on the effect of the frozen storage on changes in the FA profile, and lipid damage was considered scarce. A marked development of lipid oxidation was detected in Patagonian squid (*D. gahi*) by-products during the frozen storage (−10 and −18 °C up to 18 months) [72]; in this study, lipid deterioration increased with the time and temperature of storage. Regarding invertebrate species, the evolution of the FA profile of Chinese mitten crab (*Eriocheir sinensis*) hepatopancreas during the frozen storage at different temperatures (−20, −40, and −80 °C) was studied by Fan et al. [73]; an increase in the STFA level and a decrease in the MUFA and PUFA presence were detected with increasing temperature and time of storage.

Previous research accounts for the study of changes in the FA profile in different edible tissues of cephalopod species subjected to frozen storage conditions. Thus, Atayeter and Ercoşkun [54] analyzed the changes produced in the FA profile of the arms and mantle of European squid (*L. vulgaris*) subjected to frozen storage (−20, −40, and −80 °C). By increasing the storage time, a marked increase in the STFA presence could be detected, which was accompanied by slight decreases in the MUFA and PUFA values and remarkable increases in the ω3/ω6 ratio; additionally, marked decreases in the DHA and DPA levels were detected, but no effect on the EPA value could be proved. Gullian-Klan et al. [74] analyzed the changes produced in the FA profile of Mexican four-eyed octopus (*Octopus maya*) muscle during a 5-month storage at −18 °C. As a result, 6.07% and 9.28% decreases in the PUFA value were detected after 3 and 5 months of storage, respectively. Furthermore, the initial PI value (i.e., 3.5) decreased to values of 2.9 and 2.3 after 3 and 5 months, respectively.

3. Materials and Methods

3.1. *Raw Octopus, Sampling, Cooking and Frozen Storage*

Common octopus (*O. vulgaris*) were obtained near the Galician coast (North-West Spain) and supplied by Frigoríficos Rosa de los Vientos S. L. (Marín, Pontevedra, Spain). Three different specimen groups were considered, i.e., Group I (1–2 kg per specimen), Group II (2–3 kg per specimen), and Group III (3–4 kg per specimen). In order to carry out the study, 24 specimens of each group were used. Within each group, separation of viscera, mantle and arm was carried out.

Viscera tissue corresponding to 12 specimens was subjected to analysis (Raw viscera). For this, three independent batches were considered (n = 3; viscera corresponding to 4 specimens per batch). Viscera tissue corresponding to the other 12 specimens was subjected to frozen storage (−18 °C) for 4 months. After this period, viscera tissue was subjected to thawing and subsequent analysis (Frozen viscera). For this, three independent batches were considered (n = 3; viscera corresponding to 4 specimens per batch). In agreement with industrial practice, the viscera tissue was not subjected to the cooking process.

Mantle tissue corresponding to 6 specimens was subjected to analysis (Raw mantle). For this, three independent batches were considered (n = 3; mantles corresponding to 2 specimens per batch). Meantime, mantle tissue corresponding to 6 specimens was subjected to frozen storage (−18 °C) for 4 months. After this period, mantle tissue was subjected to thawing and subsequent analysis (Frozen mantle). For this, three independent batches were considered (n = 3; mantles corresponding to 2 specimens per batch).

Mantle tissue corresponding to the remaining 12 specimens was subjected to cooking (40 min at 90 °C). Afterwards, mantle tissue corresponding to 6 specimens was subjected to analysis (Cooked mantle). For this, three independent batches were taken into account (n = 3; mantles corresponding to 2 specimens per batch). Meantime, mantle tissue corresponding to 6 specimens was subjected to frozen storage (–18 °C) for 4 months. After this period, the mantle tissue was subjected to thawing and subsequent analysis (Cooked-Frozen mantle). For this, three independent batches were considered (n = 3; mantles corresponding to 2 specimens per batch).

For the arm tissue, the sampling procedure carried out was the same as for the mantle tissue. Thus, the following arm samples were obtained: Raw, Frozen, Cooked and Cooked-Frozen. As for the mantle tissue, three independent batches were considered (n = 3; arms corresponding to 2 specimens per batch) in all sample types.

Solvents and chemical reagents used in this study were of reagent grade (Merck, Darmstadt, Germany); otherwise, the supplier is defined.

3.2. Lipid Extraction and FA Analysis of Tissue Samples

The lipid extraction of the different tissues was carried out in agreement with the Bligh and Dyer [75] method. In a first step, this method employs a chloroform/methanol/water (1/2/0.8, *v/v/v*) mixture as an extracting solvent system. Then, the addition of chloroform and water is carried out to the mixture in order to attain a 2/2/1.8 (chloroform/methanol/water, *v/v/v*) solvent ratio so that two phases are formed. The one placed at the bottom is carefully taken and corresponds to the lipid fraction. Results on total lipid yield were calculated as g total lipids·kg^{-1} tissue. Lipid extracts were kept at −40 °C in a nitrogen atmosphere before being used.

Fatty acid methyl esters (FAMEs) were obtained from lipid extracts by employing acetyl chloride in methanol. Then, FAMEs were analyzed by gas chromatography (Perkin-Elmer 8700 chromatograph, Madrid, Spain) [76]. A fused silica capillary column SP-2330 (0.25 mm i.d. × 30 m, Supelco, Inc., Bellefonte, PA, USA) was used. The temperature program was the following: increased from 145 to 190 °C at 1.0 °C·min^{-1} and from 190 °C to 210 °C at 5.0 °C·min^{-1}, then held for 13.5 min at 210 °C. Nitrogen at 10 psig was used as carried gas and a flame ionization detector at 250 °C was used as detector. A programmed temperature vaporizer injector was used in the split mode (150:1), being heated from 45 to 275 °C at 15 °C·min^{-1}.

Identification of FAME peaks was carried out by comparison of the retention times to those of standard mixtures (Qualmix Fish, Larodan, Malmo, Sweden; Supelco 37 Component FAME Mix, Sigma-Aldrich, Laramie, WY, USA). Peak areas were automatically integrated. For quantitative purposes, C19:0 was employed as an internal standard; for that, 100 μL (i.e., 40 μg C19:0) of a 0.4 mg·mL^{-1} solution in toluene were added to each sample before the methylation reaction with acetyl chloride in methanol. Detection and quantification limits were 500 and 1,500 area units, respectively. Quantitative calibration was carried out by means of the above-mentioned Supelco FAME Mix.

The content of each FA was expressed as g·100 g^{-1} of total FAs. Results regarding FA groups (STFAs, MUFAs, PUFAs, total ω3-PUFAs and total ω6-PUFAs) and FA ratios (total ω3-PUFAs/total ω6-PUFAs, PI, and FLQ) were calculated by considering the results obtained in individual FAs. The PI and the FLQ were determined as the following ratios of FA concentrations: DHA+EPA/C16:0 and 100 × (DHA + EPA)/% total FAs, respectively.

3.3. Statistical Analysis

This research was carried out in triplicate (n = 3). For each kind of sample (raw/cooked/frozen, specimen size, and tissue), three biological replicates (three independent batches) were employed. In the case of mantle and arm, each sample analyzed was composed of tissues corresponding to two different specimens. For viscera, each sample analyzed was composed of tissues corresponding to four different specimens.

Data obtained from the FA analysis (FA groups and ratios and single ω3-PUFAs) were subjected to the ANOVA method. For it, one-way ANOVA was applied to investigate differences resulting from the following factors: tissue, cooking process and frozen storage. The effect of each factor was analyzed independently. Statistical comparisons were conducted via PASW Statistics 18 Software for Windows (SPSS Inc., Chicago, IL, USA). The least-squares difference (LSD) test was used for comparison of means. For each FA parameter, the 95% confidence interval was calculated; for this, the standard deviation of each sample and the number of replicates (n = 3) were taken into account.

4. Conclusions

A comparative study of the FA composition of non-edible (viscera) and edible (mantle and arm) tissues of octopus (*O. vulgaris*) was carried out. In agreement with current nutritional recommendations, all kinds of tissues (viscera, mantle, and arm) showed great levels of highly valuable indices related to the lipid fraction (total ω3-PUFAs, EPA, DHA, and DPA values; ω3/ω6 PI, and FLQ ratios). In most cases, the cooking process and the frozen storage led to an average decrease in the PUFA and ω3-PUFA content and to an increase in the STFA presence. In spite of such processing effects, all kinds of specimen sizes considered in the present study maintained such highly nutritional values. This result is considered especially important in the case of the viscera tissue, a substrate that is commonly discarded or employed for obtaining low-value sub-products.

This research contributes to achieve alternative sources for obtaining highly valuable constituents from waste substrates resulting from the seafood commercialization with the aim of providing healthy compounds for the food and pharmaceutical industries and increase the profitability of such by-products. Waste constituents may be used not only as food but also for other high-end applications. The results of this study could serve as a basis for the development of new functional foods enriched with ω3-PUFAs such as EPA and DHA. Such enriched foods would be likely to produce a positive and profitable impact on the health of a wide range of consumers such as infants, pregnant women, or older adults in general.

The use of the current viscera tissue would lead to the ecological upside of reducing seafood industry waste and agree with general commitments for environmental sustainability and circular economy. Further research ought to be carried out on the optimization (i.e., surface-response methodology) of the extracting conditions and considering the different process and response variables of the extraction. The use of an RSM design can provide the possibility of affording a product with, at the same time, a minimum rancidity level and maximum yield of ω3-PUFA compounds. For it, the employment of green technologies such as irradiation (microwave, ultrasound, pulsed electric fields, etc.)-assisted, supercritical fluid, or green solvent extraction ought to be developed to fulfil current ecological requirements and guarantee a high-quality ω3-PUFA extract. Before finding a subsequent use for octopus waste, international requirements regarding safety concerns (presence of heavy metals, aromatic hydrocarbons, etc.) ought to be taken into account.

Author Contributions: Conceptualization, R.P. and S.P.A.; methodology, L.F.-M. and M.T.; data curation, L.F.-M. and M.T.; writing—original draft preparation, S.P.A.; writing—review and editing, R.P. and S.P.A. All authors have read and agreed to the published version of the manuscript.

Funding: This research was funded by the Spanish Ministry of Research and Innovation, grant number PID2023-146042OB-C21 (2024-2027).

Institutional Review Board Statement: Not applicable.

Data Availability Statement: The original contributions presented in the study are included in the article; further inquiries can be directed to the corresponding author.

Acknowledgments: The authors wish to acknowledge Frigoríficos Rosa de los Vientos S. L. (Marín, Pontevedra, Spain) for kindly providing the octopus samples.

Conflicts of Interest: The authors declare no conflicts of interest.

References

1. Blanco, M.; Sotelo, C.G.; Chapela, M.J.; Pérez-Martín, R. Towards sustainable and efficient use of fishery resources: Present and future trends. *Trends Food Sci. Technol.* **2007**, *18*, 29–36. [CrossRef]
2. Tilami, S.K.; Sampels, S. Nutritional value of fish: Lipids, proteins, vitamins, and minerals. *Rev. Fish. Sci.* **2018**, *26*, 242–253.
3. Uauy, R.; Valenzuela, A. Marine oils: The health benefits of n-3 fatty acids. *Nutrition* **2000**, *16*, 680–684. [CrossRef] [PubMed]
4. Komprda, T. Eicosapentaenoic and docosahexaenoic acids as inflammation-modulating and lipid homeostasis influencing nutraceuticals: A review. *J. Funct. Foods* **2012**, *4*, 25–38. [CrossRef]

5. Xu, H.; Turchini, G.M.; Francis, D.S.; Liang, M.; Mock, T.S.; Rombenso, A.; Ai, Q. Are fish what they eat? A fatty acid's perspective. *Prog. Lipid Res.* **2020**, *80*, 101064. [CrossRef]
6. Minihane, A.M.; Armah, C.K.; Miles, E.A.; Madden, J.M.; Clark, A.B.; Caslake, M.J.; Packard, C.J.; Kofler, B.M.; Lietz, G.; Curtis, P.J.; et al. Consumption of fish oil providing amounts of eicosapentaenoic acid and docosahexaenoic acid that can be obtained from the diet reduces blood pressure in adults with systolic hypertension: A retrospective analysis. *J. Nutr.* **2016**, *146*, 516–523. [CrossRef]
7. Schunck, W.; Konkel, A.; Fischer, R.; Weylandt, K. Therapeutic potential of omega-3 fatty acid-derived epoxy eicosanoids in cardiovascular and inflammatory diseases. *Pharm. Ther.* **2018**, *183*, 177–204. [CrossRef]
8. Piclet, G. Le poisson aliment. Composition-Intérêt nutritionnel. *Cah. Nutr. Diét.* **1987**, *XXII*, 317–335.
9. Hayashi, K. Composition and distribution of lipids in different tissues of the arrow squid *Loligo bleekeri*. *Fish. Sci.* **1996**, *62*, 84–87. [CrossRef]
10. Haouas, W.; Zayene, N.; Guerbej, H.; Hammami, M.; Achour, L. Fatty acids distribution in different tissues of wild and reared *Seriola dumerili*. *Int. J. Food Sci. Technol.* **2010**, *45*, 1478–1485. [CrossRef]
11. Thammapat, P.; Raviyan, P.; Siriamornpun, S. Proximate and fatty acid composition of the muscles and viscera of Asian catfish (*Pangasius bocourti*). *Food Chem.* **2010**, *122*, 223–227. [CrossRef]
12. Rosa, R.; Nunes, L.; Sousa Reiset, C. Seasonal changes in the biochemical composition of *Octopus vulgaris* Cuvier, 1797, from three areas of Portuguese coast. *Bull. Mar. Sci.* **2002**, *71*, 739–751.
13. Quispe-Machaca, M.; Guzmán-Rivas, F.A.; Ibáñez, C.M.; Urzúa, A. Intra-individual variability in biochemical constituents and fatty acid composition of adult jumbo squid (*Dosidicus gigas*) in the southeastern Pacific Ocean. *J. Sea Res.* **2021**, *174*, 102082. [CrossRef]
14. Nakamura, Y.N.; Ando, M.; Seoka, M.; Kawasaki, K.I.; Tsukamasa, Y. Changes of proximate and fatty acid compositions of the dorsal and ventral ordinary muscles of the full-cycle cultured Pacific bluefin tuna *Thunnus orientalis* with the growth. *Food Chem.* **2007**, *103*, 234–241. [CrossRef]
15. Aubourg, S.P.; Rey-Mansilla, M.; Sotelo, C. Differential lipid damage in various muscle zones of frozen hake (*Merluccius merluccius*). *Z. Lebensm. Unters. Forsch.* **1999**, *208*, 189–193. [CrossRef]
16. Ruff, N.; Fitzgerald, R.; Cross, T.; Lynch, A.; Kerry, J. Distribution of alpha-tocopherol in fillets of turbot (*Scophthalmus maximus*) and Atlantic halibut (*Hippoglossus hippoglossus*), following dietary alpha-tocopheryl acetate supplementation. *Aquac. Nutr.* **2004**, *10*, 75–81. [CrossRef]
17. Testi, S.; Bonaldo, A.; Gatta, P.; Badiani, A. Nutritional traits of dorsal and ventral fillets from three farmed fish species. *Food Chem.* **2006**, *98*, 104–111. [CrossRef]
18. Álvarez, V.; Medina, I.; Prego, R.; Aubourg, S.P. Lipid and mineral distribution in different zones of farmed and wild blackspot seabream (*Pagellus bogaraveo*). *Eur. J. Lipid Sci. Technol.* **2009**, *111*, 957–966. [CrossRef]
19. Olsen, R.L.; Toppe, J.; Karunasagar, I. Challenges and realistic opportunities in the use of by-products from processing of fish and shellfish. *Trends Food Sci. Technol.* **2014**, *36*, 144–152. [CrossRef]
20. Ezquerra-Brauer, J.M.; Aubourg, S.P. Recent trends for the employment of jumbo squid (*Dosidicus gigas*) by-products as a source of bioactive compounds with nutritional, functional and preservative applications: A review. *Int. J. Food Sci. Technol.* **2019**, *54*, 987–998. [CrossRef]
21. Özyurt, G.; Özkütük, S. Advances in discard and by-product processing. In *Innovative Technologies in Seafood Processing*; Özoğul, Y., Ed.; CRC Press, Taylor and Francis Group: Boca Raton, FL, USA, 2020; pp. 323–350.
22. Rubio-Rodríguez, N.; Beltrán, S.; Jaime, I.; de Diego, S.M.; Sanz, M.T.; Carballido, J.R. Production of omega-3 polyunsaturated fatty acid concentrates: A review. *Innov. Food Sci. Emerg. Technol.* **2010**, *11*, 1–12. [CrossRef]
23. Atef, M.; Ojagh, M. Health benefits and food applications of bioactive compounds from fish byproducts: A review. *J. Funct. Foods* **2017**, *35*, 673–681. [CrossRef]
24. Yuan, Z.; Ye, X.; Hou, Z.; Chen, S. Sustainable utilization of proteins from fish processing by-products: Extraction, biological activities and applications. *Trend Food Sci. Technol.* **2024**, *143*, 104276. [CrossRef]
25. Ferraro, V.; Cruz, I.B.; Jorge, R.F.; Malcata, F.X.; Pintado, M.E.; Castro, P.M.L. Valorisation of natural extracts from marine source focused on marine by-products: A review. *Food Res. Int.* **2010**, *43*, 2221–2223. [CrossRef]
26. Rustad, T.; Storro, I.; Slizyte, R. Possibilities for the utilisation of marine by-products. *Int. J. Food Sci. Technol.* **2011**, *46*, 2001–2014. [CrossRef]
27. Torrinha, A.; Gomes, F.; Oliveira, M.; Cruz, R.; Mendes, E.; Delerue-Matis, C.; Casal, S.; Morais, S. Commercial squids: Characterization, assessment of potential health benefits/risks and discrimination based on mineral, lipid and vitamin E concentrations. *Food Chem. Toxicol.* **2014**, *67*, 44–56. [CrossRef]
28. Singh, A.; Benjakul, S.; Kishimura, H. Characteristics and functional properties of ovary from squid *Loligo formosana*. *J. Aquat. Food Prod. Technol.* **2017**, *26*, 1083–1092. [CrossRef]

29. Zamuz, S.; Bohrer, B.M.; Shariati, M.A.; Rebezov, M.; Kumar, M.; Pateiro, M.; Lorenzo, J.M. Assessing the quality of octopus: From sea to table. *Food Front.* **2023**, *4*, 733–749. [CrossRef]
30. Lee, S.Y.; Hur, S.J. Antihypertensive peptides from animal products, marine organisms, and plants. *Food Chem.* **2017**, *228*, 506–517. [CrossRef]
31. Lin, Y.; Tang, X.; Xua, L.; Wanga, S. Antibacterial properties and possible action mechanism of chelating peptides-zinc nanocomposite against *Escherichia coli*. *Food Cont.* **2019**, *106*, 106675. [CrossRef]
32. Sieiro, M.P.; Aubourg, S.P.; Rocha, F. Seasonal study of the lipid composition in different tissues of the common octopus (*Octopus vulgaris*). *Eur. J. Lipid Sci. Technol.* **2006**, *108*, 479–487. [CrossRef]
33. Méndez, L.; Rodríguez, A.; Aubourg, S.P.; Medina, I. Low-toxicity solvents for the extraction of valuable lipid compounds from octopus (*Octopus vulgaris*) waste. *Foods* **2023**, *12*, 3631. [CrossRef] [PubMed]
34. Aubourg, S.P.; Trigo, M.; Prego, R.; Cobelo-García, A.; Medina, I. Nutritional and healthy value of chemical constituents obtained from Patagonian squid (*Doryteuthis gahi*) by-products captured at different seasons. *Foods* **2021**, *10*, 2144. [CrossRef]
35. Shen, C.; Xie, J.; Xu, X. The components of cuttlefish (*Sepiella maindroni de Rochebruns*) oil. *Food Chem.* **2007**, *102*, 210–214. [CrossRef]
36. Kacem, M.; Sellami, M.; Kammoun, W.; Frikh, F.; Miled, N.; Rebah, F.B. Seasonal variations in proximate and fatty acid composition of viscera of *Sardinella aurita*, *Sarpa salpa*, and *Sepia officinalis* from Tunisia. *J. Aquat. Food Prod. Technol.* **2011**, *20*, 233–246. [CrossRef]
37. Rodríguez, A.; Trigo, M.; Aubourg, S.P.; Medina, I. Optimisation of healthy-lipid content and oxidative stability during oil extraction from squid (*Illex argentinus*) viscera by green processing. *Mar. Drugs* **2021**, *19*, 616. [CrossRef]
38. Ben-Youssef, S.; Selmi, S.; Ezzeddine-Najai, S.; Sadok, S. Total lipids and fatty acids composition of the coastal and the deep-sea common octopus (*Octopus vulgaris*) populations: A comparative study. *Nutr. Health* **2008**, *19*, 195–201. [CrossRef]
39. Özoğul, Y.; Duysak, O.; Özoğul, F.; Özkütük, A.S.; Türeli, C. Seasonal effects in the nutritional quality of the body structural tissue of cephalopods. *Food Chem.* **2008**, *108*, 847–852. [CrossRef]
40. Vaz-Pires, P.; Barbosa, A. Sensory, microbiological, physical and nutritional properties of iced whole common octopus (*Octopus vulgaris*). *Lebensm. Wiss. Technol.* **2004**, *37*, 105–114. [CrossRef]
41. Zlatanos, S.; Laskaridis, K.; Feist, C.; Sagredos, A. Proximate composition, fatty acid analysis and protein digestibility-corrected amino acid score of three Mediterranean cephalopods. *Mol. Nutr. Food Res.* **2006**, *50*, 967–970. [CrossRef]
42. Bonafe, E.; Aguiar, A.; Boroski, M.; Monteiro, J.; Souza, N.; Matsushita, M.; Visentainer, J. Quantification of EPA and DHA in seafood of the south coast of Brazil. *Nutr. Food Sci.* **2011**, *1*, 401–411. [CrossRef]
43. Oliveira, H.; Muniz, J.A.; Bandarra, N.M.; Castanheira, I.; Ribeiro Coelho, I.; Delgado, I.; Gonçalves, S.; Lourenço, H.M.; Motta, C.; Duarte, M.P.; et al. Effects of industrial boiling on the nutritional profile of common octopus (*Octopus vulgaris*). *Foods* **2019**, *8*, 411. [CrossRef]
44. Toyes-Vargas, E.; Robles-Romo, A.; Méndez, L.; Palacios, E.; Civera, R. Changes in fatty acids, sterols, pigments, lipid classes, and heavy metals of cooked or dried meals, compared to fresh marine by-products. *Anim. Feed Sci. Technol.* **2016**, *221*, 195–205. [CrossRef]
45. Biandolino, F.; Portacci, G.; Prato, E. Influence of natural diet on growth and biochemical composition of *Octopus vulgaris* Cuvier, 1797. *Aquac. Intern.* **2010**, *18*, 1163–1175. [CrossRef]
46. Kumari, P.; Kumar, M.; Reddy, C.R.; Jha, B. Algal lipids, fatty acids and sterols. In *Functional Ingredients from Algae for Foods and Nutraceuticals*; Domínguez, H., Ed.; Woodhead Publishing: Cambridge, UK, 2013; pp. 87–134.
47. Simopoulos, A.P. The importance of the ratio of omega-6/omega-3 essential fatty acids. *Biomed. Pharmacother.* **2002**, *56*, 365–379. [CrossRef]
48. Swanson, D.; Block, R.; Mousa, S.A. Omega-3 fatty acids EPA and DHA: Health benefits throughout life. *Adv. Nutr.* **2012**, *3*, 1–7. [CrossRef]
49. Ofosu, F.K.; Daliri, E.B.M.; Lee, B.H.; Yu, X. Current trends and future perspectives on omega-3 fatty acids. *Res. J. Biol.* **2017**, *5*, 11–20.
50. Delplanque, B.; Gibson, R.; Koletzko, B.; Lapillonne, A.; Strandvik, B. Lipid quality in infant nutrition: Current knowledge and future opportunities. *J. Pediatr. Gastroenterol. Nutr.* **2015**, *61*, 8–17. [CrossRef]
51. Li, J.; Yin, H.; Bibus, D.M.; Byelashov, O.A. The role of Omega-3 docosapentaenoic acid in pregnancy and early development. *Eur. J. Lipid Sci. Technol.* **2016**, *118*, 1692–1701. [CrossRef]
52. Byelashov, O.A.; Sinclair, A.J.; Kaur, G. Dietary sources, current intakes, and nutritional role of omega-3 docosapentaenoic acid. *Lipid Technol.* **2015**, *27*, 79–82. [CrossRef]
53. Drouin, G.; Rioux, V.; Legrand, P. The n-3 docosapentaenoic acid (DPA): A new player in the *n*-3 long chain polyunsaturated fatty acid family. *Biochimie* **2019**, *159*, 36–48. [CrossRef] [PubMed]
54. Atayeter, S.; Ercoşkun, H. Chemical composition of European squid and effects of different frozen storage temperatures on oxidative stability and fatty acid composition. *J. Food Sci. Technol.* **2011**, *48*, 83–89. [CrossRef] [PubMed]

55. Sinanoglou, V.J.; Miniadis-Meimaroglou, S. Fatty acid of neutral and polar lipids of (edible) Mediterranean cephalopods. *Food Res. Int.* **1998**, *31*, 467–473. [CrossRef]
56. Augimeri, G.; Galluccio, A.; Caparello, G.; Avolio, E.; la Russa, D.; de Rose, D.; Morelli, C.; Barone, I.; Catalano, S.; Ando, S.; et al. Potential antioxidant and anti-inflammatory properties of serum from healthy adolescents with optimal Mediterranean diet adherence: Findings from DIMENU cross-sectional study. *Antioxidants* **2021**, *10*, 1172. [CrossRef]
57. Tibbetts, S.M.; Patelakis, S.J.J. Apparent digestibility coefficients (ADCs) of intact-cell marine microalgae meal (*Pavlova* sp. 459) for juvenile Atlantic salmon (*Salmo salar* L.). *Aquaculture* **2022**, *546*, 737236. [CrossRef]
58. Rustad, T. Lipid oxidation. In *Handbook of Seafood and Seafood Products Analysis*; Nollet, L., Toldrá, F., Eds.; CRC Press, Taylor & Francis Group: Boca Raton, FL, USA, 2010; pp. 87–95.
59. Senso, L.; Suárez, M.D.; Ruiz-Cara, T.; García-Gallego, M. On the possible effects of harvesting season and chilled storage on the fatty acid profile of the fillet of farmed gilthead sea bream (*Sparus aurata*). *Food Chem.* **2007**, *101*, 298–307. [CrossRef]
60. Łuczyńska, J.; Paszczyk, B. Health risk assessment of heavy metals and lipid quality indexes in freshwater fish from Lakes of Warmia and Mazury Region, Poland. *Int. J. Environ. Res. Public Health* **2019**, *16*, 3780. [CrossRef]
61. Bazarsadueva, S.V.; Radnaeva, L.D.; Shiretorova, V.G.; Dylenova, E.P. The comparison of fatty acid composition and lipid quality indices of roach, perch, and pike of Lake Gusinoe (Western Transbaikalia). *Int. J. Environ. Res. Public Health* **2021**, *18*, 9032. [CrossRef]
62. Baldwin, D.E. Sous vide cooking: A review. *Int. J. Gastr. Food Sci.* **2012**, *1*, 15–30. [CrossRef]
63. Ling, B.; Tang, J.; Kong, F.; Mitcham, E.J.; Wang, S. Kinetics of food quality changes during thermal processing: A review. *Food Bioprocess Technol.* **2015**, *8*, 343–358. [CrossRef]
64. Undeland, I. Oxidative stability of seafood. In *Oxidative Stability and Shelf Life of Foods Containing Oils and Fats*; Hu, M., Jacobsen, C., Eds.; Academic Press and AOCS Press: London, UK, 2016; pp. 391–460.
65. Vieira, S.; Zhang, G.; Decker, E. Biological implications of lipid oxidation products. *J. Am. Oil Chem. Soc.* **2017**, *94*, 339–351. [CrossRef]
66. Czech, A.; Grela, E.R.; Ognik, K. Effect of frying on nutrients content and fatty acid composition of muscles of selected freezing seafoods. *J. Food Nutr. Res.* **2015**, *3*, 9–14. [CrossRef]
67. Sikorski, Z.E.; Kolakowski, E. Endogenous enzyme activity and seafood quality: Influence of chilling, freezing, and other environmental factors. In *Seafood Enzymes. Utilization and Influence on Postharvest Seafood Quality*; Haard, N.F., Simpson, B.K., Eds.; Marcel Dekker: New York, NY, USA, 2000; pp. 451–487.
68. Kolakowska, A. Lipid oxidation in food systems. In *Chemical and Functional Properties of Food Lipids*; Sikorski, Z., Kolakowska, A., Eds.; CRC Press: London, UK, 2003; pp. 133–165.
69. Aubourg, S.P.; Piñeiro, C.; González, M.J. Quality loss related to rancidity development during frozen storage of horse mackerel (*Trachurus trachurus*). *J. Am. Oil Chem. Soc.* **2004**, *81*, 671–678. [CrossRef]
70. Skåra, T.; Stormo, S.K.; Nilsen, H.A. Advances in Freezing and Thawing. In *Innovative Technologies in Seafood Processing*; Özoğul, Y., Ed.; CRC Press, Taylor and Francis Group: Boca Raton, FL, USA, 2020; pp. 27–44.
71. Romotowska, P.; Karlsdóttir, M.; Gudjunsdóttir, M.; Kristinsson, H.; Arason, S. Influence of feeding state and frozen storage temperature on the lipid stability of Atlantic mackerel (*Scomber scombrus*). *Int. J. Food Sci. Technol.* **2016**, *51*, 1711–1720. [CrossRef]
72. Aubourg, S.P.; Trigo, M.; González, M.J.; Lois, S.; Medina, I. Evolution of lipid damage and volatile amine content in Patagonian squid (*Doryteuthis gahi*) by-products during frozen storage. *Int. J. Food Sci. Technol.* **2022**, *57*, 5409–5418. [CrossRef]
73. Fan, L.; Xian, C.; Tang, S.; Ding, W.; Xu, C.H.; Wang, X.C. Effect of frozen storage temperature on lipid stability of hepatopancreas of *Eriocheir sinensis*. *LWT-Food Sci. Technol.* **2022**, *154*, 112513. [CrossRef]
74. Gullian-Klanian, M.; Terrats-Preciat, M.; Pech-Jiménez, E.C.; Cutz de Ocampo, J. Effect of frozen storage on protein denaturation and fatty acids profile of the red octopus (*Octopus maya*). *J. Food Proc. Preserv.* **2017**, *41*, e13072. [CrossRef]
75. Bligh, E.; Dyer, W. A rapid method of total extraction and purification. *Can. J. Biochem. Physiol.* **1959**, *37*, 911–917. [CrossRef]
76. Aubourg, S.P.; Medina, I.; Pérez-Martín, R. Polyunsaturated fatty acids in tuna phospholipids: Distribution in the *sn-2* location and changes during cooking. *J. Agric. Food Chem.* **1996**, *44*, 585–589. [CrossRef]

Disclaimer/Publisher's Note: The statements, opinions and data contained in all publications are solely those of the individual author(s) and contributor(s) and not of MDPI and/or the editor(s). MDPI and/or the editor(s) disclaim responsibility for any injury to people or property resulting from any ideas, methods, instructions or products referred to in the content.

marine drugs

Article

Polyunsaturated Fatty Acids Improved Long Term Prognosis by Reducing Oxidative Stress, Inflammation, and Endothelial Dysfunction in Acute Coronary Syndromes

Alexandru Covaciu [1,2], Theodora Benedek [1], Elena Bobescu [2,3,*], Horatiu Rus [2,3], Valentina Benza [2,3], Luigi Geo Marceanu [2], Simona Grigorescu [4] and Christian Gabriel Strempel [5]

1 M3 Department–Clinical and Medical-Surgical Disciplines, "George Emil Palade" University of Medicine, Pharmacy, Sciences and Technology from Târgu Mureș, 540139 Târgu Mureș, Romania; alexandru.covaciu92@gmail.com (A.C.); theodora.benedek@gmail.com (T.B.)
2 Department of Medical and Surgical Specialties, Faculty of Medicine, Transilvania University of Brasov, 500019 Brasov, Romania; horatiuenache@gmail.com (H.R.); valentina_stamate@yahoo.com (V.B.); marceanu@gmail.com (L.G.M.)
3 Department of Cardiology, Clinical County Emergency Hospital Brasov, 500326 Brasov, Romania
4 Department of Fundamental, Prophylactic and Clinical Disciplines, Faculty of Medicine, Transilvania University of Brasov, 500019 Brasov, Romania; simo.grigorescu@yahoo.com
5 Department of Finance, Accounting and Economic Theory, Faculty of Economic Sciences and Business Administration, Transilvania University of Brasov, 500036 Brasov, Romania; christianstrempel27@gmail.com
* Correspondence: elena.bobescu@gmail.com; Tel.: +40-722705400

Abstract: Background: Oxidative stress, inflammation, and endothelial dysfunction are important processes in the progression of atherosclerosis and the occurrence of acute coronary syndromes (ACSs). Omega-3 polyunsaturated fatty acids (Omega-3 PUFAs) are present in marine organisms and have the capacity to reduce all these processes and, at the same time, the progression of atherosclerosis and the emergence of ACSs. Aim: To evaluate the role of Omega-3 PUFAs therapy on parameters of oxidative stress, inflammatory syndrome, endothelial dysfunction, and long-term prognosis in acute coronary syndromes. Methods: One thousand one hundred forty patients were admitted to Clinic County Emergency Hospital Brasov with ACS and were enrolled in a prospective study. The study was divided into four groups related to the type of ACS and treatment with Omega-3 PUFAs added to the optimal medical therapy (OMT). The effect of Omega-3 PUFAs therapy associated with the OMT was determined by measuring the dynamics of the following parameters: (a) oxidative stress—total antioxidant status (TAS), oxidated low density lipoprotein cholesterol antibodies (Ab anti-ox-LDL), IgG anti-Myeloperoxidase antibodies (IgG type Ab anti-MPO); (b) inflammatory syndrome—C-reactive protein and fibrinogen; (c) endothelial dysfunction—flow mediated dilation (FMD) and von Willebrand factor (vWf) activity, from baseline to 6 months of follow-up. Clinical events followed at 5 years were cardiovascular and sudden death, Non-ST and ST segment elevation ACS, in stent thrombosis and restenosis, stroke, readmission in hospital for ACS and for heart failure. Results: In ACS groups, treatment with Omega-3 PUFAs added to the OMT significantly decreased the parameters of oxidative stress, inflammatory syndrome, and endothelial dysfunction at 6 months of follow-up. Regarding the clinical events, a significant reduction in the risk of cardiovascular and sudden death and a decreased incidence of Non-ST and ST segment elevation ACS, in-stent restenosis, readmission for ACS and heart failure, was observed in Omega-3 PUFA-treated groups in comparison to control groups. Conclusions: In acute coronary syndromes, therapy with Omega-3 PUFAs added to the OMT resulted in a significant decrease of parameters of oxidative stress, inflammation, and endothelial dysfunction at 6 months and also a significant improvement in the long-term prognosis.

 https://doi.org/10.3390/md23040154

Keywords: omega-3 polyunsaturated fatty acids; oxidative stress; inflammation; endothelial dysfunction; acute coronary syndrome

1. Introduction

Oxidative stress, inflammation, and endothelial dysfunction are involved in the progression of atherosclerosis and the appearance of acute coronary syndrome (ACS). Oxidative stress is the consequence of the imbalance between antioxidants and reactive oxygen species (ROS). A high concentration of ROS will cause damage to the protein and lipid structure of the arterial wall and, at the same time, DNA degradation [1–3].

The most important actions of ROS in acute coronary syndromes are an increase in the synthesis of inflammatory cytokines, thereby participating in the oxidation of LDL (a major determinant of atherogenesis); mitochondrial dysfunction leading to overproduction of ROS, which promote acute or chronic low grade inflammation, and at the same time, determine proliferation, apoptosis, and progression of atherosclerotic plaques. ROS impede the competition between polyunsaturated fatty acids (Omega-3 PUFAs) and arachidonic acid in the biosynthesis of pro-inflammatory mediators, resulting in the dysfunction of seleno-proteins, which can no longer exert their protective, antioxidant effects in the coronary arteries. It leads to the production of nitro-tyrosine, which decreases the bioavailability of nitric oxide (NO) [4–6].

Compared to native LDL, oxidated LDL is more avidly taken up by macrophages, giving rise to foam cells, forming a larger lipid core and increasing stress in the fibrous cap, which will increase the risk of plaque rupture. Hypochlorous acid (HOCl) acts with the electronic substrates of apolipoprotein B (lysine and tyrosine residues), forming 3-chloro-tyrosine myeloperoxidase, which is found in large quantities in the atherosclerotic (ATS) plaque. Moreover, the latter can generate a series of secondary products that can be involved in reactions resulting in oxidated LDL (tyrosine radicals, p-hydroxy-phenyl-acetyl-aldehyde, unsaturated glyceraldehyde, 2-hydroxypropanal, acrolein). Omega-3 polyunsaturated fatty acids (Omega-3 PUFAs) are a molecular class present in marine extracts with the capacity to reduce inflammation and oxidative stress values in the ATS plaque [7–9].

The main aim of this study was to evaluate the role of Omega-3 polyunsaturated fatty acids therapy on parameters of oxidative stress, inflammatory syndrome, and endothelial dysfunction on the long-term prognosis of acute coronary syndrome.

2. Results

2.1. Cardiovascular Risk Factors, Demographic Characteristics, and Biomarkers at Baseline

At the baseline, the cardiovascular risk factors, demographic characteristics, and the biomarkers for dyslipidemia and myocardial necrosis were analyzed (Table 1). There were no registered significant differences between the study groups. The comparison was made between groups with the same type of ACS, NSTE-ACS or STE-ACS.

Table 1. Cardiovascular risk factors, demographic characteristics, and biomarkers at baseline.

	NSTE-ACS Omega-3 PUFAs		**NSTE-ACS**			**STE-ACS Omega-3 PUFAs**		**STE-ACS**		
Total	**N = 283 Patients**	%	**N = 287 Patients**	%	*p*	**N = 293 Patients**	%	**N = 277 Patients**	%	*p*
Age > 65	113	39.93%	112	39.02%	0.825093	112	38.23%	106	38.27%	0.991793
Male	151	53.36%	153	53.31%	0.991068	155	52.90%	149	53.79%	0.831501
Smokers	127	44.88%	128	44.60%	0.946975	132	45.05%	123	44.40%	0.876634

Table 1. *Cont.*

Total	NSTE-ACS Omega-3 PUFAs N = 283 Patients	%	NSTE-ACS N = 287 Patients	%	p	STE-ACS Omega-3 PUFAs N = 293 Patients	%	STE-ACS N = 277 Patients	%	p
Arterial Hypertension	188	66.43%	188	65.51%	0.815565	195	66.55%	181	65.34%	0.76059
Diabetes Mellitus	112	39.58%	111	38.68%	0.82576	114	38.91%	109	39.35%	0.913874
BMI >25 kg/m^2	186	65.72%	184	64.11%	0.686638	194	66.21%	176	63.54%	0.503808
Total Cholesterol > 200 mg/dL	194	68.55%	197	68.64%	0.981558	202	68.94%	189	68.23%	0.854972
LDL Cholesterol > 55 mg/dL	200	70.67%	197	68.64%	0.598114	200	68.26%	195	70.40%	0.580246
HDL Cholesterol < 40 mg/dL	159	56.18%	158	55.05%	0.785738	166	56.66%	151	54.51%	0.60683
Triglycerides > 150 mg/dL	195	68.90%	129	44.95%	0.851756	135	46.08%	119	42.96%	0.454586
Troponin T > 0.1 ng/dL	278	98.23%	280	97.56%	0.576182	290	98.98%	275	99.28%	0.693338
CK-MB > 24 U/L	270	95.40%	272	94.77%	0.726689	280	95.56%	266	96.03%	0.180419

OMT = optimal medical therapy; NSTE-ACS Omega-3 PUFAs = group of patients with Non-ST elevation acute coronary syndromes treated with Omega-3 fatty acids (Omacor®) in addition to OMT; NSTE-ACS = group of patients with Non-ST elevation acute coronary syndromes with OMT without Omega-3 fatty acids (Omacor®); STE-ACS Omega-3 PUFAs = group of patients with ST elevation acute coronary syndromes treated with Omega-3 fatty acids (Omacor®) in addition to OMT; STE-ACS = group of patients with ST elevation acute coronary syndromes with OMT without Omega-3 fatty acids (Omacor®); LDL Cholesterol—Low Density Lipoprotein Cholesterol; HDL Cholesterol—High Density Lipoprotein Cholesterol; CK-MB—Creatine kinase isoenzymes MB.

2.2. Pharmacologic and Non-Pharmacologic Treatment

Optimal Medical Therapy (OMT), interventional, and surgical revascularization were defined by ACC/AHA and ESC guidelines. There were no significant differences between NSTE-ACS groups with or without Omega-3 fatty acids (Omacor®) and also in STE-ACS groups with or without Omega-3 fatty acids (Omacor®) in terms of treatment strategy (Table 2).

Table 2. Pharmacologic and non-pharmacologic treatment.

Total	NSTE-ACS Omega-3 PUFAs N = 283 Patients	%	NSTE-ACS N = 287 Patients	%	p	STE-ACS Omega-3 PUFAs N = 293 Patients	%	STE-ACS N = 277 Patients	%	p
Aspirin	283	100%	287	100%	1.0000	293	100%	277	100%	1.0000
Enoxaparin	283	100%	287	100%	1.0000	293	100%	277	100%	1.0000
Clopidogrel	283	100%	287	100%	1.0000	293	100%	277	100%	1.0000
ACEI	188	66.43%	188	65.51%	0.815565	195	66.55%	195	70.40%	0.323716
Beta-blockers	232	81.98%	232	80.84%	0.725932	141	48.12%	141	50.90%	0.507068
Calcium blockers	19	6.71%	25	8.71%	0.371765	21	7.17%	23	8.30%	0.611526
Statins	210	74.20%	210	73.17%	0.779203	218	74.40%	202	72.92%	0.688671
Intravenous nitroglycerin	90	31.80%	99	34.49%	0.494768	93	31.74%	46	16.61%	0.459779
Long-acting nitrates	126	44.52%	123	42.86%	0.688485	123	41.98%	117	42.24%	0.950137
OMT only	119	42.05%	124	43.21%	0.780202	123	41.98%	120	43.32%	0.746125
Interventional revascularization + OMT	128	45.23%	129	44.95%	0.946073	132	45.05%	125	45.13%	0.98562
Surgical revascularization + OMT	36	12.72%	34	11.85%	0.75054	38	12.97%	32	11.55%	0.606453

OMT = optimal medical therapy; NSTE-ACS Omega-3 PUFAs = group of patients with Non-ST elevation acute coronary syndromes treated with Omega-3 fatty acids (Omacor®) in addition to OMT; NSTE-ACS = group of patients with Non-ST elevation acute coronary syndromes with OMT without Omega-3 fatty acids (Omacor®); STE-ACS Omega-3 PUFAs = group of patients with ST elevation acute coronary syndromes treated with Omega-3 fatty acids (Omacor®) in addition to OMT; STE-ACS = group of patients with ST elevation acute coronary syndromes with OMT without Omega-3 fatty acids (Omacor®); ACEI = Angiotensin Converting Enzyme Inhibitors.

2.3. Results Regarding Oxidative Stress

Low values of total antioxidant status (TAS) at baseline seem to have an almost equal and high incidence in NSTE-ACS Omega-3 PUFAs (64.66%) and NSTE-ACS (63.07%), and in STE-ACS Omega-3 PUFAs (63.82%) and STE-ACS (63.90%). Also, the second markers of oxidative stress, respectively, high titers of oxidized LDL cholesterol antibodies (anti-Ox-LDL antibodies) registered similar incidences in NSTE-ACS Omega-3 PUFAs (71.02%) and NSTE-ACS (67.25%), and also in STE-ACS Omega-3 PUFAs (70.31%) and STE-ACS (67.87%). High titers of Myeloperoxidase (MPO) antibodies IgG type (Ab anti-MPO IgG), the third oxidative stress biomarker, has lower sensitivity in all the groups at baseline (Table 3).

Table 3. Oxidative stress biomarkers.

	NSTE-ACS Omega-3 PUFAs		NSTE-ACS			STE-ACS Omega-3 PUFAs		STE-ACS		
Total	N = 283 Patients	%	N = 287 Patients	%	p	N = 293 Patients	%	N = 277 Patients	%	p
Baseline										
TAS < 1.3 mmol/L	183	64.66%	181	63.07%	0.691298	187	63.82%	177	63.90%	0.984861
Ab anti-Ox LDL cholesterol >150 UI/L	201	71.02%	193	67.25%	0.32906	206	70.31%	188	67.87%	0.529031
Ab anti-MPO IgG >20 U	67	23.67%	57	19.86%	0.269773	71	24.23%	63	22.74%	0.675342
At 6 months							0.00%			
TAS < 1.3 mmol/L	59	20.85%	94	32.75%	**0.001342**	66	22.53%	98	35.38%	**0.000704**
Ab anti-Ox LDL cholesterol >150 UI/L	32	11.31%	54	18.82%	**0.012283**	31	10.58%	60	21.66%	**0.000306**
Ab anti-MPO IgG >20 U	26	9.19%	39	13.59%	0.098328	29	9.90%	36	13.00%	0.2447

OMT = optimal medical therapy; NSTE-ACS Omega-3 PUFAs = group of patients with Non-ST elevation acute coronary syndromes treated with Omega-3 fatty acids (Omacor®) in addition to OMT; NSTE-ACS = group of patients with Non-ST elevation acute coronary syndromes with OMT without Omega-3 fatty acids (Omacor®); STE-ACS Omega-3 PUFAs = group of patients with ST elevation acute coronary syndromes treated with Omega-3 fatty acids (Omacor®) in addition to OMT; STE-ACS = group of patients with ST elevation acute coronary syndromes with OMT without Omega-3 fatty acids (Omacor®); TAS = total antioxidant status; Ab anti-Ox LDL cholesterol = antibodies anti-Oxidated Low Density Lipoprotein Cholesterol; Ac anti-MPO IgG = Antibodies anti-Myeloperoxidase Immunoglobulin G type. All p values < 0.05 that are considered statistically significant are bolded.

At 6 months after the acute event, high serum levels of oxidative stress biomarkers registered a low incidence in all groups, which once again confirms the correlations between high levels of oxidative stress and acute coronary ischemic events. Moreover, the oxidative stress persisted at 6 months in a significantly lower percentage of patients in the groups treated with Omega-3 PUFAs compared to the groups that did not receive this therapy as follows: Low values of total antioxidant status (TAS) in NSTE-ACS Omega-3 PUFAs at 20.85% and NSTE-ACS at 32.75% ($p < 0.01$) and also in STE-ACS Omega-3 PUFAs at 22.53% and STE-ACS at 35.38% ($p < 0.001$). Similar results about the second oxidative stress biomarker were obtained: high titers of anti-Ox LDL antibodies in NSTE-ACS Omega-3 PUFAs at 11,31% and NSTE-ACS at 18.82% ($p < 0.05$) and also in STE-ACS Omega-3 PUFAs at 10.58% and STE-ACS at 21.66% ($p < 0.05$). The third biomarker of oxidative stress, Ab anti-MPO IgG, registered lower titers in Omega-3 PUFAs groups but not significantly lower, mostly because of low titers at baseline (Table 3).

2.4. Results Regarding Inflammation

At baseline, a high incidence of inflammation was registered in all NSTE-ACS and STE-ACS groups; high serum levels of C-reactive protein (CRP) > 0.5 mg/dL were obtained in NSTE-ACS Omega-3 PUFAs (56.89%) and NSTE-ACS (55.05%) and also in STE-ACS Omega-3 PUFAs (55.97%) and STE-ACS (56.92%), and high plasma values of fibrinogen > 400 mg/dL were obtained in NSTE-ACS Omega-3 PUFAs (48.41%) and NSTE-ACS (49.83%) and also in STE-ACS Omega-3 PUFAs (51.19%) and STE-ACS (50.54%). At baseline, a high incidence of the inflammatory syndrome can be observed in more than half of the patients with acute coronary syndrome, if we refer to the increased serum level of C-reactive protein, more sensitive than fibrinogen, with increased plasma values in almost half of the patients (Table 4).

Table 4. Biomarkers of inflammation.

	NSTE-ACS Omega-3 PUFAs		NSTE-ACS			STE-ACS Omega-3 PUFAs		STE-ACS		
Total	N = 283 Patients	%	N = 287 Patients	%	*p*	N = 293 Patients	%	N = 277 Patients	%	*p*
Baseline										
C-reactive protein > 0.5 mg/dL	161	56.89%	158	55.05%	0.658483	164	55.97%	156	56.32%	0.933882
Fibrinogen > 400 mg/dL	137	48.41%	143	49.83%	0.735701	150	51.19%	140	50.54%	0.876136
At 6 months										
C-reactive protein > 0.5 mg/dL	64	22.61%	91	31.71%	**0.014716**	53	18.09%	82	29.60%	**0.001231**
Fibrinogen > 400 mg/dL	31	10.95%	51	17.77%	**0.020429**	40	13.65%	72	25.99%	**0.00021**

OMT = optimal medical therapy; NSTE-ACS Omega-3 PUFAs = group of patients with Non-ST elevation acute coronary syndromes treated with Omega-3 fatty acids (Omacor®) in addition to OMT; NSTE-ACS = group of patients with Non-ST elevation acute coronary syndromes with OMT without Omega-3 fatty acids (Omacor®); STE-ACS Omega-3 PUFAs = group of patients with ST elevation acute coronary syndromes treated with Omega-3 fatty acids (Omacor®) in addition to OMT; STE-ACS = group of patients with ST elevation acute coronary syndromes with OMT without Omega-3 fatty acids (Omacor®). All p values < 0.05 that are considered statistically significant are bolded.

At 6 months of follow-up, inflammation was registered as having an important reduction in all groups. Inflammation persisted at 6 months in a significantly lower percentage of patients in the groups treated with Omega-3 PUFAs compared to the control groups as follows: High serum levels of CRP in NSTE-ACS Omega-3 PUFAs at 22.61% and NSTE-ACS at 31.71% ($p < 0.05$) and also in STE-ACS Omega-3 PUFAs at 18.09% and STE-ACS at 29,60% ($p < 0.01$). Similar results about high plasma levels of fibrinogen were obtained in NSTE-ACS Omega-3 PUFAs at 10.95% and NSTE-ACS at 17.77% ($p < 0.05$) and also in STE-ACS Omega-3 PUFAs at 13.65% and STE-ACS at 25.99% ($p < 0.001$) (Table 4).

2.5. Results Regarding Endothelial Dysfunction

At baseline, in all ACS patients, endothelial dysfunction parameters registered an increased incidence as follows: Low value of flow-mediated dilation (FMD) < 4.5%—in NSTE-ACS Omega-3 PUFAs (53%) and NSTE-ACS (50.87%) and also in STE-ACS Omega-3 PUFAs (52.56%) and STE-ACS (51.26%); high level of von Willebrand factor activity (vWf activity) > 169.7%—in NSTE-ACS Omega-3 PUFAs (55.12%) and NSTE-ACS (52.61%) and also in STE-ACS Omega-3 PUFAs (54.61%) and STE-ACS (54.51%). All of these data illustrate the pivotal role of endothelial dysfunction in triggering the acute coronary ischemic event (Table 5).

After 6 months of follow-up, the parameters of endothelial dysfunction were more reduced in Omega-3 PUFA-treated groups and registered a significantly lower percentage of patient with endothelial dysfunction in these groups as follows: Low value of flow-mediated dilation (FMD) in NSTE-ACS Omega-3 PUFAs at 19.79% and NSTE-ACS at 29.27% ($p < 0.05$) and also in STE-ACS Omega-3 PUFAs at 14.68% and STE-ACS at 27.80% ($p < 0.001$). Similar results about high plasma level of vWf activity were obtained in NSTE-ACS Omega-3 PUFAs at 18.73% and NSTE-ACS at 28.22% ($p < 0.05$) and also in STE-ACS Omega-3 PUFAs at 15.70% and STE-ACS at 29.60% ($p < 0.01$) (Table 5).

Table 5. Endothelial dysfunction parameters.

	NSTE-ACS Omega-3 PUFAs		NSTE-ACS			STE-ACS Omega-3 PUFAs		STE-ACS		
Total	N = 283 Patients	%	N = 287 Patients	%	*p*	N = 293 Patients	%	N = 277 Patients	%	*p*
Baseline										
Flow-Mediated Vasodilatation (FMV) < 4.5%	150	53.00%	146	50.87%	0.61043	154	52.56%	142	51.26%	0.756888
Von Willebrand factor (vWf) activity > 169.7%	156	55.12%	151	52.61%	0.54779	160	54.61%	151	54.51%	0.98186
At 6 months										
Flow-Mediated Dilation (FMD) < 4.5%	56	19.79%	84	29.27%	**0.008563**	43	14.68%	77	27.80%	**0.000123**
Von Willebrand factor (vWf) activity > 169.7%	53	18.73%	81	28.22%	**0.007521**	46	15.70%	82	29.60%	**0.00007**

OMT = optimal medical therapy; NSTE-ACS Omega-3 PUFAs = group of patients with Non-ST elevation acute coronary syndromes treated with Omega-3 fatty acids (Omacor®) in addition to OMT; NSTE-ACS = group of patients with Non-ST elevation acute coronary syndromes with OMT without Omega-3 fatty acids (Omacor®); STE-ACS Omega-3 PUFAs = group of patients with ST elevation acute coronary syndromes treated with Omega-3 fatty acids (Omacor®) in addition to OMT; STE-ACS = group of patients with ST elevation acute coronary syndromes with OMT without Omega-3 fatty acids (Omacor®). All *p* values < 0.05 that are considered statistically significant are bolded.

2.6. Long-Term Clinical Results

The results regarding patient monitoring and the occurrence of cardiovascular event components individually evaluated—cardiovascular death, sudden death, NST/ST elevation acute coronary syndrome (NSTE/STE-ACS), in-stent thrombosis, in-stent restenosis, stroke, readmission for ACS, and readmission for heart failure—were reported at 5 years of follow-up. In groups with Omega-3 PUFAs added to the optimal medical therapy, with or without interventional and/or surgery revascularization, a significantly reduced incidence of most cardiovascular events was observed. The risk of cardiovascular death was significantly lower at 6.71% ($p < 0.05$) in the NSTE-ACS Omega-3 PUFAs group in comparison with 12.89% in the NSTE-ACS group, and also significantly reduced to 7.17% ($p < 0.01$) in STE-ACS Omega-3 PUFAs versus 15.88% in STE-ACS (Table 6).

Similar results about sudden death were significantly lower at 3.89% ($p < 0.05$) in the NSTE-ACS Omega-3 PUFAs group versus 8.71% in the NSTE-ACS group and also 4.10% ($p < 0.01$) in the STE-ACS Omega-3 PUFAs group versus 10.11% in the NSTE-ACS group. It can be observed that the results are almost similar regarding cardiovascular and sudden death as a confirmation of a bad prognosis of ACS, even in NSTE or STE (Table 6).

Regarding the occurrence of reinfarction (recurrence of NSTE/STE-ACS), the reduction of oxidative stress, inflammatory syndrome, and endothelial dysfunction was associated with a significantly lower risk of these coronary ischemic events in the groups treated with Omega-3 PUFAs compared to the conventionally treated groups (Table 6).

In-stent acute thrombosis, a trend has been registered regarding the reduction of risk in the Omega-3 PUFAs groups in comparison with the control groups, without reaching statistical significance (Table 6).

Occurrence of in-stent restenosis—the intravascular event highly dependent on the activity of the atheroma plaque—was shown to occur in a significantly reduced percentage in patients treated with Omega-3 PUFAs compared to control groups that did not receive this treatment, proving the anti-inflammatory and antioxidant effects of Omega-3 PUFAs at 2.47% ($p < 0.00001$) in the NSTE-ACS Omega-3 PUFAs group versus 13.24% in the NSTE-ACS group and also 3.07% ($p < 0.00001$) in the STE-ACS Omega-3 PUFAs group versus 14.80% in the NSTE-ACS group (Table 6).

Table 6. Clinical results at 5 years.

	NSTE-ACS Omega-3 PUFAs		NSTE-ACS			STE-ACS Omega-3 PUFAs		STE-ACS		
Total	N = 283 Patients	%	N = 287 Patients	%	p	N = 293 Patients	%	N = 277 Patients	%	p
Cardiovascular death	19	6.71%	37	12.89%	**0.013221**	21	7.17%	44	15.88%	**0.001066**
Sudden death	11	3.89%	25	8.71%	**0.01722**	12	4.10%	28	10.11%	**0.004973**
Recurrence of NSTE/STE-ACS	32	11.31%	58	20.21%	**0.003568**	39	13.31%	67	24.19%	**0.00085**
In-stent acute thrombosis	6	2.12%	12	4.18%	0.159467	8	2.73%	14	5.05%	0.150016
In-stent restenosis	7	2.47%	38	13.24%	**<0.00001**	9	3.07%	41	14.80%	**<0.00001**
Stroke	6	2.12%	17	5.92%	**0.02105**	8	2.73%	15	5.42%	0.103518
Readmission for ACS	46	16.25%	75	26.13%	**0.003932**	49	16.72%	84	30.32%	**0.000124**
Readmission for heart failure	39	13.78%	67	23.34%	**0.003343**	41	13.99%	71	25.63%	**0.000474**
Loss from follow-up	10	3.53%	10	3.48%	0.974512	9	3.07%	11	3.97%	0.559693

OMT = optimal medical therapy; NSTE-ACS Omega-3 PUFAs = group of patients with Non-ST elevation acute coronary syndromes treated with Omega-3 fatty acids (Omacor®) in addition to OMT; NSTE-ACS = group of patients with Non-ST elevation acute coronary syndromes with OMT without Omega-3 fatty acids (Omacor®); STE-ACS Omega-3 PUFAs = group of patients with ST elevation acute coronary syndromes treated with Omega-3 fatty acids (Omacor®) in addition to OMT; STE-ACS = group of patients with n ST elevation acute coronary syndromes with OMT without Omega-3 fatty acids (Omacor®). All *p* values < 0.05 that are considered statistically significant are bolded.

The risk of stroke was significantly lower in 2.12% ($p < 0.05$) only in the NSTE-ACS Omega-3 PUFAs group in comparison with 5.92% in the NSTE-ACS group. There was also a trend regarding the reduction of stroke risk in the STE-ACS Omega-3 PUFAs group in comparison with the STE-ACS group, without reaching statistical significance (Table 6).

Regarding the soft endpoints represented by readmission for ACS and readmission for heart failure, the incidence of both of these were significantly lower in groups treated with Omega-3 PUFAs in comparison with control groups, as can be seen in Table 6.

Loss from follow-up was registered in similar proportions—under 4%—in all studied groups (Table 6).

Multiple linear regression analysis for predictors of cardiovascular death, sudden death, recurrence of NSTE/STE-ACS, in-stent restenosis, stroke, readmission for ACS, and heart failure in acute coronary syndrome patients was conducted. Multiple linear regression analysis was used to adjust for potential confounders, including risk factors of age > 65 years, male gender, diabetes mellitus, smokers, arterial hypertension, and body mass index > 25 kg/m^2. Omega-3 PUFA therapy remained independently associated with a lower risk of cardiovascular death in patients with acute coronary syndromes included in the present study (Table 7).

Table 7. The results of multiple linear regression analysis for predictors of cardiovascular death.

Predictor Variable	Coefficient beta (ß)	95% Confidence Interval (CI)	Standard Error	*p*-Value
Intercept	**0.161**	0.010 to 0.303	**0.069**	**0.02**
Omega-3 PUFA Therapy	**−0.112**	−0.216 to −0.025	**0.059**	**0.03**
Age > 65	0.211	0.091 to 0.332	0.052	**0.005**
Male	0.069	0.058 to 0.175	0.042	**0.02**
Smokers	0.108	0.026 to 0.189	0.042	**0.01**
Arterial Hypertension	0.154	0.061 to 0.232	0.045	**0.001**
Diabetes Mellitus	0.114	0.034 to 0.188	0.045	**0.009**
BMI > 25 kg/m^2	0.076	−0.158 to −0.060	0.051	0.05

BMI = Body Mass Index; Omega-3 PUFAs = Omega-3 fatty acids (Omacor®). All *p* values < 0.05 that are considered statistically significant are bolded.

Similar results were obtained regarding the other long term endpoints: Omega-3 PUFAs treatment remained independently associated also with a lower risk of sudden death ($p = 0.04$), recurrence of NSTE/STE-ACS ($p = 0.009$), in-stent restenosis ($p = 0.002$), readmission for ACS ($p = 0.001$) and heart failure ($p = 0.004$) in patients with acute coronary syndromes included in the present study. For in-stent acute thrombosis and for stroke, Omega-3 PUFA therapy is not independently associated with lower risk in our study.

3. Discussion

The results of the present study regarding the significant decrease at 6 months of the levels of oxidative stress, inflammation, and endothelial dysfunction parameters in patients with Non-ST elevation ACS (NSTE-ACS) and ST elevation ACS (STE-ACS) treated with Omega-3 PUFAs added to OMT were in accordance with most of the data from the literature about the effectiveness of treatment with Omega-3 PUFAs in cardiovascular disease. Epidemiological and therapeutic studies have shown that a diet rich in fish significantly reduces coronary artery disease risk [10]. Marine Omega-3 PUFAs reduced the infiltration of inflammatory cells and pro-inflammatory activity into the plaque [11]. In other interventional and therapeutic studies, treatment with marine Omega-3 PUFAs in patients with advanced atherosclerotic plaques was followed by stabilization of the atheroma plaques and reduction of inflammation [12]. In the OCEAN study, the levels of matrix metalloproteinases were lower in atheroma plaques from patients treated with marine Omega-3 PUFAs, and the plaque vulnerability was thus reduced [13]. In 5480 patients, the dietary intake of marine Omega-3 PUFAs was inversely correlated with subclinical atherosclerosis [14]. The JELIS study demonstrated a reduction in major acute coronary events using EPA in patients with hypercholesteremia [15].

More than that, marine Omega-3 PUFAs were demonstrated to be anti-atherogenic agents [16,17]. In addition, in a cohort of 600 men with cardiovascular disease, fish oil supplement administration was followed by reduced parameters of atherothrombotic risk [18]. Omega-3 PUFAs also decreased atherosclerotic plaque dimensions and stabilized plaques, preventing plaque rupture and acute coronary syndrome [19,20].

With regard to the antioxidant effect, it was demonstrated that the EPA neutralized extracellular reactive oxygen species and reduced oxidized LDL-C levels in plasma [21]. Moreover, Omega-3 PUFAs had antioxidant properties, improved endothelial function, and contributed to anti-atherosclerotic benefits [19,22].

Omega 3 PUFAs act through several mechanisms including: the stabilization of vulnerable plaques, reducing platelet aggregation and inflammation.. At the same time, Omega 3 PUFAs are involved in reducing the risk of arrhythmias and promoting vasodilation. [19,20].

Moreover, a retrospective study in Japan in patients with cardiovascular disease demonstrated that low DHA levels were correlated with reduced flow-mediated dilation (FMD) as a measure of endothelial dysfunction [21]. Similar data about improvement of endothelial function and anti-inflammatory effects were reported in patients with metabolic syndrome after treatment with Omega-3 PUFAs [23].

The most important results of the present study are those related to the long-term prognosis, respectively, at 5 years of follow-up, which highlighted the effectiveness of the association of treatment with Omega-3 PUFAs to OMT in patients with ACS. Thus, the risk for cardiovascular and sudden death, reinfarction, stroke, and readmissions for ACS and heart failure were significantly reduced in the groups treated with Omega-3 PUFAs compared to the control groups.

Previous studies have various limitations regarding this issue according to demographic characteristics, comorbidities, the lack of homogeneity of study groups in meta-analyses, and the predominance of short-term results in most studies in comparison with long-term follow-up, which could obtain consistent results with high statistical value. In this context, a systematic review reported that Omega-3 PUFAs administrated daily were associated with a significant reduction in sudden cardiac death. A meta-analysis of 13 studies also demonstrated that marine Omega-3 PUFAs supplementation significantly reduced the risk of myocardial infarction and cardiovascular morbidity [22]. A meta-analysis of 38 studies involving 149,051 participants demonstrated that Omega-3 PUFAs reduce cardiovascular mortality, the risk of non-fatal myocardial infarction, and coronary heart disease events [22]. In the GISSI Prevenzione trial, the risk of the primary endpoint (death, non-fatal myocardial infarction, and stroke) was significantly decreased by treatment with Omega-3 PUFAs at 3.5 years of follow-up [24]. In a short-term follow-up study, the reduction in the risk of sudden death was statistically significant at 4 months [25].

Even though some negative trials of EPA + DHA were published, the debate and meta-analyses demonstrated the value of Omega-3 PUFAs in preventing cardiovascular events and reducing residual risk [22].

A narrative review of publications published in the last 10 years highlighted the important role of dietary Omega-3 PUFAs in reducing oxidative stress related to mitochondrial dysfunction, the apoptosis of endothelial cells, and increasing endogenous antioxidant enzyme activity. At the same time, Omega-3 polyunsaturated fatty acids reduce the level of pro-inflammatory cytokines in the heart muscle and vessels [26].

Higher levels of plasma EPA are associated with a lower risk of major cardiovascular events (MACE), such as any cause mortality, acute myocardial infarction, stroke, and readmission for heart failure. Additionally, it reduces the risk of sudden cardiovascular events caused by plaque rupture by significant thickening of the fibrous cap and increasing the stability of the atherosclerotic plaque. DHA is an Omega-3 fatty acid that has anti-arrhythmic and anti-thrombotic effects administrated in fish oil pills [26,27].

In accordance with these data, Omega-3 PUFA supplementation has been suggested as a therapeutic strategy in primary and secondary prevention of cardiovascular disease [28].

The limitations of this study are represented by the relatively small number of patients included compared to the large clinical trials, but these limitations are partially compensated for by the long-term follow-up of the patients—5 years, not very often met in the large clinical trials.

Another limitation of this research is the absence of the Omega-3 index at baseline and in dynamics among the biomarkers. A value of this index above 8% is demonstrated to be protective in terms of the risk of sudden death, an aspect demonstrated by retrospective and prospective epidemiological studies. The results of fundamental research on animal models and cell cultures have demonstrated the action of Omega-3 polyunsaturated fatty acids at the level of the myocardial cell membrane, reducing the risk of fatal arrhythmias and sudden death. These results led to the recommendation of the Guidelines of the European and American Societies of Cardiology of the ingestion of an average of 1 g of Omega-3 polyunsaturated fatty acids/day, especially in the secondary prevention of sudden death. A large individual variability depending on age, individual genetic substrate, health status, and geographical region habits determine the Omega-3 index [29].

In this study, Omega-3 PUFA therapy remained independently associated with a lower risk of cardiovascular death, sudden death, recurrence of NSTE/STE-ACS, in-stent restenosis, and readmission for ACS and heart failure in patients with acute coronary syndromes included in the present study, after adjusting for potential confounders, including risk factors of age > 65 years, male gender, diabetes mellitus, smokers, arterial hypertension, and body mass index > 25 kg/m^2 by multiple linear regression analysis.

4. Materials and Methods

One thousand one hundred forty patients were admitted to Clinic County Emergency Hospital Brasov with ACS and were enrolled in a prospective study. The study was divided into 4 groups related to the type of ACS and Omega-3 PUFAs treatment added to the optimal medical therapy (OMT). In ESC Guidelines for the management of acute coronary syndromes, the 2 types of ACS were defined as Non-ST segment elevation acute coronary syndromes (NSTE-ACSs) and ST segment elevation acute coronary syndromes (STE-ACSs), related to changes on a 12-lead electrocardiogram (ECG).

According to epidemiological studies, people who consume fish, due to the rich Omega-3 PUFA content, have a reduced risk of cardiovascular disease. The most important Omega-3 PUFAs recovered in marine extracts are eicosapentaenoic acid (EPA) and docosahexaenoic acid (DHA). Therapy with 1 g of Omega-3 PUFAs daily is the most recommended dose for secondary prevention. Omacor® was the commercial form of Omega-3 PUFAs used [10,24,25,30,31]. Omacor®—1 g capsule contains 465 mg of EPA and 375 mg of DHA—is most commonly administered in Europe [7]. In this context, Omacor® was administrated to all patients in the Omega-3 PUFAs groups.

The study protocol and informed consent were approved by the Ethics Committee of the County Emergency Hospital of Brasov. The patients received the informed consent that they read and signed before inclusion in the study, according to the Declaration of Helsinki. All data were anonymized during the analysis [32].

The patients were divided in four groups in relation to type of ACS—Non-ST segment elevation acute coronary syndrome (NSTE-ACS) or ST segment elevation acute coronary syndrome (STE-ACS)—and with addition of Omega-3 polyunsaturated fatty acids (Omacor®) to optimal medical therapy (OMT): Group 1: NSTE-ACS Omega-3 PUFAs—282 patients with Non-ST segment elevation acute coronary syndromes treated with Omega-3 polyunsaturated fatty acids (Omacor®) in addition to OMT; Group 2 (NSTE-ACS control group): NSTE-ACS—287 patients with Non-ST segment elevation acute coronary syndromes with OMT, without Omega-3 polyunsaturated fatty acids (Omacor®); Group 3: STE-ACS Omega-3 PUFAs—293 patients with ST segment elevation acute coronary syndromes treated with Omega-3 polyunsaturated fatty acids (Omacor®) in addition to OMT; Group 4 (STE-ACS control group): STE-ACS—277 patients with ST segment elevation acute coronary syndromes with OMT without Omega-3 polyunsaturated fatty acids (Omacor®). Acute coronary syndrome (ACS) was diagnosed by clinical and paraclinical criteria.

At admission to the hospital, the biomarkers to confirm NSTE-ACS and STE-ACS diagnostic and specific parameters for oxidative stress, inflammatory syndrome, and endothelial dysfunction were analyzed. Biomarkers of myocardial necrosis were evaluated: Creatine kinase isoenzymes MB (CK-MB) with Immunologic assay—normal value < 24 U/L; Troponin T with ECLIA assay—normal value < 0.1 ng/dL.

Among the oxidative stress biomarkers, the following biomarkers were chosen for determination in this study: Anti-Myeloperoxidase antibodies (MPO) type IgG—turbidimetric method—ELISA INOVA kits—normal value < 20 U [4,5,33–36]; Anti-Ox-LDL antibody—ELISA INOVA technique—normal value < 150U/L [9,30,35]; Total antioxidant status (TAS)—ABTS ® Method—RANDOX kits normal value > 1.3 mmol/L. The principle of the method is the incubation of ABTS® (2,2-azinobis (3-ethylbenzothiazoline-6-sulfonate)) with a peroxidase-met-myoglobin and H_2O_2 to produce the ABTS® + cation radical measured at 600 nm. The total antioxidant status is a measure of all components with antioxidant activity [35,36].

Biomarkers of inflammation were also analyzed: C-reactive protein—Immunoturbodimetric Assay—normal values 0–0.5 mg/dL; fibrinogen—Continuous sequential photo-optical method—normal values 200–400 mg/dL [37–41].

Parameters of endothelial dysfunction were determined. Flow-mediated dilation (FMD) was analyzed using brachial artery ultrasound measurement. Endothelial dysfunction was defined as a percentage variation of brachial artery basal diameter after 60 s post-hyperemia. Endothelial dysfunction was considered if flow-mediated dilation was less than 4.5% [40,41]. The second endothelial dysfunction analyzed was Von Willebrand factor (vWf) activity, using ELISA kits—normal values < 169.7% [42,43].

All these parameters were measured at baseline and at 6 months to evaluate the role of treatment with Omega-3 PUFAs (Omacor®) added to the OMT in patients with acute coronary syndromes.

The results regarding patient monitoring and the occurrence of cardiovascular events —cardiovascular death and sudden death, Non-ST/ST elevation acute coronary syndrome (NSTE-ACS/STE-ACS), in-stent thrombosis and restenosis, stroke, and readmission for ACS and for heart failure—were reported at 5 years of follow-up.

Statistical Analysis: The research methodology and the statistical analysis methods were chosen to be appropriate to the research profile. Statistical analysis was performed with the following tests: GraphPad Prism 10 version 10.4.1 and Excel Office 2019; demographic data and clinical characteristics of the study population were analyzed using descriptive statistics, expressed as percentages for categorical variables; Fisher's exact test to evaluate percentage differences in categorical variables; Pearson's correlation to evaluate the relationship between variables; multiple linear regression analysis to define independent variables. Statistical significance was defined as a value of $p < 0.05$ for all tests.

5. Conclusions

In patients with acute coronary syndromes treated with Omega-3 polyunsaturated fatty acids, at 6 months from the index event, the parameters of total antioxidant status, inflammation, and endothelial dysfunction normalized in a significantly larger number of patients in comparison with the control group. The improvement of long-term prognosis at 5 years of follow-up represents the strongest evidence of the effectiveness of Omega-3 polyunsaturated fatty acid supplementation. Regarding long-term prognosis, in Non-ST and ST segment elevation acute coronary syndromes, treatment with Omega-3 polyunsaturated fatty acids added to the optimal medical therapy significantly decreased the risk of cardiovascular death, sudden death, reinfarction, and readmissions for acute coronary syndrome or heart failure in comparison with Omega-3 polyunsaturated fatty acids untreated groups.

At the same time, multiple linear regression analysis proved that Omega-3 PUFA therapy remained independently associated with a lower risk of major acute cardiovascular events in patients with acute coronary syndromes after adjusting for potential confounders, including risk factors.

Analyzing the results of this clinical study, which are in accordance with the already published results, we concluded that it is important to individually evaluate the patients regarding the presence of oxidative stress and inflammation that play an important role in the unfavorable prognosis in acute coronary syndromes and to improve oxidative and inflammatory status by adding Omega-3 polyunsaturated fatty acids to the specific treatment.

Author Contributions: Conceptualization, A.C. and E.B.; methodology, L.G.M.; software, C.G.S. and L.G.M.; validation, A.C., T.B. and E.B.; formal analysis, A.C., V.B., H.R. and L.G.M.; investigation, S.G., V.B. and H.R.; resources, A.C., V.B., H.R. and S.G.; data curation, A.C., C.G.S., V.B., H.R. and S.G.; writing—original draft preparation, A.C, V.B., H.R. and S.G.; writing—review and editing, E.B. and C.G.S.; visualization, E.B. and L.G.M.; supervision, T.B. and E.B.; project administration, C.G.S. All authors have read and agreed to the published version of the manuscript.

Funding: This research received no external funding.

Institutional Review Board Statement: The study was conducted in accordance with the Declaration of Helsinki and approved by the Ethics Committee of Clinic County Emergency Hospital of Brasov, Romania (protocol code: PUFA-ACS-OSIED; registration number/date of approval: Nr. 512/5 January 2014), for studies involving humans.

Informed Consent Statement: Informed consent was obtained from all subjects involved in the study.

Data Availability Statement: The data presented in this study are available on request from the corresponding author.

Conflicts of Interest: The authors declare no conflicts of interest.

Abbreviations

The following abbreviations are used in this manuscript:

ATS	Atherosclerotic
ECG	Electrocardiogram
OMT	Optimal Medical Therapy
BMI	Body Mass Index
NSTE-ACS Omega-3 PUFAs	group of patients with Non-ST elevation acute coronary syndromes treated with Omega-3 fatty acids (Omacor®) in addition to OMT
NSTE-ACS	group of patients with Non-ST elevation acute coronary syndromes with OMT without Omega-3 fatty acids (Omacor®)
STE-ACS Omega-3 PUFAs	group of patients with ST elevation acute coronary syndromes treated with Omega-3 fatty acids (Omacor®) in addition to OMT
STE-ACS	group of patients with n ST elevation acute coronary syndromes with OMT without Omega-3 fatty acids (Omacor®)
LDL Cholesterol	Low Density Lipoprotein Cholesterol
HDL Cholesterol	High Density Lipoprotein Cholesterol
CK-MB ACEI	Creatine Kinase isoenzymes MB Angiotensin Converting Enzymes Inhibitors
FMD	Flow-Mediated Dilation
vWf	von Willebrand factor
TAS	Total Antioxidant Status
Ab anti Ox LDL cholesterol	antibodies anti-Oxidated Low Density Lipoprotein Cholesterol
Ac anti MPO IgG	Antibodies anti-Myeloperoxidase Immunoglobulin G type

References

1. Kotlyarov, S.; Kotlyarova, A. Clinical significance of polyunsaturated fatty acids in the prevention of cardiovascular diseases. *Front. Nutr.* **2022**, *9*, 998291. [CrossRef]
2. Herrington, W.; Lacey, B.; Sherliker, P.; Armitage, J.; Lewington, S. Epidemiology of Atherosclerosis and the Potential to Reduce the Global Burden of Atherothrombotic Disease. *Circ. Res.* **2016**, *118*, 535–546. [CrossRef] [PubMed]
3. Visseren, F.L.J.; Mach, F.; Smulders, Y.M.; Carballo, D.; Koskinas, K.C.; Bäck, M.; Benetos, A.; Biffi, A.; Boavida, J.-M.; Capodanno, D.; et al. 2021 ESC Guidelines on cardiovascular disease prevention in clinical practice. *Eur. Heart J.* **2021**, *42*, 3227–3337. [CrossRef] [PubMed]
4. D'oria, R.; Schipani, R.; Leonardini, A.; Natalicchio, A.; Perrini, S.; Cignarelli, A.; Laviola, L.; Giorgino, F. The Role of Oxidative Stress in Cardiac Disease: From Physiological Response to Injury Factor. *Oxid. Med. Cell Longev.* **2020**, *2020*, 1–29. [CrossRef]
5. Balan, A.I.; Halațiu, V.B.; Scridon, A. Oxidative Stress, Inflammation, and Mitochondrial Dysfunction: A Link between Obesity and Atrial Fibrillation. *Antioxidants* **2024**, *13*, 117. [CrossRef]
6. Zong, Y.; Li, H.; Liao, P.; Chen, L.; Pan, Y.; Zheng, Y.; Zhang, C.; Liu, D.; Zheng, M.; Gao, J. Mitochondrial dysfunction: Mechanisms and advances in therapy. *Signal Transduct. Target. Ther.* **2024**, *9*, 124. [CrossRef]
7. Frangie, C.; Daher, J. Role of myeloperoxidase in inflammation and atherosclerosis (Review). *Biomed. Rep.* **2022**, *16*, 53. [CrossRef] [PubMed] [PubMed Central]

8. Hong, C.G.; Florida, E.; Li, H.; Parel, P.M.; Mehta, N.N.; Sorokin, A.V. Oxidized low-density lipoprotein associates with cardiovascular disease by a vicious cycle of atherosclerosis and inflammation: A systematic review and meta-analysis. *Front. Cardiovasc. Med.* **2023**, *9*, 1023651. [CrossRef]
9. Munno, M.; Mallia, A.; Greco, A.; Modafferi, G.; Banfi, C.; Eligini, S. Radical Oxygen Species, Oxidized Low-Density Lipoproteins, and Lectin-like Oxidized Low-Density Lipoprotein Receptor 1: A Vicious Circle in Atherosclerotic Process. *Antioxidants* **2024**, *13*, 583. [CrossRef]
10. Haque, N.; Parveen, S.; Tang, T.; Wei, J.; Huang, Z. Marine Natural Products in Clinical Use. *Mar. Drugs* **2022**, *20*, 528. [CrossRef]
11. Calder, P.C. The role of marine omega-3 (n-3) fatty acids in inflammatory processes, atherosclerosis and plaque stability. *Mol. Nutr. Food Res.* **2012**, *56*, 1073–1080. [CrossRef] [PubMed]
12. Thies, F.; MC Garry, J.; Yaqoob, P.; Rerkasem, K.; Williams, J.; Shearman, C.P.; Gallagher, P.J.; Calder, P.C.; Grimble, R.F. Association of n-3 polyunsaturated fatty acids with stability of atherosclerotic plaques: A randomised controlled trial. *Lancet* **2003**, *361*, 477–485. [CrossRef] [PubMed]
13. Cawood, A.L.; Ding, R.; Napper, F.L.; Young, R.H.; Williams, J.A.; Ward, M.J.; Gudmundsen, O.; Vige, R.; Payne, S.P.; Ye, S.; et al. Eicosapentaenoic acid (EPA) from highly concentrated n−3 fatty acid ethyl esters is incorporated into advanced atherosclerotic plaques and higher plaque EPA is associated with decreased plaque inflammation and increased stability. *Atherosclerosis* **2010**, *212*, 252–259. [CrossRef] [PubMed]
14. He, K.; Liu, K.; Daviglus, M.L.; Mayer-Davis, E.; Jenny, N.S.; Jiang, R.; Ouyang, P.; Steffen, L.M.; Siscovick, D.; Wu, C.; et al. Intakes of long-chain n–3 polyunsaturated fatty acids and fish in relation to measurements of subclinical atherosclerosis. *Am. J. Clin. Nutr.* **2008**, *88*, 1111–1118. [CrossRef]
15. Yokoyama, M.; Origasa, H.; Matsuzaki, M.; Matsuzawa, Y.; Saito, Y.; Ishikawa, Y.; Oikawa, S.; Sasaki, J.; Hishida, H.; Itakura, H.; et al. Effects of eicosapentaenoic acid on major coronary events in hypercholesterolaemic patients (JELIS): A randomised open-label, blinded endpoint analysis. *Lancet* **2007**, *369*, 1090–1098. [CrossRef]
16. Casas, R.; Estruch, R.; Sacanella, E. Influence of Bioactive Nutrients on the Atherosclerotic Process: A Review. *Nutrients* **2018**, *10*, 1630. [CrossRef]
17. Massaro, M.; Scoditti, E.; Carluccio, M.A.; De Caterina, R. Nutraceuticals and Prevention of Atherosclerosis: Focus on ω-3 Polyunsaturated Fatty Acids and Mediterranean Diet Polyphenols. *Cardiovasc. Ther.* **2010**, *28*, e13–e19. [CrossRef]
18. Franzese, C.J.; Bliden, K.P.; Gesheff, M.G.; Pandya, S.; Guyer, K.E.; Singla, A.; Tantry, U.S.; Toth, P.P.; Gurbel, P.A. Relation of Fish Oil Supplementation to Markers of Atherothrombotic Risk in Patients With Cardiovascular Disease Not Receiving Lipid-Lowering Therapy. *Am. J. Cardiol.* **2015**, *115*, 1204–1211. [CrossRef]
19. Kar, S. Omacor and Omega-3 Fatty Acids for Treatment of Coronary Artery Disease and the Pleiotropic Effects. *Am. J. Ther.* **2014**, *21*, 56–66. [CrossRef]
20. Mason, R.P.; Libby, P.; Bhatt, D.L. Emerging Mechanisms of Cardiovascular Protection for the Omega-3 Fatty Acid Eicosapentaenoic Acid. *Arterioscler. Thromb. Vasc. Biol.* **2020**, *40*, 1135–1147. [CrossRef]
21. Yagi, S.; Aihara, K.; Fukuda, D.; Takashima, A.; Hara, T.; Hotchi, J.; Ise, T.; Yamaguchi, K.; Tobiume, T.; Iwase, T.; et al. Effects of Docosahexaenoic Acid on the Endothelial Function in Patients with Coronary Artery Disease. *J. Atheroscler. Thromb.* **2015**, *22*, 447–454. [CrossRef] [PubMed]
22. Khan, S.U.; Lone, A.N.; Khan, M.S.; Virani, S.S.; Blumenthal, R.S.; Nasir, K.; Miller, M.; Michos, E.D.; Ballantyne, C.M.; Boden, W.E.; et al. Effect of omega-3 fatty acids on cardiovascular outcomes: A systematic review and meta-analysis. *eClinicalMedicine* **2021**, *38*, 100997. [CrossRef] [PubMed]
23. Tousoulis, D.; Plastiras, A.; Siasos, G.; Oikonomou, E.; Verveniotis, A.; Kokkou, E.; Maniatis, K.; Gouliopoulos, N.; Miliou, A.; Paraskevopoulos, T.; et al. Omega-3 Omega-3 PUFAs improved endothelial function and arterial stiffness with a parallel antiinflammatory effect in adults with metabolic syndrome. *Atherosclerosis* **2014**, *232*, 10–16. [CrossRef] [PubMed]
24. GISSI-Prevenzione Inormal valueestigators (Gruppo Italiano per lo Studio della Sopravvivenza nell'Infarto miocardico). Dietary supplementation with n-3 polyunsaturated fatty acids and vitamin E after myocardial infarction: Results of the GISSI-Prevenzione trial. *Lancet* **1999**, *354*, 447–455. [CrossRef]
25. Marchioli, R.; Barzi, F.; Bomba, E.; Chieffo, C.; Di Gregorio, D.; Di Mascio, R.; Franzosi, M.G.; Geraci, E.; Levantesi, G.; Maggioni, A.P.; et al. Early Protection Against Sudden Death by n-3 Polyunsaturated Fatty Acids After Myocardial Infarction. *Circulation* **2002**, *105*, 1897–1903. [CrossRef]
26. Banaszak, M.; Dobrzyńskaa, M.; Kawkac, A.; Górnaa, I.; Woźniaka, D.; Przysławskia, J.; Drzymała-Czyż, S. Role of Omega-3 fatty acids eicosapentaenoic (EPA) and docosahexaenoic (DHA) as modulatory and anti-inflammatory agents in noncommunicable diet-related diseases—Reports from the last 10 years. *Clin. Nutr. ESPEN* **2024**, *63*, 240–258.
27. Veenstra, J.; Kalsbeek, A.; Westra, J.; Disselkoen, C.; Smith, C.E.; Tintle, N. Genome-Wide Interaction Study of Omega-3 PUFAs and Other Fatty Acids on Inflammatory Biomarkers of Cardiovascular Health in the Framingham Heart Study. *Nutrients* **2017**, *9*, 900. [CrossRef]

28. Zanetti, M.; Gortan Cappellari, G.; Barbetta, D.; Semolic, A.; Barazzoni, R. Omega 3 Polyunsaturated Fatty Acids Improve Endothelial Dysfunction in Chronic Renal Failure: Role of eNOS Activation and of Oxidative Stress. *Nutrients* **2017**, *9*, 895. [CrossRef]
29. Schuchardt, J.P.; Beinhorn, P.; Hu, X.F.; Chan, H.M.; Roke, K.; Bernasconi, A.; Hahn, A.; Sala-Vila, A.; Stark, K.D.; Harris, W.S. Omega-3 world map: 2024 update. *Prog. Lipid Res.* **2024**, *95*, 101286. [CrossRef]
30. Zhang, B.; Xiong, K.; Cai, J.; Ma, A. Fish Consumption and Coronary Heart Disease: A Meta-Analysis. *Nutrients* **2020**, *12*, 2278. [CrossRef]
31. Dolecek, T.; Granditis, G. Dietary Polyunsaturated Fatty Acids and Mortality in the Multiple Risk Factor Intervention Trial (MRFIT). *World Rev. Nutr. Diet.* **1991**, *66*, 205–216. [CrossRef] [PubMed]
32. Purcaru, D.; Preda, A.; Popa, D.; Moga, M.A.; Rogozea, L. Informed consent: How much awareness is there? *PLoS ONE* **2014**, *9*, e110139. [CrossRef] [PubMed]
33. Rodrigo, R.; Retamal, C.; Schupper, D.; Vergara-Hernández, D.; Saha, S.; Profumo, E.; Buttari, B.; Saso, L. Antioxidant Cardioprotection against Reperfusion Injury: Potential Therapeutic Roles of Resveratrol and Quercetin. *Molecules* **2022**, *27*, 2564. [CrossRef] [PubMed] [PubMed Central]
34. Madaudo, C.; Coppola, G.; Parlati, A.L.M.; Corrado, E. Discovering Inflammation in Atherosclerosis: Insights from Pathogenic Pathways to Clinical Practice. *Int. J. Mol. Sci.* **2024**, *25*, 6016. [CrossRef]
35. Bobescu, E.; Covaciu, A.; Rus, H.; Rogozea, L.M.; Badea, M.; Marceanu, L.G. Low Response to Clopidogrel in Coronary Artery Disease. *Am. J. Ther.* **2020**, *27*, e133–e141. [CrossRef]
36. Halliwell, B.; Gutteridge, J.M.C. *Free Radicals in Biology and Medicine*, 5th ed.; Oxford University Press Inc.: New York, NY, USA, 2015; Volume 348–353, pp. 539–549.
37. Sánchez, A.; Mirabel, J.L.; Barrenechea, E.; Eugui, J.; Puelles, A.; Castañeda, A. Evaluation of an improved immunoturbidimetic assay for serum C-reactive protein on a COBAS INTEGRA 400 Analyzer. *Clin. Lab.* **2002**, *48*, 313–317. [PubMed]
38. Louka, M.; Kaliviotis, E. Development of an Optical Method for the Evaluation of Whole Blood Coagulation. *Biosensors* **2021**, *11*, 113. [CrossRef] [PubMed] [PubMed Central]
39. Mackie, I.; Casini, A.; Pieters, M.; Pruthi, R.; Reilly-Stitt, C.; Suzuki, A. International council for standardisation in haematology recommendations on fibrinogen assays, thrombin clotting time and related tests in the investigation of bleeding disorders. *Int. J. Lab. Hematol.* **2024**, *46*, 20–32. [CrossRef]
40. Bobescu, E.; Marceanu, L.G.; Dima, L.; Balan, A.; Strempel, C.G.; Covaciu, A. Trimetazidine Therapy in Coronary Artery Disease: The Impact on Oxidative Stress, Inflammation, Endothelial Dysfunction, and Long-Term Prognosis. *Am. J. Ther.* **2021**, *28*, e540–e547. [CrossRef]
41. Williams, J.S.; Wiley, E.; Cheng, J.L.; Stone, J.C.; Bostad, W.; Cherubini, J.M.; Gibala, M.J.; Tang, A.; MacDonald, M.J. Differences in cardiovascular risk factors associated with sex and gender identity, but not gender expression, in young, healthy cisgender adults. *Front. Cardiovasc. Med.* **2024**, *11*, 1374765. [CrossRef]
42. Kozlov, S.; Okhota, S.; Avtaeva, Y.; Melnikov, I.; Matroze, E.; Gabbasov, Z. Von Willebrand factor in diagnostics and treatment of cardiovascular disease: Recent advances and prospects. *Front. Cardiovasc. Med.* **2022**, *9*, 1038030. [CrossRef]
43. Kawecki, C.; Lenting, P.J.; Denis, C.V. von Willebrand factor and inflammation. *J. Thromb. Haemost.* **2017**, *15*, 1285–1294. [CrossRef]

Disclaimer/Publisher's Note: The statements, opinions and data contained in all publications are solely those of the individual author(s) and contributor(s) and not of MDPI and/or the editor(s). MDPI and/or the editor(s) disclaim responsibility for any injury to people or property resulting from any ideas, methods, instructions or products referred to in the content.

marine drugs

MDPI

Article

Unveiling the Lipid Features and Valorization Potential of Atlantic Salmon (*Salmo salar*) Heads

João Pedro Monteiro [1,2,3,*], Tiago Sousa [1,2], Tânia Melo [1,2], Carla Pires [4,5], António Marques [4,5], Maria Leonor Nunes [5], Ricardo Calado [6,*] and M. Rosário Domingues [1,2,*]

1 Centro de Espetrometria de Massa & LAQV-REQUIMTE & Departamento de Química, Universidade de Aveiro, Campus Universitário de Santiago, 3810-193 Aveiro, Portugal; tmms@ua.pt (T.S.); taniamelo@ua.pt (T.M.)

2 CESAM & Departamento de Química, Universidade de Aveiro, Campus Universitário de Santiago, 3810-193 Aveiro, Portugal

3 CIVG—Vasco da Gama Research Center/EUVG—Vasco da Gama University School, 3020-210 Coimbra, Portugal

4 Division of Aquaculture, Upgrading and Bioprospection, Portuguese Institute for the Sea and Atmosphere (IPMA, I.P.), Av. Doutor Alfredo Magalhães Ramalho 6, 1495-165 Algés, Portugal; cpires@ipma.pt (C.P.); amarques@ipma.pt (A.M.)

5 Interdisciplinary Centre of Marine and Environmental Research (CIIMAR/CIMAR-LA), University of Porto, Terminal de Cruzeiros do Porto de Leixões, Av. General Norton de Matos s/n, 4450-208 Matosinhos, Portugal; nunes.leonor@gmail.com

6 ECOMARE & CESAM & Departamento de Biologia, Universidade de Aveiro, Campus Universitário de Santiago, 3810-193 Aveiro, Portugal

* Correspondence: jpspmonteiro@yahoo.com (J.P.M.); rjcalado@ua.pt (R.C.); mrd@ua.pt (M.R.D.)

Abstract: The sustainable utilization of co-products derived from the salmon processing industry is crucial for enhancing the viability and decreasing the environmental footprint of both capture and aquaculture operations. Salmon (*Salmo salar*) is one of the most consumed fish worldwide and a major species produced in aquaculture. As such, significant quantities of salmon co-products are produced in pre-commercialization processing/steaking procedures. The present study characterized a specific co-product derived from the processing of salmon: minced salmon heads. More specifically, this work aimed to reveal the nutritional profile of this co-product, with a special focus on its lipid content, including thoroughly profiling fatty acids and fully appraising the composition in complex lipids (polar lipids and triglycerides) for the first time. The antioxidant potential of lipid extracts from this salmon co-product was also studied in order to bioprospect lipid functional properties and possibly unveil new pathways for added-value applications. Our analysis indicated that these minced salmon heads are exceptionally rich in lipids. Oleic acid is the most prevalent fatty acid in this co-product, followed by palmitic acid, stearic acid, and linoleic acid. Moreover, relevant lipid indexes inferred from the fatty acid composition of this co-product revealed good nutritional traits. Lipidome analysis revealed that triglycerides were clearly the predominant lipid class present in this co-product while phospholipids, as well as ceramides, were also present, although in minimal quantities. The bioprospecting of antioxidant activity in the lipid extracts of the minced salmon heads revealed limited results. Given the high concentration of triglycerides, minced salmon heads can constitute a valuable resource for industrial applications from the production of fish oil to biodiesel (as triglycerides can be easily converted into fatty acid methyl esters), as well as possible ingredients for cosmetics, capitalizing on their alluring emollient properties. Overall, the valorization of minced salmon heads, major co-products derived from the processing of one of the most intensively farmed fish in the world, not only offers economic benefits but also contributes to the sustainability of the salmon processing industry by reducing waste and promoting a more efficient use of marine bioresources.

Keywords: aquaculture; omega-3 fatty acids; fish co-products; lipidomics; nutritional quality; salmon; sustainability

1. Introduction

Atlantic salmon (*Salmo salar*) is a critical resource for human nutrition, normally recognized for its high-quality protein content and for the presence of beneficial fatty acids (FAs), namely omega-3 [1–4]. As one of the most consumed fish globally, its production is both indispensable and substantial. In 2023, global salmon production reached approximately 3 million metric tons [5], with aquaculture accounting for about 70% of this volume [6]. This high level of production generates significant quantities of co-products including heads, bones, skin, and viscera. Estimates suggest that the fillet yield could represent only 50–70% of the whole salmonid weight [7], implying a production of roughly up to 1.5 million metric tons of salmon co-products annually.

A considerable number of studies have studied the lipid content of *S. salar* muscle/edible tissue, including characterizations at a molecular level and the use of lipidomic approaches [8–14], which is proof of the interest and relevance of this species. Regarding salmon co-products in particular, the phospholipid composition of the head, roe, and skin (although in a different salmon species, *Oncorhynchus tshawytscha*) was studied using nuclear magnetic resonance spectroscopy [15]. This study showed that salmon heads contained less phospholipid and omega-3 fatty acid content (including eicosapentaenoic (EPA) and docosahexaenoic (DHA) acids) than roe and skin [15]. The other available report using lipidomics to study salmon co-products characterized the phospholipid extracts of the heads of an unspecified salmon species using more standard liquid chromatography coupled with a mass spectrometry (LC-MS) approach [16]. This study reported high percentages of DHA and EPA in the fatty acid chains of phosphatidylcholine (PC) and phosphatidylethanolamine (PE), the two main phospholipid classes present [16]. Other studies using lipidomics tools to characterize the lipid content of the heads of other fish species included those conducted on the silver carp (*Hypophthalmichthys molitrix*) [16], Pacific blue mackerel (*Scomber australasicus*) [17], and king salmon (*Oncorhynchus tshawytscha*) [15], in the later species using a nuclear magnetic resonance (NMR), with all focusing specifically on the content in omega-3-containing phospholipids. This interest in the content in omega-3-containing phospholipids is justified by the wide range of health benefits ascribed to these compounds when acquired through the diet [18,19]. Other studies simply focused on the characterization of the fatty acid profiles of *S. salar* co-products in specific, unveiling desirable features in terms of the omega-3 fatty content in heads [20,21], as well as in frames, skin trimmings, and viscera [21]. Despite these reportedly enticing characteristics of the lipid content of salmon co-products, the bioprospecting of bioactive lipids in salmon co-products with a perspective of valorizing these resources for higher-end applications in the industry still represents an unexplored path.

The exploitation of salmon co-products as prospective sources of valuable compounds was the subject of numerous studies, mostly focused specifically on the protein content and related bioprospecting of the biological activity of protein fractions and hydrolysates [22–34]. Despite the lipid content of salmon co-products remaining rather less explored in terms of the bioprospecting of biological activity and the isolation of bioactive lipids, a few preliminary assessments assigned antioxidant [35], anti-inflammatory [35,36], antithrombotic [37], cardioprotective [37], and antimicrobial activity [35,38] to lipid fractions and oils obtained from salmon heads. Also, orally administered phospholipid extracts from salmon heads were linked to ameliorating effects upon rodent models used to study a metabolic syndrome [16]. Therefore, further bioprospecting and characterizing, in detail, the lipid fractions of salmon co-products could substantiate the pertinence of recovering bioactive lipids or lipid fractions from these bioresources, envisioning biorefinery approaches that allow taking full advantage of both protein and lipid contents [39].

The valorization of fish co-products is paramount for the sustainability and economic viability of the seafood industry [39]. Using these co-products can significantly reduce waste, lower the environmental impact, and create new revenue streams [40,41]. Making use of state-of-the-art technical resources including gas chromatography–mass spectrometry (GC-MS) and liquid chromatography–mass spectrometry (LC-MS), this study aimed to achieve an in-depth characterization of the whole lipid content of Atlantic salmon heads. Moreover, since the available studies characterizing the lipid content in salmon co-products at a molecular level only focused on phospholipids, we aimed to extend the characterization of minced salmon heads to their whole lipidome. The bioprospecting of biological activity in the lipid extracts of minced salmon heads could also pave the way for novel, higher-end applications for these bioresources in the industry. We believe that this type of characterization of fish co-products is instrumental to unveiling new applications and/or directing these resources to the most suitable industrial uses, contributing to a more sustainable and economically viable farmed fish industry framed by circularity and contributing more efficiently to a blue bioeconomy.

2. Results

2.1. Elemental Composition and Biochemical Characterization

The results of the elemental analysis quantifying the contents of carbon, hydrogen, nitrogen, and sulfur are depicted in Supplementary Figure S1. In terms of the biochemical characterization of the minced salmon heads, the most striking result was the remarkable content of lipids (23.97 ± 0.72% WW), which was much higher than the content of protein (16.20 ± 20% WW; Table 1). This salmon co-product also displayed an abundant moisture content (58.04 ± 0.99% WW) and very limited contents of ash (0.81 ± 0.22% WW) and carbohydrates and other compounds (0.97 ± 0.80% WW; Table 1), which were determined as the remaining component percentages. Another remarkable feature of this co-product was the considerably low phospholipid content in the total lipid fractions obtained from these resources (1.16 ± 0.17% of total lipid; Table 1).

Table 1. Proximate composition of the minced salmon heads. Data are shown as means ± standard deviations (SDs) for 5 samples of salmon head homogenate (n = 5).

Component (WW)	%
Moisture	58.04 ± 0.99
Ash	0.81 ± 0.22
Protein	16.20 ± 1.02
Carbohydrates and other compounds	0.97 ± 0.80
Lipid	23.97 ± 0.72
Phospholipid [1]	1.16 ± 0.17

[1] Phospholipid content is presented as percentage of total lipid content.

2.2. Fatty Acid Profiles

The fatty acid profile of the minced salmon heads was characterized by a very significant content of fatty acids with 18-carbon chains. Oleic acid (C18:1*n*-9) was clearly the most abundant fatty acid present (34.66 ± 1.26%, Figure 1; 116.0 ± 15.4 μg/mg DW, Supplementary Figure S2) while stearic (C18:0) and linoleic acids (C18:2*n*-6) were also among the major fatty acids present (11.64 ± 1.14% and 12.10 ± 0.41%, respectively; Figure 1). Palmitic acid (C16:0) was the other fatty acid that was present in more noticeable amounts, representing 13.24 ± 1.44% of all the fatty acid content (Figure 1). Omega-3 fatty acids were present at modest amounts and included linolenic acid (C18:3*n*-3; 3.70 ± 0.12%, Figure 1; 15.8 ± 2.2 μg/mg DW, Supplementary Figure S2), eicosapentaenoic acid (EPA, C20:5*n*-3, 2.70 ± 0.12%, Figure 1; 12.6 ± 1.7 μg/mg DW, Supplementary Figure S2), and docosahexaenoic acid (DHA, C22:6*n*-3, 2.86 ± 0.11%, Figure 1; 12.7 ± 1.4 μg/mg DW, Supplementary Figure S2). In Supplementary Figure S2, the results of the fatty acid profiling are presented in absolute quantities.

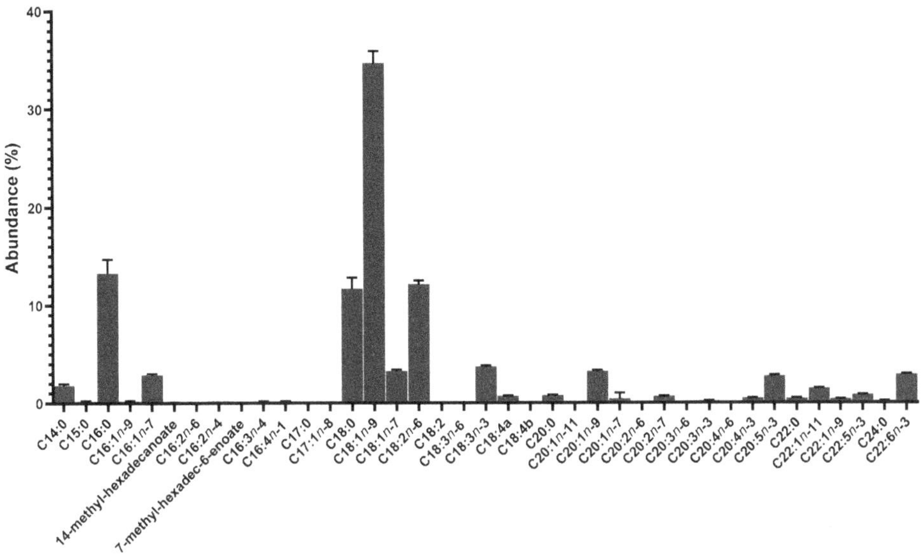

Figure 1. Fatty acid profile of the minced salmon heads (results are presented as percentages of total fatty acid content). Data are shown as means ± standard deviations (SDs) for 5 samples of salmon head homogenate (n = 5).

Monounsaturated fatty acids (MUFAs) represented almost half of the total fatty acid content (46.59 ± 1.98%; Table 2) while the contents of saturated (SFA) and polyunsaturated fatty acids (PUFA) were lower (28.29 ± 2.56% and 25.12 ± 0.79%, respectively). The minced salmon heads contained more omega-6 (or n-6, 12.44 ± 0.38%) than omega-3 (or n-3, 10.68 ± 0.35%) fatty acids, although the resulting n-6/n-3 ratio was still reasonably low (1.17 ± 0.02), as was the saturated-to-unsaturated (SFA/UFA) ratio (0.39 ± 0.05; Table 2). The average fatty acid chain length was very close to 18 (18.01 ± 0.04), confirming the significant contributions of fatty acids like oleic (the most abundant), stearic, and linoleic acids to the total content of fatty acid. This salmon co-product was also shown to present interestingly low atherogenic (AI) and thrombogenic (TI) indexes (0.28 ± 0.03 and 0.43 ± 0.06, respectively), along with a remarkably high hypocholesterolemic/hypercholesterolemic (h/H) ratio (0.42 ± 0.05; Table 2).

Table 2. Indexes/factors derived from the fatty acid profiles of minced salmon heads. Data are shown as means ± standard deviations (SDs) for 5 samples of salmon head homogenate (n = 5).

Index/Factor	Value
n-3	10.68 ± 0.35%
n-6	12.44 ± 0.38%
n-6/n-3	1.17 ± 0.02
SFA	28.29 ± 2.56%
MUFA	46.59 ± 1.98%
PUFA	25.12 ± 0.79%
SFA/PUFA	1.13 ± 0.12
TI	0.43 ± 0.06
AI	0.28 ± 0.03

Table 2. *Cont.*

Index/Factor	Value
h/H	4.24 ± 0.49
PoI	0.42 ± 0.05
UI	125.59 ± 3.93
ACL	18.01 ± 0.04

ACL: average chain length; AI: atherogenic index; h/H: hypocholesterolemic/hypercholesterolemic index; MUFA: monounsaturated fatty acid; PoI: polienic index; PUFA: polyunsaturated fatty acid; SFA: saturated fatty acid; UI: unsaturation index; TI: thrombogenic index.

2.3. Lipidome Characterization

Analysis of the lipidome of the minced salmon heads by LC-MS allowed the identification and semi-quantification of 150 different lipid species (Figure 2). This analysis revealed that triglycerides (TGs) were not only the most abundant but also the most diverse class of lipids present (Figure 2; Supplementary Tables S1 and S2). In fact, 108 different TG species were identified, including one oxidized TG form (TG 50:2;O). Other acylglycerides, namely diglycerides (DGs) were also present, with 14 different species being detected, all in low amounts. In terms of the polar lipid content and phospholipids in particular, the detected species were also present in residual amounts, as was already somewhat inferred from the quantification of phospholipids in lipid total extracts (Table 1). Nevertheless, fifteen PC species and five lysophosphatidylcholines (LPCs) were detected. Other phospholipid classes, namely molecular PEs (2), lysophosphatidylethanolamine (LPE, 1) phosphatidylglycerol (PG, 1) and phosphatidylserine (PS, 2), were barely represented (Figure 2; Supplementary Tables S1 and S2). Two sphingolipids, one ceramide (Cer 42:2;2O), and one hexosylceramide (HexCer 42:2;2O) were also detected.

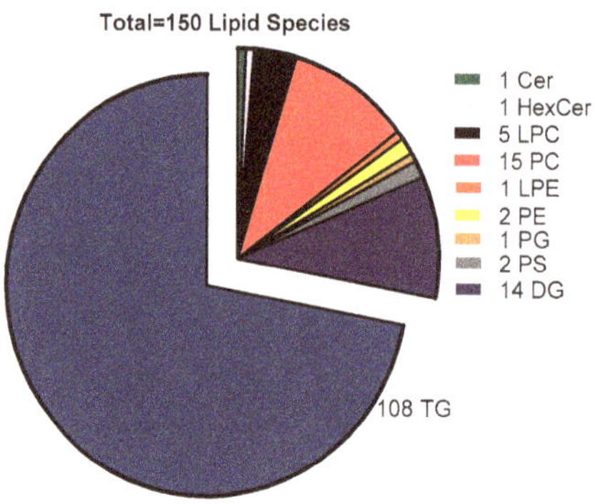

Figure 2. Number of molecular lipid species detected in minced salmon heads according to lipid class.

In terms of the specific profiles of TGs, the most prevalent lipid class present in samples of minced salmon heads (Supplementary Figure S3), 108 different molecular species were recorded, with total fatty acyl carbon chains ranging from 37 to 66 total carbons and with the most unsaturated species being TG 62:13 (Figure 3), corresponding to a TG specifically containing one oleic and two DHA esterified fatty acids. Further analysis of the TG profile revealed that TG 54:3 (containing three esterified oleic acids) was the most abundant TG species, representing 16.20 ± 0.35% of the content of the class (Figures 3 and 4). The only other two TG species with contents representing over 10% of the class were TG 52:2 (with the major form containing one palmitic and two oleic esterified fatty acids, representing 11.54 ± 0.32% of total TGs) and TG 52:3 (with the major form containing one palmitic, one oleic, and one linoleic esterified fatty acid, amounting to 10.04 ± 0.15% of the class; Figure 4).

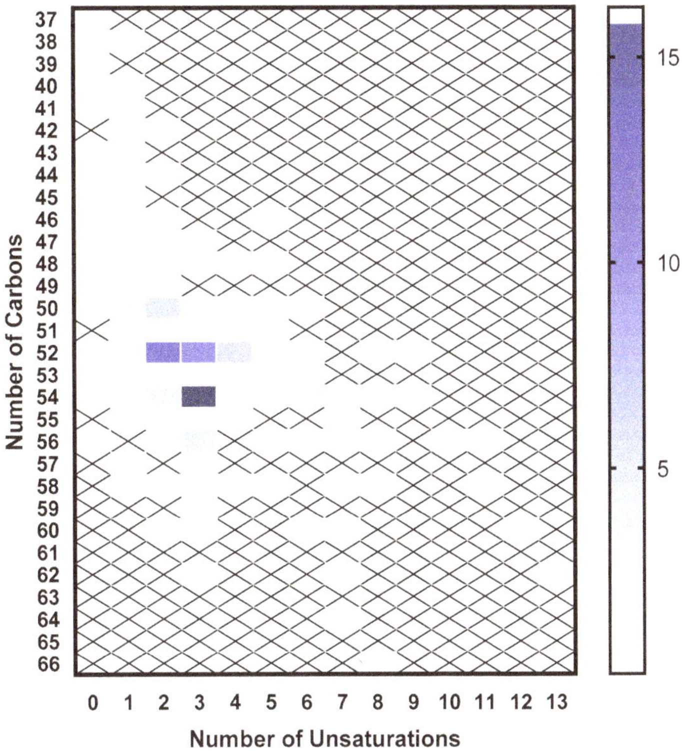

Figure 3. Heatmap depicting the percentages of triglycerides present in minced salmon heads as distributed according to total acyl chain length and unsaturation degree.

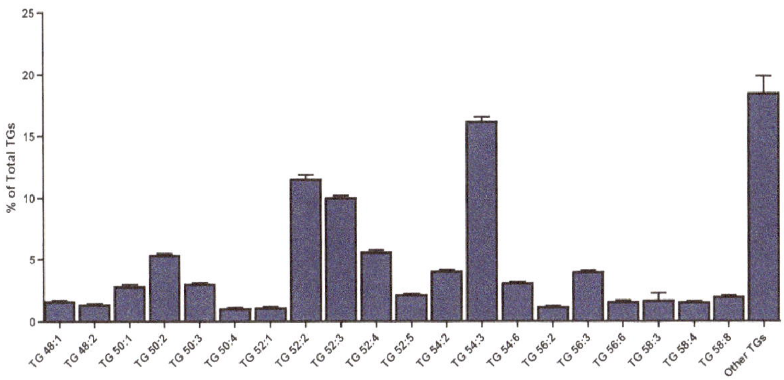

Figure 4. Percentages of molecular lipid species of triglycerides present in minced salmon heads in terms of the total content of the class (only the top 20 most abundant species are presented).

The percentage of marine TGs (as defined as the ones containing EPA and/or DHA) was also ascertained, representing 12.71 ± 0.22% of the total lipid content of the class.

2.4. Screening of Antioxidant Activity

The antioxidant activity of total lipid extracts of minced salmon heads was assessed by the use of both DPPH$^{\bullet}$ and ABTS$^{\bullet+}$ assays (Table 3). Lipid extracts from minced salmon heads were shown to display a limited free radical scavenging capacity. In fact, it was only possible to determine 10.76% of the inhibition for the lipid extracts of minced salmon heads in the DPPH• assays and 15.41% in the ABTS•+ assay using 250 μg mL^{-1} of the lipid extract (Table 3). The DPPH• assay also resulted in low values in terms of TE for the minced salmon head lipid extract, with a value of 4.39 for the DPPH• assay and a value of 1.67 in the ABTS•+ assay.

Table 3. Free radical scavenging capacity of the total lipid extracts of minced salmon heads expressed as inhibition percentages for 250 μg mL^{-1} of lipid extract for DPPH• and ABTS•+ assays and Trolox equivalents (TE, as μmol g^{-1}) for each assay also. Data are shown as means ± standard deviations (SDs) for 3 samples of lipid extracts from minced salmon heads (*n* = 5).

DPPH• Assay		ABTS•+ Assay	
Inhibition (%)	TE (μmol g^{-1})	Inhibition (%)	TE (μmol g^{-1})
10.76% ± 0.69	4.39 ± 0.28	15.41% ± 0.62	1.67 ± 0.06

3. Discussion

The proximate composition of the minced salmon heads showed a remarkable amount of lipid content, certainly one of the highest reported for fish and other seafood co-products [39]. Atlantic salmon is considered a high fat fish species, according to tentative classifications [42], and the lipid content of their heads seems to corroborate that assumption. Thus, salmon heads represent an obvious possibility for exploring this naturally abundant content in lipids for novel applications, for which a great first step would always be a thorough characterization of lipid composition at molecular level.

The results presented here for the proximate composition, in general, agree with results reported in previous studies. Vásquez et al. [23] reported values of 62.6% of moisture (WW) and contents of 54.3% of lipids and 30.1% of protein (DW), as well as a very restricted ash content (2.7%), in *S. salar* heads. Another study by Malcorps and colleagues reported the moisture content to range between 50 and 55% and confirmed the remarkable quantity of lipids (>40% DW), which was higher than the protein content (≈30%), describing, however, a significantly higher ash content in salmon heads (≈8%) [21]. It is important to mention that reports studying the proximate composition of salmon flesh/fillets highlight very meaningful compositional features regarding salmon heads, underscoring nutritional quality [1,3,4] and nutraceutical potential [20,36,37]. These reports systematically propose that in salmon flesh (unlike in the case of minced salmon heads), the composition of protein is higher than that of lipids (sometimes three times larger in relative terms) [1,3,4]. Other studies directly comparing the fat (lipid) content in salmon heads to that in salmon flesh/fillets corroborated that the first displayed a content much higher in terms of lipids than proteins [20,36]. Another remarkable feature of salmon heads was the notably low content of total phospholipids present in total lipid extracts. This overwhelming predominance of neutral lipids and TGs in particular (accounting for well over 90% of lipids) in salmon heads had already been highlighted in previous studies [20,37].

Regarding the fatty acid profile of salmon head homogenate, our study showed fatty acids with 18-carbon chains to be especially enriched in this co-product, especially oleic acid, representing more than a third of total fatty acid content. Other studies characterizing the fatty acid profile of Atlantic salmon heads were in line with our findings, namely confirming oleic acid to consistently be the main fatty acid present, and highlighted linoleic and palmitic acids as the main components of fatty acid profiles [20,21,35,36,38]. However, these studies also generally reported lower stearic acid (C18:0) levels than the ones reported in the present study (2.3–5.4%) [20,21,35,36,38]. When considering the contents of marine omega-3 fatty acids in salmon heads, the literature reports vary significantly, with some pointing to quantities of EPA similar to the ones reported in the present work but higher contents of DHA [21,36]. Nonetheless, other studies referred higher relative quantities of both EPA and DHA [20,35] while other works completely failed to detect/report EPA or DHA [38]. Methodological differences, both in lipid extraction methods and in the derivatization and determination of fatty acid content, may account at least for some of the discrepancies between these reports. Moreover, growth conditions, especially the diet provided to fish during grow-out, may also contribute to the disparities reported in salmon head fatty acid profiles [43–45]. Also, the fatty acid profile of commercially available *S. salar* was reported to vary significantly according to several production specifications, namely ranges (value, standard, premium, or organic) and origins [3], making it particularly challenging to compare the fatty acid profiles of their heads without any background information on diet and other grow-out conditions (e.g., water temperature). A study characterizing the fatty acid profiles of Atlantic salmon fillets from ten different retailers clearly documented this variability, with oleic acid, despite being consistently the most abundant fatty acid (as in *S. salar* heads), varying between 24.3 and 42.0% [3] while other reports described an even lower content (17.7% in wild *S. salar*) [1]. Linoleic acid ranged from 8.3 to 15.1% and palmitic acid from 8.7 to 14.1% as examples of other main fatty acid components [3]. The contents of marine omega-3 FAs were found to be very variable, with the EPA ranging from 2.6 to 8.5% and DHA from 3.1 to 9.4% [1,3,4].

Omega-3 fatty acids collectively, and EPA and DHA in particular, have been ascribed a wide range of health benefits [46–51], with a combined intake of EPA + DHA from 250 to 500 mg/day in adults being recommended by health organizations [52]. Our calculations showed that less than 50 g of salmon head homogenate (DW) would be enough to meet those requirements. Therefore, even though the content in salmon heads of omega-3 fatty acids, namely EPA and DHA, is somewhat restricted in relative terms when compared to that of other fatty acids, the richness of salmon heads in lipid contents makes them good prospective sources of these fatty acids in qualitative or absolute terms. Moreover, despite this aforementioned restricted content of EPA and DHA, minced salmon heads still present a very enticing *n*-6/*n*-3 ratio (of 1.17). An *n*-6/*n*-3 ratio lower than 4–5 is recommended in the diet [53,54] as presenting several health benefits [55,56] and actively mitigating the prejudicial effects of the high *n*-6/*n*-3 ratios characteristic of Western diets [57,58]. Moreover, the saturated-to-polyunsaturated (SFA/PUFA) fatty acid index, one of the other most commonly used indexes for appraising the nutritional value of dietary foods, is also in line with that described for the edible parts of many other fish species, or even generally lower [59]. Additionally, minced salmon heads also present low thrombogenic indexes (TIs) and remarkably high hypocholesterolemic/hypercholesterolemic (h/H) indexes when generally compared to other fish species [59], which represent additional beneficial features in nutritional terms.

The analysis of the lipidome of the minced Atlantic salmon heads allowed detecting TGs containing small-chain fatty acids (8–12 carbons), as well as other odd-chain fatty acids in their fragmentation profiles (MS/MS) that were not detected by GC-MS. This fact is explained by a lower sensitivity of the GC-MS technique with regard to high resolution LC-MS, and by the fact that the TG species containing those fatty acids were vestigial with regard to the majority TG species. This study also highlighted the predominance of TGs as the most well-represented lipid class in this co-product. A study also characterizing the phospholipid content in salmon heads also reported residual contents of phospholipids, such as phosphatidylinositol (PI) and sphingomyelin (SM), in salmon heads, in addition to the phospholipid classes that were detected in our study [20]. Another study using lipidomics focused specifically on the characterization of phospholipid extracts from salmon heads [16], detecting molecular species of the following lipid classes: LPA, LPC, LPE, LPG, LPI, LPS, PA, PC, PE, PG, PI, and PS [16]. However, in our case, we were not able to detect lipids from the LPA, LPG, LPI, LPS, PA, and PG and PI classes given the overwhelming predominance of TGs in our total lipid extracts [16]. This previous study reported PC 34:1 as being the most abundant PC in salmon heads and PE O-34:2 to be the most abundant molecular species from the PE class [16]. In our study, PC 34:1 was, in fact, one of the most abundant PCs in the minced salmon heads, along with PC 42:6 and PC

38:6, while we were not able to detect ether lipids belonging to the PE lipid class. In the same previous study, the percentage of phospholipids containing EPA and DHA was larger than 20%, therefore being higher than the percentage of TGs containing those fatty acids reported here [16]. This enrichment of the polar lipid components of Atlantic salmon heads with regard to the results presented here for TGs was confirmed in another study, reporting EPA and DHA contents in polar lipid extracts to attain 10.1 and 16.8%, respectively [37]. Moreover, in king salmon (*Oncorhynchus tshawytscha*), this trend was also recorded, with polar lipid fractions of salmon heads being reported to be especially enriched in EPA and DHA with regard to the neutral lipid content [15].

A study specifically characterizing the TG profile in salmon muscle tissue through lipidomics identified TG 58:7 (16.4% of the class total) as the major TG present in the lipid extracts of these co-products, followed by TG 54:5 (11.6%) [11]. These results mean that the TG composition in the salmon muscle differs considerably from that of the head. Indeed, the most abundant TG in the head is TG 54:3 (16.2%) while the main TG in the muscle (TG 58:7) only occurs residually (≈0.02%) in the head. Two other different studies also employing a lipidomic approach to characterize the fillets/muscle of Atlantic salmon and focusing only on its phospholipid content identified PC 38:6 and PE 38:6 as the most abundant molecular species for each phospholipid class [8,10]. In salmon heads, PC 38:6 was among the most abundant PC present, along with PC 42:6, while we were not able to detect PE 38:6.

With regard to works characterizing the heads of other fish species by lipidomics tools, the available studies are difficult to compare to ours because they focused specifically on the content in omega-3-containing phospholipids. A study characterizing silver carp heads phospholipids detected lipid species from the LPC, LPE, LPG, LPI, LPS, phosphatidic acid (PA), PC, PE, PG, PI, and PS classes [16]. Here, we were not able to detect lipids from the LPG, LPI, LPS, PA, and PI classes, probably because of the fact that the polar lipid content is almost negligible in total lipid extracts of minced salmon heads. With regard to the classes that we were able to detect, this study reported LPC 16:0, LPE 22:6, PC 34:1, PE-O 34:2, PG 34:1, and PS 36:1 to be the most abundant lipid species of their class [16]. In our study, although we were only able to detect a restricted number of phospholipid species, we were able to detect LPC 16:0 and PC 34:1, and LPE 22:6 was, in fact, the only member of its class that we detected.

While the analysis of the lipid content and profile of salmon heads revealed a somewhat limited content in omega-3 fatty acids, and the marine fatty acids EPA and DHA, when compared to other animal seafood co-products [39], that content, in absolute terms, ended up being very considerable given the substantial content of lipids present in this salmon co-product. The global demand for omega-3 fatty acids was estimated to attain 3.61 billion by 2028 [60]. This projection intensified demand and consolidated the nutritional importance of omega-3 fatty acids, mostly due to the plethora of beneficial health-promoting benefits generally ascribed to the consumption of these fatty acids (namely EPA and DHA). These biomolecules have been reported to yield beneficial effects in the mitigation of neurodegenerative and cardiovascular diseases and cancer and protection against inflammation specifically [61–69]. However, the production of fish oils containing marine omega-3 fatty acids is under pressure, both from the side of demand, because of current demographics, and from the side of production, due to climate change impacts and overfishing [70,71]. Therefore, the search for new alternative sources of omega-3 fatty acids (namely EPA and DHA) is now more intensive than ever [72]. Fish co-products have already been profusely explored for the production of fish oils containing the marine omega-3 fatty acids EPA and DHA [73–78]. Given the volume of production and consumption of salmon at a worldwide scale, along with the remarkable level of lipid content in salmon heads, the situation can make these resources very enticing for food/feed applications. In this sense, given the current demand for omega-3 fatty acids, lipid fractions obtained from salmon heads could be potentially more readily explored [79–81]. Moreover, the similarities between salmon heads and olive oil, namely the high content of oleic acid [82–84], also overwhelmingly predominant in the form of triglycerides [85–87], may also be promising for future applications given the range of health benefits already reported for these biomolecules, especially in the context of cardiovascular disease and metabolic syndrome [88–92]. It is important to notice, however, that eventual uses of these resources for food/feed should take into account that lipids, particularly those containing PUFA, may be readily degraded by lipid oxidation reactions with the production of secondary oxidation products contributing to flavor deterioration and the occurrence of off-flavors [93–95]. Moreover, some of these products may present toxicity risks [96–98]. Therefore, the development of new approaches to detect and quantify oxidized lipids in resources like these co-products is important, and studies like this, characterizing the lipid components at a molecular level, represent a first step in that direction.

In the cosmetics industry, TGs from salmon heads may also represent valuable resources taking advantage of the emollient qualities of TGs [99]. In this case, this salmon co-product may represent an alternative not only to vegetable fat but also to other animal fat sources that have already been exploited in the industry [100–103]. Rendered poultry fat, tallow, and lard are widely used in the cosmeceutical industry for their occlusive, emulsion-stabilizing, emollient, surfactant, and viscosity-increasing properties [104]. The natural origin of these lipids can be a significant marketing point, catering to the increasing consumer demand for natural and sustainably sourced cosmetic ingredients. However, the purposeful use of salmon head lipids for cosmeceutical applications is still constrained by a general lack of bioprospecting of biological activity and bioactive lipids in the lipid fractions of these co-products.

This salmon co-product has, therefore, valorization potential in the food/feed and cosmetics industry. But another opportunity for its valorization while those applications are implemented and developed comes from an increasingly challenged energy industry. In fact, another alternative suitable application could include the production of biodiesel [105,106] and the production of skin care products. In biodiesel production, TGs represent primary materials due to their high energy content and efficient conversion to FAMEs [107–109]. The production of biodiesel from waste animal fat is rather established and presents great potential as these resources do not compete with the final production of food products and also contribute to a global reduction in waste produced by the food industry [106,110]. Reports have shown that the use of animal fat for the production of biodiesel has approximately doubled in the past decade [111]. Moreover, the demand for animal fat for biofuel production is projected to triple by 2030 with regard to the beginning of this decade [111]. The use of salmon-head-derived TGs may represent a valid alternative to vegetable oils and other animal fat sources [112–114] for biodiesel production, enabling an alternative and renewable energy source that can reduce reliance on fossil fuels and lower greenhouse gas emissions. The importance of salmon for human nutrition and the production volume of this fish species [115] make salmon co-products readily available bioresources. Interestingly, oleic acid in particular, by far the most abundant fatty acid in salmon heads, is a major component in almost all biodiesel formulations and was reported to present highly favorable features for biodiesel production, namely a convenient balance between oxidative stability and low-temperature operability [116–120]. Moreover, the very low levels of phospholipids in the lipid extracts of salmon heads may represent an additional advantage from an industrial point of view, simplifying the processing and refinement of these resources to highly concentrated TG fractions.

Here, we reported the lipid extracts of salmon heads to display moderate antioxidant activity. The lipid content in salmon heads and other salmon co-products remains, for the most part, rather unexplored in terms of the bioprospecting of biological activities and bioactive lipids, with the exception of a handful of studies. Salmon backbones, heads, and viscera oils obtained by Soxhlet and microwave-assisted extractions were previously tested for cytotoxic, antioxidant, anti-inflammatory, and antimicrobial activities with promising results [35]. As also reported here, the antioxidant activity of salmon head lipids was rather limited, and even only detected when salmon head oil was obtained specifically by one of the techniques employed (microwave-assisted extraction) [35]. The anti-inflammatory potential of lipid extracts from the heads of *S. salar* was confirmed in another study [36], as were the antimicrobial properties of oils produced from *S. salar* co-products: in this case, oils from the heads, but also discarded soft tissues [38]. Another study reported lipid extracts heads of *S. salar* to display antithrombotic activity and potential cardioprotective properties [37]. Lastly, phospholipid extracts from salmon heads were shown to display general ameliorating effects in rodent models of metabolic syndrome [16]. However, despite the promise of these bioprospecting studies and the remarkable lipid content of salmon heads, much remains to be explored in terms of screening for biological activity and bioactive lipids in salmon heads in particular and salmon co-products in general.

Other than lipids, the protein content of salmon co-products has also been explored for novel added-value applications. Not only the heads of salmon but also the trimmings, backbones/frames, viscera, tailfins, and the skin were all explored for the recovery of valuable protein fractions, gelatin, collagen, and protein hydrolysates [22–27,30–34]. Moreover, protein extracts and hydrolysates from salmon co-products (including the heads) have been reported to yield multiple bioactive properties [22–24,26–30]. These promising bioactivities and applications targeting the protein content of salmon co-products, including heads, along with the rather untapped potential of their lipids, may sustain and justify an integrated biorefinery approach that optimally recovers both proteins and lipids and directs these biomolecules for selected high-end uses. Such an approach would not only increase the profitability of the salmon industry by creating multiple new revenue streams but also contribute to its sustainability by minimizing waste and promoting a circular economy. In this study, the priority was to characterize, as deeply and thoroughly as possible, the lipid content in these resources to fully unveil their potential in terms of their lipid content. For this purpose, we used the traditional extraction methods (the Bligh and Dyer method, more specifically [121]) known to guarantee the highest yield and coverage of lipid classes. It is obvious that if food/feed applications are to be prioritized, lipid extraction, either alone or in biorefinery setups, must be explored using green extraction methods and solvents. In this case, our study offers a good baseline for comparison and for an eventual optimization of other methods.

The limitations of this work include the fact that the lipidomics approach involved a semi-quantitative procedure, and therefore, it is not possible to calculate, quantitatively, the amounts of each lipid species and lipid classes per quantity of biomass of the co-product. Moreover, since the objective of this work was to perform a characterization of the lipid content as thoroughly as possible, we used the standard methods (Bligh and Dyer) known to assure an optimized extraction of total lipid and lipid class coverage. However, these classical methods use solvents that are not compatible with uses in the food/feed industry and, therefore, the next steps of the valorization of these co-products include testing the use of green methods/solvents for the lipid extraction in these resources. Another limitation was the fact that we were only able to bioprospect biological activity in lipid extracts from these resources in terms of antioxidant activity. In the future, it will be important to extend this screening to other possible manifestations of biological activity in order to improve the possibilities of unveiling other innovative uses for these resources in possible higher-end applications. Finally, future work also includes the planning and testing of biorefinery setups able to optimally recover and guarantee the quality of both lipid and protein contents from these co-products.

4. Materials and Methods

4.1. Chemicals

High-grade liquid chromatography (HPLC)-grade dichloromethane (CH_2Cl_2), 96% absolute ethanol (CH_3CH_2OH), and methanol (CH_3OH) were obtained from Fisher Scientific Ltd. (Loughborough, UK). Milli-Q purified water was obtained using the Synergy® system from Millipore Corporation (Billerica, MA, USA). Whatman No. 1 filter paper was acquired from Sigma-Aldrich, St. Louis, MO, USA. The 2,2′-azino-bis(3-ethylbenzothiazoline-6-

sulfonic acid) radical cation ($ABTS^{•+}$) was obtained from Fluka (Buchs, Switzerland) while the α,α-diphenyl-β-picrylhydrazyl radical ($DPPH^{•}$) was obtained from Sigma-Aldrich (St. Louis, MO, USA). The 37 Component FAME Mix was purchased from Supelco (Sigma-Aldrich, St. Louis, MO, USA) along with the internal standard methyl nonadecanoate (≥99% purity) from Sigma-Aldrich (St. Louis, MO, USA). Lipid internal standards for lipidomic analyses were acquired from Avanti Polar Lipids, Inc. (Alabaster, AL, USA), including 1,2-dimyristoyl-*sn*-glycero-3-phosphate (dMPA), 1,2-dimyristoyl-*sn*-glycero-3-phospho-(1′-rac-glycerol) (dMPG), 1,2-dimyristoyl-*sn*-glycero-3-phosphocholine (dMPC), 1,2-dipalmitoyl-*sn*-glycero-3-phosphatidylinositol (dPPI), 1,2-dimyristoyl-sn-glycero-3-phosphoethanolamine (dMPE), 1,2-dimyristoyl-*sn*-glycero-3-phosphatidylserine (dMPS), 1-nonadecanoyl-2-hydroxy-*sn*-glycero-3-phosphocholine (LPC), N-heptadecanoyl-D-erythro-sphingosine (Cer), 1′,3′-bis[1,2-di-tetradecanoyl-*sn*-glycero-3-phospho]-*sn*-glycerol (CL), and N-heptadecanoyl-D-erythro-sphingosylphosphorylcholine (SM). All other reagents were purchased from leading commercial suppliers guaranteeing product quality.

4.2. Samples

Fresh heads of farmed Atlantic salmon (*Salmo salar*) were purchased at a local supermarket and transported to the laboratory on ice. Heads were ground and minced in an industrial meat grinder (HOBART, Troy, OH, USA) and stored at −20 °C until analysis. Heads were mechanically ground and minced in an industrial meat grinder (HOBART, Troy, OH, USA) and stored at −20 °C until analysis. These co-products did not include any additive since only mechanical means are used in the industrial processing. Five different portions of minced heads were subsequently subjected to all the analyses described in this section.

4.3. Biochemical and Elemental Composition

Moisture and ash contents were assessed using standard analytical methods. Five portions of minced salmon heads (250 mg) were placed in ceramic crucibles and dried overnight in an oven set at 105 °C. Following a 24 h period, the crucibles were cooled to room temperature in a desiccator containing silica gel and weighed to determine moisture content. Ash determination followed a two-step procedure. First, the salmon-head mince was pre-incinerated by placing the crucibles on a heated plate for 20 min. Subsequently, the crucibles were transferred to a muffle furnace and maintained at 575 °C for 6 h. Once cooled, the crucibles were weighed to determine the ash content.

The elemental composition of lyophilized co-product samples, specifically carbon (C), hydrogen (H), nitrogen (N), and sulfur (S) contents, was analyzed using a Leco Truspec-Micro CHNS 630-200-200 elemental analyzer. The combustion furnace temperature was set at 1075 °C, with the afterburner set to 850 °C. Approximately 2 mg of minced salmon heads was combusted in an oxygen/carrier gas mixture, ensuring complete combustion and conversion of co-products to water vapor, carbon dioxide, and nitrogen for gas analysis. Carbon, hydrogen, and sulfur were detected via infrared absorption while nitrogen was measured using thermal conductivity. This elemental analysis enabled the estimation of protein content using a nitrogen-to-protein conversion factor. The standard conversion factor of 6.25 was employed as it had commonly been used in previous studies to determine protein content in salmon [3,20,21,23]. Lipid content was measured by gravimetry after lipid extraction, and carbohydrates (with other compounds) represented the remaining wet weight percentage after accounting for moisture, ash, lipid, and protein contents.

4.4. Lipid Extraction

Total lipid extracts from salmon head homogenate were obtained using the Bligh and Dyer method, with slight adaptations [121,122]. Approximately 10 mg of freeze-dried co-product samples were thoroughly minced and homogenized with a mortar and pestle and placed in glass tubes. To these tubes, 2.5 mL of methanol and 1.25 mL of dichloromethane were added. The mixture was vortexed for 1 min and another 1.25 mL was added to each sample-containing tube. After vortexing for another minute, the mixtures were incubated for 30 min in an orbital shaker on ice (4 °C) and then centrifuged at 626× *g* for 10 min at room temperature. The supernatant was transferred to a new glass tube and the precipitate (biomass) was re-extracted two times (adding 2.5 mL methanol and 1.25 mL dichloromethane each time, followed by 1 min vortexing and centrifugation at 626× *g* for 10 min at room temperature, with subsequent collection of the supernatant into the same tube). The combined organic phases were completely dried under a nitrogen stream, re-dissolved in 2 mL of dichloromethane and 2 mL of methanol, and vortexed for 1 min. Ultra-pure water (1.8 mL) was added to each tube; the mixtures were vortexed for 1 min and centrifuged at 626× *g* for 10 min at room temperature. Organic (lower) phases were collected into new tubes and the aqueous (upper) phases were re-extracted with 2 mL of dichloromethane, followed by 2 min of vortexing and a final centrifugation at 626× *g* for 10 min at room temperature. The lower organic phases were combined in the same tube, filtered (Whatman No. 1 filter paper), dried under a nitrogen stream, and preserved at −20 °C for further analysis. Finally, the total lipid content in salmon heads was estimated by gravimetric analysis.

4.5. Phospholipid Quantification in Total Lipid Extracts

The phospholipid quantification in total lipid extracts of minced salmon heads was performed using a modified version of Bartlett and Lewis method [122–124]. Briefly, samples were dissolved in 300 µL of dichloromethane, and small aliquots (10 µL, in duplicates) were transferred to acid-washed glass tubes and evaporated under a nitrogen stream. Each sample-containing tube was then added 125 µL of perchloric acid ($HClO_4$, 70% m/V) and incubated for 60 min at 180 °C in a steel heating block. Following the hydrolysis step, 850 µL of water, 125 µL of ammonium molybdate (2.5%, m/V), and 125 µL of ascorbic acid (10%, m/V) were added to the content of each tube. The mixture was then vortexed and incubated for 10 min at 100 °C in a water bath. A calibration curve was prepared using standards with known phosphorus concentrations ranging from 0.1 to 2.0 µg (using a standard solution of $NaH_2PO_4 \cdot 2H_2O$ containing 100 µg of phosphorus per mL). Absorbance measurements were performed at 797 nm using a Thermo Scientific Multiskan Go microplate UV-vis spectrophotometer (Thermo Scientific, Hudson, NH, USA).

4.6. Gas Chromatography–Mass Spectrometry (GC-MS)

The fatty acid content in lipid extracts of minced salmon heads was analyzed by GC-MS following transmethylation. Thirty microgram aliquots of the lipid extracts were transferred to glass tubes and dried under a nitrogen stream. The lipid films were then dissolved in 1 mL of *n*-hexane containing C19:0 as an internal standard (1 µg mL^{-1}, CAS number 1731-94-8, Merck, Darmstadt, Germany). Each tube was added 200 µL of a potassium hydroxide (KOH) 2 M solution in methanol and the mixture was vortexed for 2 min. Next, 2 mL of a saturated sodium chloride (NaCl) solution was added and the mixture was centrifuged for 5 min at 626× *g* to promote phase separation. Cholesterol in the upper (organic) phase was removed using a protocol available on the Lipid Web "https://lipidhome.co.uk/ms/basics/msmeprep/index.htm (accessed on 5 May 2024). A 10 mm silica column in a pipette tip with wool was pre-conditioned with 5 mL of hexane. Methyl esters were added to the top of the column and eluted with a hexane ether (95:5, *v*/*v*, 3 mL) mixture, then completely dried under a nitrogen stream. Finally, fatty acid methyl esters (FAMEs) were dissolved in 100 µL of *n*-hexane, and 2 µL of the resulting solution was injected into an Agilent Technologies 8860 GC System (Santa Clara, CA, USA) equipped with a DB-FFAP column (30 m length, 0.32 mm internal diameter, and 0.25 µm film thickness, J&W Scientific, Folsom, CA, USA). The gas chromatograph was connected to an Agilent 5977B Network Mass Selective Detector, operating with electron impact ionization at 70 eV, scanning the mass range of m/z 50–550 in a 1 s cycle in full scan mode. The oven temperature program started at 58 °C for 2 min and increased by 25 °C min^{-1} to 160 °C, by 2 °C min^{-1} to 210 °C, and by 30 °C min^{-1} to 250 °C, holding for

10 min. The injector and detector temperatures were set at 220 °C and 280 °C, respectively. Helium was used as the carrier gas at a flow rate of 1.4 mL min^{-1}. Fatty acids were identified by comparing retention times to those of commercial FAME standards in the Supelco 37 Component FAME Mix (ref. 47885-U, Sigma-Aldrich, Darmstadt, Germany) and by MS-spectrum comparison with chemical databases (Wiley 275 library and AOCS lipid library). The relative percentages of fatty acids were calculated using the percent relative area method and quantification was performed by normalization with the included internal standard methyl nonadecanoate (19:0). Various indexes, including average chain length (ACL), double bond index (DBI), peroxidizability index (PI), atherogenic index (AI), thrombogenic index (TI), hypocholesterolemic/hypercholesterolemic index (h/H), and polienic index (PoI), were calculated as previously described [122,125].

4.7. Reverse-Phase Liquid Chromatography–Mass Spectrometry (C18–LC–MS)

Total lipid extracts from minced salmon heads were analyzed by reverse-phase liquid chromatography (C18-LC-MS) using a Dionex Ultimate 3000 (Thermo Fisher Scientific, Bremen, Germany) with an Ascentis® Express 90 Å C18 column (Sigma-Aldrich®, 2.1 × 150 mm, 2.7 μm) coupled to a Q-Exactive® hybrid quadrupole Orbitrap mass spectrometer (Thermo Fisher, Scientific, Bremen, Germany). The mobile phases used for the gradient during RP-LC-MS determinations consisted of mobile phase A (Milli-Q water/acetonitrile (40/60%) with 10 mM ammonium formate (NH_4HCO_2) and 0.1% formic acid (CH_2O_2)) and mobile phase B (isopropanol/acetonitrile (90/10%) with 10 mM ammonium formate and 0.1% formic acid). The gradient defined for the experiments was as follows: 32% B at 0 min, 45% B at 1.5 min, 52% B at 4 min, 58% B at 5 min, 66% B at 8 min, 70% B at 11 min, 85% B at 14 min, 97% B at 18 min, 97% B at 25 min, 32% B at 25.01 min, and 32% B at 33 min. A mixture containing 1 μg of lipid extract from salmon head homogenate was prepared in 91 μL of a solvent system consisting of 50% isopropanol/50% methanol and 8 μL of a mixture of phospholipid standards (dMPC—0.04 μg, SM d18:1/17:0—0.04 μg, dMPE—0.04 μg, LPC—0.04 μg, dPPI—0.08 μg, CL(14:0)4—0.16 μg; dMPG—0.024 μg, Cer 17:0/d18:1—0.08 μg, dMPS—0.08 μg, and dMPA—0.16 μg) and loaded into the C18 column at 50 °C with a flow rate of 260 μL min^{-1}. The mass spectrometer operated simultaneously in positive (3.0 kV) and negative (−2.7 kV) modes. The capillary temperature was set to 320 °C and the sheath gas flow to 35 U. Data acquisition was performed in full-scan mode with a high resolution of 70,000 and automatic gain control (AGC) target of 3×10^6, in an m/z range of 300–1600, with 2 micro scans and a maximum injection time (IT) of 100 ms. Tandem mass spectra (MS/MS) resolution was 17,500, with an AGC target of 1×10^5, 1 micro scan, and a maximum IT of 100 ms. The cycles included a full-scan mass spectrum and 10 data-dependent MS/MS scans, continuously repeated throughout the experiments with a dynamic exclusion of 30 s and an intensity threshold of 8×10^4. The normalized collision energy (CE) ranged between 20, 24, and 28 eV in the negative mode and 25 and 30 eV in the positive mode. Data acquisition was performed using the Xcalibur data system (V3.3, Thermo Fisher Scientific, Bremen, Germany). Molecular lipid species were identified using the Lipostar software version 2.1.5 (Molecular Discovery Ltd., Borehamwood, UK) [126]. This software is capable of processing raw-data importing, peak detection, integration, and identification. Lipid assignment and identification were performed against a database created from the LIPID MAPS structure database (version of June 2024). The database was fragmented using the DB Manager Module of Lipostar, following Lipostar fragmentation rules. The raw files were directly imported and aligned according to the settings defined by Lange et al. [127]. Automatic peak picking was carried out with the SDA smoothing level set to high and a minimum signal-to-noise ratio of 3. Automatic isotope clustering settings included a tolerance of 7 ppm and a retention time tolerance of 0.2 min. The MS/MS filter was applied to retain features with MS/MS spectra for identification. Lipid identification was performed using the following parameters: 5 ppm precursor ion mass tolerance and 10 ppm product ion mass tolerance. Lipostar annotations and assignments were manually confirmed based on the analysis of the fragmentation (MS/MS spectra) and the presence of the m/z signaling the presence of the fatty acids and class-characteristic fragments. The areas of the peaks of each lipid species were normalized by calculating the ratio against the area of the respective class internal lipid standard included at a known concentration. The relative abundance of each lipid species was estimated by dividing the normalized peak areas of each lipid species by the sum of the total normalized peak areas.

4.8. Antioxidant Activity

The antioxidant scavenging activities of minced salmon heads total lipid extracts against α,α-diphenyl-β-picrylhydrazyl ($DPPH^{\bullet}$) and 2,2′-azino-bis-3-ethylbenzothiazoline-6-sulfonic acid radical cation ($ABTS^{\bullet+}$) were evaluated using previously described methods [125,128]. Briefly, 150 µL of an ethanolic dilution of the lipid extracts (50, 250, and 500 µg mL^{-1}) was combined with 150 µL of $DPPH^{\bullet}$ or $ABTS^{\bullet+}$ working solution in ethanol (absorbance ≈ 0.9). The samples were incubated for 120 min and the absorbance was measured at 517 nm for $DPPH^{\bullet}$ and 734 nm for $ABTS^{\bullet+}$ every 5 min using a UV-vis spectrophotometer (Multiskan GO 1.00.38, Thermo Scientific, Hudson, NH, USA). Controls were prepared by substituting the radical solution with ethanol. To ensure the stability of the radicals, solutions with the radical plus ethanol were also prepared. All measurements were performed in triplicate. The same procedure was applied to the Trolox standard solution (12.5, 62.5, 125, 250 µg mL^{-1} in ethanol). The antioxidant activity, as expressed as the percentage of inhibition of the $DPPH^{\bullet}$ (or $ABTS^{\bullet+}$), was calculated using Equation (1), which is as follows:

$$\text{Inhibition (\%)} = ((\text{Abs}_{\text{Radical}} - (\text{Abs}_{\text{Sample}} - \text{Abs}_{\text{Control}}))/\text{Abs}_{\text{Radical}}) \times 100 \quad (1)$$

Here, $\text{Abs}_{\text{Radical}}$ is the absorbance of the radical ($DPPH^{\bullet}$ or $ABTS^{\bullet+}$), $\text{Abs}_{\text{Sample}}$ is the absorbance of the sample with radical ($DPPH^{\bullet}$ or $ABTS^{\bullet+}$), and $\text{Abs}_{\text{Control}}$ is the absorbance of the sample with ethanol.

The antioxidant activity expressed in Trolox equivalents (TE) was calculated according to Equation (2):

$$\text{TE } (\mu\text{mol g}^{-1}) = \text{IC Trolox } (\mu\text{mol g}^{-1}) \times 1000/\text{IC of samples } (\mu\text{g mL}^{-1}) \quad (2)$$

Here, IC is the concentration of lipid extract per sample and of Trolox, which promotes inhibition at the tested concentrations of the extract in the radicals $DPPH^{\bullet}$ or $ABTS^{\bullet+}$.

4.9. Statistical Analysis

Figures were produced using GraphPad Prism version 7.00 for Windows (GraphPad Software, La Jolla, CA, USA). All experimental data are shown as means ± standard deviations (SDs) for 5 samples of minced salmon heads (n = 5).

5. Conclusions

Given its significant role in human nutrition and its importance as one of the most produced and preferred fish species for human consumption worldwide, the salmon industry will continue to generate considerable quantities of co-products that cannot be regarded as waste. Therefore, the efficient use of salmon co-products is paramount to foster sustainability in fisheries and aquaculture. One of the most important findings of this study was the remarkably rich lipid content of minced salmon heads (23.97 ± 0.72 WW). Such a substantial lipid content should warrant further study, particularly targeting bioprospecting and valorization pathways. The lipid content of minced salmon heads occurs predominantly in the form of triglycerides (TGs), paving the way for multiple uses (Figure 5). A biorefinery setup, which optimizes the extraction of both lipids and proteins contents, could represent a promising approach to handle these resources. This dual extraction strategy could lead to a more comprehensive utilization of salmon heads, contributing to the economic and environmental sustainability of the salmon industry. This study proved that the biochemical characterization of co-products, in particular the elucidation of the lipid content, can play a key role in identifying novel uses for these bioresources, which, in the case of these particular resources, may include food/feed and cosmeceutical applications. Overall, the present study ultimately highlighted the importance of a thorough characterization of fish co-products for the optimization of their use and value, ultimately supporting a more sustainable and economically viable fish industry under a blue bioeconomy framework.

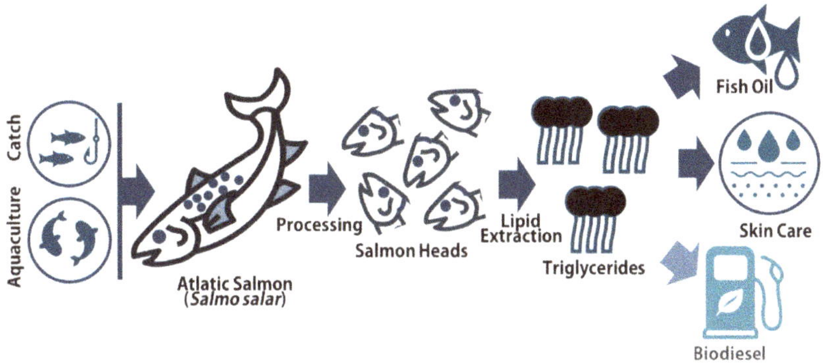

Figure 5. Salmon heads as source of concentrated triglyceride fractions and potential industry applications as pathways or valorization.

Supplementary Materials: The following supporting information can be downloaded at: https://www.mdpi.com/article/10.3390/md22110518/s1, Figure S1: Elemental composition of the grinded salmon head; Figure S2: Fatty acid profile of the grinded salmon heads as presented in absolute terms; Figure S3: Total ion count chromatogram of a grinded salmon sample in the positive mode; Table S1: Molecular species identified by mass accuracy HPLC–MS and characterized by MS/MS analyses in minced salmon heads; Table S2: Normalized areas of the peaks of the molecular species identified by mass accuracy HPLC–MS in minced salmon heads.

Author Contributions: Conceptualization, J.P.M., R.C. and M.R.D.; methodology and formal analysis, J.P.M., T.S. and T.M.; data curation, J.P.M. and T.S.; investigation, and writing—original draft preparation, J.P.M. and T.S.; writing—review and editing, C.P., A.M., M.L.N., R.C. and M.R.D.; statistics, software, and validation, J.P.M.; resources. C.P., A.M., M.L.N., R.C. and M.R.D.; funding acquisition, R.C. and M.R.D. All authors have read and agreed to the published version of the manuscript.

Funding: This study was performed within the scope of the project "BLUE BIOECONOMY PACT" (Project N°. C644915664-00000026), co-funded by the Next Generation EU European Fund, under the incentive line "Agendas for Business Innovation" within Component 5—Capitalization and Business Innovation of the Portuguese Recovery and Resilience Plan (RRP): specifically, under the transversal Co-Products Valorization Platform (WP8).

Institutional Review Board Statement: Not applicable.

Data Availability Statement: The data presented in this study are available on request from the corresponding author.

Acknowledgments: This is a contribution of the Marine Lipidomics Laboratory at the University of Aveiro. The authors are thankful to the project "BLUE BIOECONOMY PACT" (Project N°. C644915664-00000026), co-funded by the Next Generation EU European Fund, under the incentive line "Agendas for Business Innovation" within Component 5—Capitalization and Business Innovation of the Portuguese Recovery and Resilience Plan (RRP): specifically, under the transversal Co-Products Valorization Platform (WP8). The authors also acknowledge the COST Action EpiLipidNET, CA19105-Pan-European Network in Lipidomics and EpiLipidomics. This work is a contribution of the Marine Lipidomics Laboratory. The authors thank the University of Aveiro, Fundação para a Ciência e Tecnologia (FCT), and Ministério da Ciência Tecnologia e Ensino Superior (MCTES) for the financial support to the research units CESAM (UIDB/50017/2020 + UIDP/50017/2020 + LA/P/0094/2020) and LAQV-REQUIMTE (UIDP/50006/2020 + UIDB/50006/2020) through national funds and co-funded by the ERDF, within the Portugal 2020 Partnership Agreement and Compete

2020, and the Portuguese Mass Spectrometry Network (RNEM, LISBOA-01-0145-FEDER-402-022125). The authors also acknowledge Mass Analytica (https://mass-analytica.com/) for making Lipostar available for the data analysis in this study.

Conflicts of Interest: The authors declare no conflict of interest.

References

1. Jensen, I.J.; Mæhre, H.K.; Tømmerås, S.; Eilertsen, K.E.; Olsen, R.L.; Elvevoll, E.O. Farmed Atlantic salmon (*Salmo salar* L.) is a good source of long chain omega-3 fatty acids. *Nutr. Bull.* **2012**, *37*, 25–29. [CrossRef]
2. Kousoulaki, K.; Østbye, T.K.; Krasnov, A.; Torgersen, J.S.; Mørkøre, T.; Sweetman, J. Metabolism, health and fillet nutritional quality in Atlantic salmon (Salmo salar) fed diets containing n-3-rich microalgae. *J. Nutr. Sci.* **2015**, *4*, e24. [CrossRef] [PubMed]
3. Sprague, M.; Fawcett, S.; Betancor, M.B.; Struthers, W.; Tocher, D.R. Variation in the nutritional composition of farmed Atlantic salmon (Salmo salar L.) fillets with emphasis on EPA and DHA contents. *J. Food Compos. Anal.* **2020**, *94*, 103618. [CrossRef]
4. Głuchowski, A.; Czarniecka-Skubina, E.; Rutkowska, J. Salmon (*Salmo salar*) Cooking: Achieving Optimal Quality on Select Nutritional and Microbiological Safety Characteristics for Ready-to-Eat and Stored Products. *Molecules* **2020**, *25*, 5661. [CrossRef] [PubMed]
5. Food and Agriculture Organization of the United Nations, Salmon-Main Producers See Record-Breaking Exports. Available online: https://www.fao.org/in-action/globefish/news-events/news/news-detail/Salmon---Main-producers-see-record-breaking-exports/en (accessed on 20 June 2024).
6. Fund, W.W. Farmed Seafood-Farmed Salmon: Overview. Available online: https://www.worldwildlife.org/industries/farmed-salmon (accessed on 20 June 2024).
7. Station, T.R. *Yield and Nutritional Value of the Commercially More Important Fish Species*; Food & Agriculture Org.: Rome, Italy, 1989.
8. Yu, X.; Li, L.; Wang, H.; Song, G.; Wang, J.; Li, S.; Wang, Y.; Shen, Q. Lipidomics study of rainbow trout (*Oncorhynchus mykiss*) and salmons (*Oncorhynchus tshawytscha* and *Salmo salar*) using hydrophilic interaction chromatography and mass spectrometry. *LWT-Food Sci. Technol.* **2020**, *121*, 108988. [CrossRef]
9. Ellisor, D.L.; Bayless, A.L.; Schock, T.B.; Davis, W.C.; Knott, B.T.; Seghers, J.; Leys, H.; Emteborg, H. Multi-omics characterization of NIST seafood reference materials and alternative matrix preparations. *Anal. Bioanal. Chem.* **2024**, *416*, 773–785. [CrossRef]
10. Yeo, J.; Parrish, C.C. Shotgun Lipidomics for the Determination of Phospholipid and Eicosanoid Profiles in Atlantic Salmon (*Salmo salar* L.) Muscle Tissue Using Electrospray Ionization (ESI)-MS/MS Spectrometric Analysis. *Int. J. Mol. Sci.* **2021**, *22*, 2272. [CrossRef]
11. Yeo, J.; Parrish, C.C. Evaluation of triacylglycerol (TAG) profiles and their contents in salmon muscle tissue using ESI-MS/MS spectrometry with multiple neutral loss scans. *Food Chem.* **2020**, *324*, 126816. [CrossRef]
12. Broughton, R.; Tocher, D.R.; Napier, J.A.; Betancor, M.B. Profiling Phospholipids within Atlantic Salmon Salmo salar with Regards to a Novel Terrestrial Omega-3 Oil Source. *Metabolites* **2022**, *12*, 851. [CrossRef]
13. Jin, Y.; Harvey, T.N.; Bartosova, Z.; Hassani, S.; Bruheim, P.; Sandve, S.R.; Vik, J.O. Diet and Life Stage-Associated Lipidome Remodeling in Atlantic Salmon. *J. Agric. Food Chem.* **2021**, *69*, 3787–3796. [CrossRef]
14. Fiorino, G.M.; Losito, I.; De Angelis, E.; Arlorio, M.; Logrieco, A.F.; Monaci, L. Assessing fish authenticity by direct analysis in real time-high resolution mass spectrometry and multivariate analysis: Discrimination between wild-type and farmed salmon. *Food Res. Int.* **2019**, *116*, 1258–1265. [CrossRef] [PubMed]
15. Ahmmed, M.K.; Carne, A.; Ahmmed, F.; Stewart, I.; Sabrina Tian, H.; Bekhit, A.E.A. Positional distribution of fatty acids and phospholipid composition in King salmon (*Oncorhynchus tshawytscha*) head, roe and skin using nuclear magnetic resonance spectroscopy. *Food Chem.* **2021**, *363*, 130302. [CrossRef]
16. Wang, Q.; Wang, R.; Zhao, X.; Lu, H.; Zhang, P.; Dong, X.; Wang, Y. Comparison of the Effect of Phospholipid Extracts from Salmon and Silver Carp Heads on High-Fat-Diet-Induced Metabolic Syndrome in C57BL/6J Mice. *Mar. Drugs* **2023**, *21*, 409. [CrossRef]
17. Ahmmed, M.K.; Ahmmed, F.; Stewart, I.; Carne, A.; Tian, H.S.; Bekhit, A.E.A. Omega-3 phospholipids in Pacific blue mackerel (*Scomber australasicus*) processing by-products. *Food Chem.* **2021**, *353*, 129451. [CrossRef] [PubMed]
18. Zhang, H.; Lu, Y.; Zhang, Y.; Dong, J.; Jiang, S.; Tang, Y. DHA-enriched phosphatidylserine ameliorates cyclophosphamide-induced liver injury via regulating the gut-liver axis. *Int. Immunopharmacol.* **2024**, *140*, 112895. [CrossRef]
19. Zhang, T.-T.; Xu, J.; Wang, Y.-M.; Xue, C.-H. Health benefits of dietary marine DHA/EPA-enriched glycerophospholipids. *Prog. Lipid Res.* **2019**, *75*, 100997. [CrossRef]
20. Gbogouri, G.; Linder, M.; Fanni, J.; Parmentier, M. Analysis of lipids extracted from salmon (*Salmo salar*) heads by commercial proteolytic enzymes. *Eur. J. Lipid Sci. Technol.* **2006**, *108*, 766–775. [CrossRef]
21. Malcorps, W.; Newton, R.W.; Sprague, M.; Glencross, B.D.; Little, D.C. Nutritional Characterisation of European Aquaculture Processing By-Products to Facilitate Strategic Utilisation. *Front. Sustain. Food Syst.* **2021**, *5*, 720595. [CrossRef]
22. Idowu, A.T.; Benjakul, S.; Sinthusamran, S.; Sookchoo, P.; Kishimura, H. Protein hydrolysate from salmon frames: Production, characteristics and antioxidative activity. *J. Food Biochem.* **2019**, *43*, e12734. [CrossRef]

23. Vázquez, J.A.; Sotelo, C.G.; Sanz, N.; Pérez-Martín, R.I.; Rodríguez-Amado, I.; Valcarcel, J. Valorization of Aquaculture By-Products of Salmonids to Produce Enzymatic Hydrolysates: Process Optimization, Chemical Characterization and Evaluation of Bioactives. *Mar. Drugs* **2019**, *17*, 676. [CrossRef]
24. Taroncher, M.; Rodríguez-Carrasco, Y.; Aspevik, T.; Kousoulaki, K.; Barba, F.J.; Ruiz, M.J. Cytoprotective Effects of Fish Protein Hydrolysates against H(2)O(2)-Induced Oxidative Stress and Mycotoxins in Caco-2/TC7 Cells. *Antioxidants* **2021**, *10*, 975. [CrossRef] [PubMed]
25. de la Fuente, B.; Pallarés, N.; Berrada, H.; Barba, F.J. Salmon (Salmo salar) Side Streams as a Bioresource to Obtain Potential Antioxidant Peptides after Applying Pressurized Liquid Extraction (PLE). *Mar. Drugs* **2021**, *19*, 323. [CrossRef] [PubMed]
26. Henaux, L.; Pereira, K.D.; Thibodeau, J.; Pilon, G.; Gill, T.; Marette, A.; Bazinet, L. Glucoregulatory and Anti-Inflammatory Activities of Peptide Fractions Separated by Electrodialysis with Ultrafiltration Membranes from Salmon Protein Hydrolysate and Identification of Four Novel Glucoregulatory Peptides. *Membranes* **2021**, *11*, 528. [CrossRef]
27. Liu, H.; Li, B. Separation and identification of collagen peptides derived from enzymatic hydrolysate of Salmo salar skin and their anti-inflammatory activity in lipopolysaccharide (LPS)-induced RAW264.7 inflammatory model. *J. Food Biochem.* **2022**, *46*, e14122.
28. Harnedy, P.A.; Parthsarathy, V.; McLaughlin, C.M.; O'Keeffe, M.B.; Allsopp, P.J.; McSorley, E.M.; O'Harte, F.P.M.; FitzGerald, R.J. Atlantic salmon (Salmo salar) co-product-derived protein hydrolysates: A source of antidiabetic peptides. *Food Res. Int.* **2018**, *106*, 598–606. [CrossRef]
29. Daskalaki, M.G.; Axarlis, K.; Tsoureki, A.; Michailidou, S.; Efraimoglou, C.; Lapi, I.; Kolliniati, O.; Dermitzaki, E.; Venihaki, M.; Kousoulaki, K.; et al. Fish-Derived Protein Hydrolysates Increase Insulin Sensitivity and Alter Intestinal Microbiome in High-Fat-Induced Obese Mice. *Mar. Drugs* **2023**, *21*, 343. [CrossRef]
30. Lapi, I.; Kolliniati, O.; Aspevik, T.; Deiktakis, E.E.; Axarlis, K.; Daskalaki, M.G.; Dermitzaki, E.; Tzardi, M.; Kampranis, S.C.; Marsni, Z.E.; et al. Collagen-Containing Fish Sidestream-Derived Protein Hydrolysates Support Skin Repair via Chemokine Induction. *Mar. Drugs* **2021**, *19*, 396. [CrossRef] [PubMed]
31. Wu, H.; Abdollahi, M.; Undeland, I. Effect of recovery technique, antioxidant addition and compositional features on lipid oxidation in protein enriched products from cod- salmon and herring backbones. *Food Chem.* **2021**, *360*, 129973. [CrossRef] [PubMed]
32. Vázquez, J.A.; Hermida-Merino, C.; Hermida-Merino, D.; Piñeiro, M.M.; Johansen, J.; Sotelo, C.G.; Pérez-Martín, R.I.; Valcarcel, J. Characterization of Gelatin and Hydrolysates from Valorization of Farmed Salmon Skin By-Products. *Polymers* **2021**, *13*, 2828. [CrossRef]
33. Buscaglia, M.; Guérard, F.; Roquefort, P.; Aubry, T.; Fauchon, M.; Toueix, Y.; Stiger-Pouvreau, V.; Hellio, C.; Le Blay, G. Mechanically Enhanced Salmo salar Gelatin by Enzymatic Cross-linking: Premise of a Bioinspired Material for Food Packaging, Cosmetics, and Biomedical Applications. *Mar. Biotechnol.* **2022**, *24*, 801–819. [CrossRef]
34. Jeong, H.; Yoon, S.; Yang, N.E.; Youn, M.Y.; Hong, S.J.; Jo, S.M.; Kim, K.S.; Jeong, E.J.; Kim, H.W.; Shin, E.C. Chemometric approach for an application of Atlantic salmons (*Oncorhynchus keta*) by-product for potential food sources. *Food Sci. Biotechnol.* **2024**, *33*, 855–876. [CrossRef] [PubMed]
35. de la Fuente, B.; Pinela, J.; Mandim, F.; Heleno, S.A.; Ferreira, I.C.F.R.; Barba, F.J.; Berrada, H.; Caleja, C.; Barros, L. Nutritional and bioactive oils from salmon (*Salmo salar*) side streams obtained by Soxhlet and optimized microwave-assisted extraction. *Food Chem.* **2022**, *386*, 132778. [CrossRef] [PubMed]
36. Ahmad, T.B.; Rudd, D.; Kotiw, M.; Liu, L.; Benkendorff, K. Correlation between Fatty Acid Profile and Anti-Inflammatory Activity in Common Australian Seafood by-Products. *Mar. Drugs* **2019**, *17*, 155. [CrossRef]
37. Tsoupras, A.; O'Keeffe, E.; Lordan, R.; Redfern, S.; Zabetakis, I. Bioprospecting for Antithrombotic Polar Lipids from Salmon, Herring, and Boarfish By-Products. *Foods* **2019**, *8*, 416. [CrossRef]
38. Inguglia, L.; Chiaramonte, M.; Di Stefano, V.; Schillaci, D.; Cammilleri, G.; Pantano, L.; Mauro, M.; Vazzana, M.; Ferrantelli, V.; Nicolosi, R.; et al. *Salmo salar* fish waste oil: Fatty acids composition and antibacterial activity. *PeerJ* **2020**, *8*, e9299. [CrossRef]
39. Monteiro, J.P.; Domingues, M.R.; Calado, R. Marine Animal Co-Products-How Improving Their Use as Rich Sources of Health-Promoting Lipids Can Foster Sustainability. *Mar. Drugs* **2024**, *22*, 73. [CrossRef] [PubMed]
40. Cooney, R.; de Sousa, D.B.; Fernández-Ríos, A.; Mellett, S.; Rowan, N.; Morse, A.P.; Hayes, M.; Laso, J.; Regueiro, L.; Wan, A.H.L.; et al. A circular economy framework for seafood waste valorisation to meet challenges and opportunities for intensive production and sustainability. *J. Clean. Prod.* **2023**, *392*, 136283. [CrossRef]
41. Coppola, D.; Lauritano, C.; Palma Esposito, F.; Riccio, G.; Rizzo, C.; de Pascale, D. Fish Waste: From Problem to Valuable Resource. *Mar. Drugs* **2021**, *19*, 116. [CrossRef]
42. Oğuz, T.; Mahmut Ali, G. Fatty Acids in Fish. In *Fatty Acids*; Angel, C., Ed.; IntechOpen: Rijeka, Croatia, 2017; Chapter 8.
43. Glencross, B.D.; Arnott, S.; Carr, I.; Santigosa, E. Modelling changes in whole body and fillet fatty acid composition of Atlantic salmon with changing levels of dietary omega-3 intake. *Aquaculture* **2024**, *593*, 741289. [CrossRef]
44. Xu, H.; Turchini, G.M.; Francis, D.S.; Liang, M.; Mock, T.S.; Rombenso, A.; Ai, Q. Are fish what they eat? A fatty acid's perspective. *Prog. Lipid Res.* **2020**, *80*, 101064. [CrossRef]
45. Ng, W.-K.; Sigholt, T.; Bell, J.G.B. The influence of environmental temperature on the apparent nutrient and fatty acid digestibility in Atlantic salmon (*Salmo salar* L.) fed finishing diets containing different blends of fish oil, rapeseed oil and palm oil. *Aquac. Res.* **2004**, *35*, 1228–1237. [CrossRef]

46. Bae, J.H.; Lim, H.; Lim, S. The Potential Cardiometabolic Effects of Long-Chain ω-3 Polyunsaturated Fatty Acids: Recent Updates and Controversies. *Adv. Nutr.* **2023**, *14*, 612–628. [CrossRef]
47. Djuricic, I.; Calder, P.C. Beneficial Outcomes of Omega-6 and Omega-3 Polyunsaturated Fatty Acids on Human Health: An Update for 2021. *Nutrients* **2021**, *13*, 2421. [CrossRef] [PubMed]
48. DiNicolantonio, J.J.; O'Keefe, J.H. The Importance of Marine Omega-3s for Brain Development and the Prevention and Treatment of Behavior, Mood, and Other Brain Disorders. *Nutrients* **2020**, *12*, 2333. [CrossRef]
49. Liao, J.; Xiong, Q.; Yin, Y.; Ling, Z.; Chen, S. The Effects of Fish Oil on Cardiovascular Diseases: Systematical Evaluation and Recent Advance. *Front. Cardiovasc. Med.* **2021**, *8*, 802306. [CrossRef] [PubMed]
50. Mone, P.; Varzideh, F.; Kansakar, U.; Infante, C.; Lombardi, A.; de Donato, A.; Frullone, S.; Santulli, G. Omega-3 fatty acids coordinate glucose and lipid metabolism in diabetic patients. *Lipids Health Dis.* **2022**, *21*, 31. [CrossRef]
51. Shahidi, F.; Ambigaipalan, P. Omega-3 Polyunsaturated Fatty Acids and Their Health Benefits. *Annu. Rev. Food Sci. Technol.* **2018**, *9*, 345–381. [CrossRef] [PubMed]
52. Richard, C.; Monk, J.M. Docosahexaenoic acid. *Adv. Nutr.* **2024**, *15*, 100161. [CrossRef]
53. Gonzalez-Becerra, K.; Barron-Cabrera, E.; Muñoz-Valle, J.F.; Torres-Castillo, N.; Rivera-Valdes, J.J.; Rodriguez-Echevarria, R.; Martinez-Lopez, E. A Balanced Dietary Ratio of n-6:n-3 Polyunsaturated Fatty Acids Exerts an Effect on Total Fatty Acid Profile in RBCs and Inflammatory Markers in Subjects with Obesity. *Healthcare* **2023**, *11*, 2333. [CrossRef]
54. Simopoulos, A.P.; DiNicolantonio, J.J. The importance of a balanced ω-6 to ω-3 ratio in the prevention and management of obesity. *Open Heart* **2016**, *3*, e000385. [CrossRef]
55. Liput, K.P.; Lepczyński, A.; Ogłuszka, M.; Nawrocka, A.; Poławska, E.; Grzesiak, A.; Ślaska, B.; Pareek, C.S.; Czarnik, U.; Pierzchała, M. Effects of Dietary n-3 and n-6 Polyunsaturated Fatty Acids in Inflammation and Cancerogenesis. *Int. J. Mol. Sci.* **2021**, *22*, 6965. [CrossRef] [PubMed]
56. Mukhametov, A.; Yerbulekova, M.; Aitkhozhayeva, G.; Tuyakova, G.; Dautkanova, D. Effects of ω-3 fatty acids and ratio of ω-3/ω-6 for health promotion and disease prevention. *Food Sci. Technol.* **2022**, *42*, e58321. [CrossRef]
57. Hibbeln, J.R.; Nieminen, L.R.; Blasbalg, T.L.; Riggs, J.A.; Lands, W.E. Healthy intakes of n-3 and n-6 fatty acids: Estimations considering worldwide diversity. *Am. J. Clin. Nutr.* **2006**, *83* (Suppl. S6), 1483s–1493s. [CrossRef] [PubMed]
58. Simopoulos, A.P. The importance of the ratio of omega-6/omega-3 essential fatty acids. *Biomed. Pharmacother.* **2002**, *56*, 365–379. [CrossRef]
59. Chen, J.; Liu, H. Nutritional Indices for Assessing Fatty Acids: A Mini-Review. *Int. J. Mol. Sci.* **2020**, *21*, 5695. [CrossRef]
60. Qin, J.; Kurt, E.; LBassi, T.; Sa, L.; Xie, D. Biotechnological production of omega-3 fatty acids: Current status and future perspectives. *Front. Microbiol.* **2023**, *14*, 1280296. [CrossRef]
61. Jesionowska, M.; Ovadia, J.; Hockemeyer, K.; Clews, A.C.; Xu, Y. EPA and DHA in microalgae: Health benefits, biosynthesis, and metabolic engineering advances. *J. Am. Oil Chem. Soc.* **2023**, *100*, 831–842. [CrossRef]
62. Kousparou, C.; Fyrilla, M.; Stephanou, A.; Patrikios, I. DHA/EPA (Omega-3) and LA/GLA (Omega-6) as Bioactive Molecules in Neurodegenerative Diseases. *Int. J. Mol. Sci.* **2023**, *24*, 10717. [CrossRef]
63. Şimşek, H.; Uçar, A. Polyunsaturated fatty acids as a nutraceutical for age-related neurodegenerative diseases: Current knowledge and future directions. *Clin. Nutr. Open Sci.* **2024**, *56*, 65–73. [CrossRef]
64. Nomali, M.; Heidari, M.E.; Ayati, A.; Tayebi, A.; Shevchuk, O.; Mohammadrezaei, R.; Navid, H.; Khayyatzadeh, S.S.; Palii, S.; Valizade Shiran, F.; et al. Omega-3 supplementation and outcomes of heart failure: A systematic review of clinical trials. *Medicine* **2024**, *103*, e36804. [CrossRef]
65. Sherratt, S.C.R.; Mason, R.P.; Libby, P.; Steg, P.G.; Bhatt, D.L. Do patients benefit from omega-3 fatty acids? *Cardiovasc. Res.* **2024**, *119*, 2884–2901. [CrossRef] [PubMed]
66. Szlendak, M.; Kapała, A. Does the ratio of eicosapentaenoic acid to docosahexaenoic acid matter in cancer treatment? A systematic review of their effects on cachexia-related inflammation. *Nutrition* **2024**, *124*, 112466. [CrossRef]
67. de Freitas Rodrigues, J.; Philippsen, H.K.; Dolabela, M.F.; Nagamachi, C.Y.; Pieczarka, J.C. The Potential of DHA as Cancer Therapy Strategies: A Narrative Review of In Vitro Cytotoxicity Trials. *Nutrients* **2023**, *15*, 2006. [CrossRef] [PubMed]
68. Blaauw, R.; Calder, P.C.; Martindale, R.G.; Berger, M.M. Combining proteins with n-3 PUFAs (EPA + DHA) and their inflammation pro-resolution mediators for preservation of skeletal muscle mass. *Crit. Care* **2024**, *28*, 38. [CrossRef] [PubMed]
69. Poggioli, R.; Hirani, K.; Jogani, V.G.; Ricordi, C. Modulation of inflammation and immunity by omega-3 fatty acids: A possible role for prevention and to halt disease progression in autoimmune, viral, and age-related disorders. *Eur. Rev. Med. Pharmacol. Sci.* **2023**, *27*, 7380–7400.
70. Shepon, A.; Makov, T.; Hamilton, H.A.; Müller, D.B.; Gephart, J.A.; Henriksson, P.J.G.; Troell, M.; Golden, C.D. Sustainable optimization of global aquatic omega-3 supply chain could substantially narrow the nutrient gap. *Resour. Conserv. Recycl.* **2022**, *181*, 106260. [CrossRef]
71. Glencross, B.D. Mind the (supply) gap. *Nat. Food* **2020**, *1*, 26. [CrossRef]
72. Venegas-Calerón, M.; Napier, J.A. New alternative sources of omega-3 fish oil. *Adv. Food Nutr. Res.* **2023**, *105*, 343–398. [PubMed]
73. Shepherd, C.J.; Jackson, A.J. Global fishmeal and fish-oil supply: Inputs, outputs and markets. *J. Fish Biol.* **2013**, *83*, 1046–1066. [CrossRef]
74. Šimat, V.; Vlahović, J.; Soldo, B.; Generalić Mekinić, I.; Čagalj, M.; Hamed, I.; Skroza, D. Production and characterization of crude oils from seafood processing by-products. *Food Biosci.* **2020**, *33*, 100484. [CrossRef]

75. Rodrigues, M.; Rosa, A.; Almeida, A.; Martins, R.; Ribeiro, T.â.; Pintado, M.; Gonçalves, R.F.S.; Pinheiro, A.C.; Fonseca, A.J.M.; Maia, M.R.G.; et al. Omega-3 fatty acids from fish by-products: Innovative extraction and application in food and feed. *Food Bioprod. Process.* **2024**, *145*, 32–41. [CrossRef]
76. Pinela, J.; Fuente, B.D.L.; Rodrigues, M.; Pires, T.C.S.P.; Mandim, F.; Almeida, A.; Dias, M.I.; Caleja, C.; Barros, L. Upcycling Fish By-Products into Bioactive Fish Oil: The Suitability of Microwave-Assisted Extraction. *Biomolecules* **2023**, *13*, 1. [CrossRef] [PubMed]
77. Byun, H.-G.; Taekil, E.; Jung, W.-K.; Kim, S.-K. Characterization of Fish Oil Extracted from Fish Processing By-products. *J. Food Sci. Nutr.* **2008**, *13*, 7–11. [CrossRef]
78. Soldo, B.; Šimat, V.; Vlahović, J.; Skroza, D.; Ljubenkov, I.; Generalić Mekinić, I. High Quality Oil Extracted from Sardine By-Products as an Alternative to Whole Sardines: Production and Refining. *Eur. J. Lipid Sci. Technol.* **2019**, *121*, 1800513. [CrossRef]
79. Ahmad, S.; Ali, M.; Khardali, A.; Ali, M.; Khan, G.; Alam, N.; Thaibah, H.; Khatun, A.; Anwer, T.; Alam, M. Incredible Use of Omega-3 Fatty Acids: A Review on Current Use and Future Prospective. *J. Young Pharm.* **2024**, *16*, 177–186. [CrossRef]
80. Baker, E.J. Alternative sources of bioactive omega-3 fatty acids: What are the options? *Curr. Opin. Clin. Nutr. Metab. Care* **2024**, *27*, 106–115. [CrossRef]
81. Derbyshire, E.J.; Birch, C.S.; Bonwick, G.A.; English, A.; Metcalfe, P.; Li, W. Optimal omegas–barriers and novel methods to narrow omega-3 gaps. A narrative review. *Front. Nutr.* **2024**, *11*, 1325099. [CrossRef] [PubMed]
82. Hernández, M.L.; Sicardo, M.D.; Belaj, A.; Martínez-Rivas, J.M. The Oleic/Linoleic Acid Ratio in Olive (*Olea europaea* L.) Fruit Mesocarp Is Mainly Controlled by OeFAD2-2 and OeFAD2-5 Genes Together With the Different Specificity of Extraplastidial Acyltransferase Enzymes. *Front. Plant Sci.* **2021**, *12*, 653997. [CrossRef]
83. Lopez, S.; Bermúdez, B.; Montserrat-de la Paz, S.; Pacheco, Y.; Ortega Gómez, A.; Varela, L.; Lemus, A.; Millan-Linares, M.; Rosillo, M.; Abia, R.; et al. Oleic acid—The main component of olive oil on postprandial metabolic processes. In *Olives and Olive Oil in Health and Disease Prevention*; Elsevier: Amsterdam, The Netherlands, 2021; pp. 639–649.
84. Revelou, P.-K.; Xagoraris, M.; Alexandropoulou, A.; Kanakis, C.D.; Papadopoulos, G.K.; Pappas, C.S.; Tarantilis, P.A. Chemometric Study of Fatty Acid Composition of Virgin Olive Oil from Four Widespread Greek Cultivars. *Molecules* **2021**, *26*, 4151. [CrossRef]
85. Viola, P.; Viola, M. Virgin olive oil as a fundamental nutritional component and skin protector. *Clin. Dermatol.* **2009**, *27*, 159–165. [CrossRef]
86. Mariotti, M.; Peri, C. The composition and nutritional properties of extra-virgin olive oil. In *The Extra-Virgin Olive Oil Handbook*; John Wiley & Sons Ltd.: Hoboken, NJ, USA, 2014; pp. 21–34.
87. Boskou, D. *Olive Oil: Chemistry and Technology*; AOCS Publishing: Champaign, IL, USA, 2006.
88. Lu, Y.; Zhao, J.; Xin, Q.; Yuan, R.; Miao, Y.; Yang, M.; Mo, H.; Chen, K.; Cong, W. Protective effects of oleic acid and polyphenols in extra virgin olive oil on cardiovascular diseases. *Food Sci. Hum. Wellness* **2024**, *13*, 529–540. [CrossRef]
89. Gaforio, J.J.; Visioli, F.; Alarcón-de-la-Lastra, C.; Castañer, O.; Delgado-Rodríguez, M.; Fitó, M.; Hernández, A.F.; Huertas, J.R.; Martínez-González, M.A.; Menendez, J.A.; et al. Virgin Olive Oil and Health: Summary of the III International Conference on Virgin Olive Oil and Health Consensus Report, JAEN (Spain) 2018. *Nutrients* **2019**, *11*, 2039. [CrossRef] [PubMed]
90. Jiménez-Sánchez, A.; Martínez-Ortega, A.; Remón-Ruiz, P.; Piñar-Gutierrez, A.; Pereira, J.; García-Luna, P. Therapeutic Properties and Use of Extra Virgin Olive Oil in Clinical Nutrition: A Narrative Review and Literature Update. *Nutrients* **2022**, *14*, 1440. [CrossRef] [PubMed]
91. Pastor, R.; Bouzas, C.; Tur, J.A. Beneficial effects of dietary supplementation with olive oil, oleic acid, or hydroxytyrosol in metabolic syndrome: Systematic review and meta-analysis. *Free Radic. Biol. Med.* **2021**, *172*, 372–385. [CrossRef] [PubMed]
92. Seidita, A.; Soresi, M.; Giannitrapani, L.; Di Stefano, V.; Citarrella, R.; Mirarchi, L.; Cusimano, A.; Augello, G.; Carroccio, A.; Iovanna, J.L.; et al. The clinical impact of an extra virgin olive oil enriched mediterranean diet on metabolic syndrome: Lights and shadows of a nutraceutical approach. *Front. Nutr.* **2022**, *9*, 980429. [CrossRef]
93. de Oliveira, D.A.S.B.; Minozzo, M.G.; Licodiedoff, S.; Waszczynskyj, N. Physicochemical and sensory characterization of refined and deodorized tuna (*Thunnus albacares*) by-product oil obtained by enzymatic hydrolysis. *Food Chem.* **2016**, *207*, 187–194. [CrossRef]
94. Miyashita, K.; Uemura, M.; Hosokawa, M. Effective Prevention of Oxidative Deterioration of Fish Oil: Focus on Flavor Deterioration. *Annu. Rev. Food Sci. Technol.* **2018**, *9*, 209–226. [CrossRef]
95. Song, G.; Li, L.; Wang, H.; Zhang, M.; Yu, X.; Wang, J.; Shen, Q. Electric Soldering Iron Ionization Mass Spectrometry Based Lipidomics for in Situ Monitoring Fish Oil Oxidation Characteristics during Storage. *J. Agric. Food Chem.* **2020**, *68*, 2240–2248. [CrossRef]
96. Foret, M.K.; Lincoln, R.; Do Carmo, S.; Cuello, A.C.; Cosa, G. Connecting the "Dots": From Free Radical Lipid Autoxidation to Cell Pathology and Disease. *Chem. Rev.* **2020**, *120*, 12757–12787. [CrossRef]
97. Kubow, S. Routes of formation and toxic consequences of lipid oxidation products in foods. *Free Radic. Biol. Med.* **1992**, *12*, 63–81. [CrossRef]
98. Schaich, K.M. Toxicity of Lipid Oxidation Products Consumed in the Diet. In *Bailey's Industrial Oil and Fat Products*; John Wiley & Sons Ltd.: Hoboken, NJ, USA, 2020; pp. 1–88.
99. Franco-Gil, M.E.; Graça, A.; Martins, A.; Marto, J.; Ribeiro, H.M. Emollients in dermatological creams: Early evaluation for tailoring formulation and therapeutic performance. *Int. J. Pharm.* **2024**, *653*, 123825. [CrossRef] [PubMed]

100. Fiume, M.M.; Bergfeld, W.F.; Belsito, D.V.; Hill, R.A.; Klaassen, C.D.; Liebler, D.C.; Marks, J.G., Jr.; Shank, R.C.; Slaga, T.J.; Snyder, P.W.; et al. Amended Safety Assessment of Triglycerides as Used in Cosmetics. *Int. J. Toxicol.* **2022**, *41* (Suppl. S3), 22–68. [CrossRef] [PubMed]
101. Rahman, F.C.K.; Farooq, T.; Khan, S.; Talukder, S.; Kumar, R. Rendered animal fat: A boon to the cosmetic industry. *Indian Vet. J.* **2022**, *99*, 20–26.
102. Russell, M.F.; Sandhu, M.; Vail, M.; Haran, C.; Batool, U.; Leo, J. Tallow, Rendered Animal Fat, and Its Biocompatibility with Skin: A Scoping Review. *Cureus* **2024**, *16*, e60981. [CrossRef] [PubMed]
103. Cristiano, L.; Guagni, M. Zooceuticals and Cosmetic Ingredients Derived from Animals. *Cosmetics* **2022**, *9*, 13. [CrossRef]
104. Dhanavel, N.; Nandakrishnan, M.H. A Review of Animal Fat: A Great Source for Industrial Applications. *J. Chem. Rev.* **2024**, *6*, 115–137.
105. Cheng, L.-H.; Cheng, Y.-F.; Yen, S.-Y.; Chen, J. Ultrafiltration of triglyceride from biodiesel using the phase diagram of oil–FAME–MeOH. *J. Membr. Sci.* **2009**, *330*, 156–165. [CrossRef]
106. Sarker, S. By-products of fish-oil refinery as potential substrates for biogas production in Norway: A preliminary study. *Results Eng.* **2020**, *6*, 100137. [CrossRef]
107. Andreo-Martínez, P.; Ortiz-Martínez, V.M.; Salar-García, M.J.; Veiga-del-Baño, J.M.; Chica, A.; Quesada-Medina, J. Waste animal fats as feedstock for biodiesel production using non-catalytic supercritical alcohol transesterification: A perspective by the PRISMA methodology. *Energy Sustain. Dev.* **2022**, *69*, 150–163. [CrossRef]
108. Long, F.; Liu, W.; Jiang, X.; Zhai, Q.; Cao, X.; Jiang, J.; Xu, J. State-of-the-art technologies for biofuel production from triglycerides: A review. *Renew. Sustain. Energy Rev.* **2021**, *148*, 111269. [CrossRef]
109. Hasan, N.; Ratnam, M.V. Biodiesel Production from Waste Animal Fat by Transesterification Using H_2SO_4 and KOH Catalysts: A Study of Physiochemical Properties. *Int. J. Chem. Eng.* **2022**, *2022*, 6932320. [CrossRef]
110. Sander, A.; Antonije Košćak, M.; Kosir, D.; Milosavljević, N.; Parlov Vuković, J.; Magić, L. The influence of animal fat type and purification conditions on biodiesel quality. *Renew. Energy* **2018**, *118*, 752–760. [CrossRef]
111. Malins, C. *The Fat of the Land*; Transport and Environment: Ixelles, Belgium, 2023.
112. Ramos, M.; Dias, A.P.S.; Puna, J.F.; Gomes, J.; Bordado, J.C. Biodiesel Production Processes and Sustainable Raw Materials. *Energies* **2019**, *12*, 4408. [CrossRef]
113. Toldrá-Reig, F.; Mora, L.; Toldrá, F. Trends in Biodiesel Production from Animal Fat Waste. *Appl. Sci.* **2020**, *10*, 3644. [CrossRef]
114. Banković-Ilić, I.B.; Stojković, I.J.; Stamenković, O.S.; Veljkovic, V.B.; Hung, Y.-T. Waste animal fats as feedstocks for biodiesel production. *Renew. Sustain. Energy Rev.* **2014**, *32*, 238–254. [CrossRef]
115. FAO. *The State of World Fisheries and Aquaculture, 2022*; Food & Agriculture Organization: Rome, Italy, 2022; Volume 3.
116. Knothe, G. Dependence of biodiesel fuel properties on the structure of fatty acid alkyl esters. *Fuel Process. Technol.* **2005**, *86*, 1059–1070. [CrossRef]
117. Moser, B.R. Biodiesel production, properties, and feedstocks. *Vitr. Cell. Dev. Biol.-Plant* **2009**, *45*, 229–266. [CrossRef]
118. Amran, N.A.; Bello, U.; Hazwan Ruslan, M.S. The role of antioxidants in improving biodiesel's oxidative stability, poor cold flow properties, and the effects of the duo on engine performance: A review. *Heliyon* **2022**, *8*, e09846. [CrossRef] [PubMed]
119. Lanjekar, R.D.; Deshmukh, D. A review of the effect of the composition of biodiesel on NOx emission, oxidative stability and cold flow properties. *Renew. Sustain. Energy Rev.* **2016**, *54*, 1401–1411. [CrossRef]
120. Aparamarta, H.W.; Gunawan, S.; Husin, H.; Azhar, B.; Tri Aditya, H. The effect of high oleic and linoleic fatty acid composition for quality and economical of biodiesel from crude Calophyllum inophyllum oil (CCIO) with microwave-assisted extraction (MAE), batchwise solvent extraction (BSE), and combination of MAE–BSE methods. *Energy Rep.* **2020**, *6*, 3240–3248.
121. Bligh, E.G.; Dyer, W.J. A rapid method of total lipid extraction and purification. *Can. J. Biochem. Physiol.* **1959**, *37*, 911–917. [CrossRef] [PubMed]
122. Monteiro, J.P.; Maciel, E.; Maia, R.; Pereira, A.T.; Calado, R.; Domingues, P.; Melo, T.; Eira, C.; Domingues, M.R. Characterization of the cardiac phospholipidome of small cetaceans provides adaptational insight and a foundation for indirect population health screening. *Mar. Mammal Sci.* **2021**, *37*, 1406–1427. [CrossRef]
123. Bartlett, E.M.; Lewis, D.H. Spectrophotometric determination of phosphate esters in the presence and absence of orthophosphate. *Anal. Biochem.* **1970**, *36*, 159–167. [CrossRef]
124. Rey, F.; Lopes, D.; Maciel, E.; Monteiro, J.; Skjermo, J.; Funderud, J.; Raposo, D.; Domingues, P.; Calado, R.; Domingues, M.R. Polar lipid profile of Saccharina latissima, a functional food from the sea. *Algal Res.* **2019**, *39*, 101473. [CrossRef]
125. Maurício, T.; Couto, D.; Lopes, D.; Conde, T.; Pais, R.; Batista, J.; Melo, T.; Pinho, M.; Moreira, A.S.P.; Trovão, M.; et al. Differences and Similarities in Lipid Composition, Nutritional Value, and Bioactive Potential of Four Edible Chlorella vulgaris Strains. *Foods* **2023**, *12*, 1625. [CrossRef] [PubMed]
126. Goracci, L.; Tortorella, S.; Tiberi, P.; Pellegrino, R.M.; Di Veroli, A.; Valeri, A.; Cruciani, G. Lipostar, a Comprehensive Platform-Neutral Cheminformatics Tool for Lipidomics. *Anal. Chem.* **2017**, *89*, 6257–6264. [CrossRef]
127. Lange, M.; Angelidou, G.; Ni, Z.; Criscuolo, A.; Schiller, J.; Blüher, M.; Fedorova, M. AdipoAtlas: A reference lipidome for human white adipose tissue. *Cell Rep. Med.* **2021**, *2*, 100407. [CrossRef]
128. Santos, F.; Monteiro, J.P.; Duarte, D.; Melo, T.; Lopes, D.; da Costa, E.; Domingues, M.R. Unraveling the Lipidome and Antioxidant Activity of Native Bifurcaria bifurcata and Invasive Sargassum muticum Seaweeds: A Lipid Perspective on How Systemic Intrusion May Present an Opportunity. *Antioxidants* **2020**, *9*, 642. [CrossRef]

Disclaimer/Publisher's Note: The statements, opinions and data contained in all publications are solely those of the individual author(s) and contributor(s) and not of MDPI and/or the editor(s). MDPI and/or the editor(s) disclaim responsibility for any injury to people or property resulting from any ideas, methods, instructions or products referred to in the content.

marine drugs

MDPI

Article

Reversed-Phase Medium-Pressure Liquid Chromatography Purification of Omega-3 Fatty Acid Ethyl Esters Using AQ-C18

Mingxin Sang [1,2], Nan Pan [2,*], Jingna Wu [3], Xiaoting Chen [2], Shuilin Cai [2], Huan Fang [2,4], Meitian Xiao [1], Xiaoming Jiang [5] and Zhiyu Liu [6,*]

1 College of Chemical Engineering, Huaqiao University, Xiamen 361021, China; 22014087026@stu.hqu.edu.cn (M.S.); mtxiao@hqu.edu.cn (M.X.)
2 Fisheries Research Institute of Fujian, Xiamen 361013, China; xtchen@jmu.edu.cn (X.C.); caishuilin@hqu.edu.cn (S.C.); 220820081@fzu.edu.cn (H.F.)
3 Xiamen Key Laboratory of Marine Medicinal Natural Products Resources, Fujian Universities and Colleges Engineering Research Center of Marine Biopharmaceutical Resources, Xiamen Medical College, Xiamen 361023, China; wjn@xmmc.edu.cn
4 College of Biological Science and Engineering, Fuzhou University, Fuzhou 350108, China
5 Quanzhou Institute of Marine Bioresources Industry, Quanzhou 362000, China; jxm@ouc.edu.cn
6 Key Laboratory of Cultivation and High-Value Utilization of Marine Organisms in Fujian Province, National Research and Development Center for Marine Fish Processing (Xiamen), Xiamen 361013, China
* Correspondence: npan01@qub.ac.uk (N.P.); 13906008638@163.com (Z.L.)

Abstract: Omega-3 fatty acids are in high demand due to their efficacy in treating hypertriglyceridemia and preventing cardiovascular diseases. However, the growth of the industry is hampered by low purity and insufficient productivity. This study aims to develop an efficient RP-MPLC purification method for omega-3 fatty acid ethyl esters with high purity and capacity. The results indicate that the AQ-C18 featuring polar end-capped silanol groups outperformed C18 and others in retention time and impurity separation. By injecting pure fish oil esters with a volume equivalent to a 1.25% bed volume on an AQ-C18 MPLC column using a binary isocratic methanol–water (90:10, *v:v*) mobile phase at 30 mL/min, optimal omega-3 fatty acid ethyl esters were obtained, with the notable purity of 90.34% and a recovery rate of 74.30%. The total content of EPA and DHA produced increased from 67.91% to 85.27%, meeting the acceptance criteria of no less than 84% set by the 2020 edition of the Pharmacopoeia of the People's Republic of China. In contrast, RP-MPLC significantly enhanced the production efficiency per unit output compared to RP-HPLC. This study demonstrates a pioneering approach to producing omega-3 fatty acid ethyl esters with high purity and of greater quantity using AQ-C18 RP-MPLC, showing this method's significant potential for use in industrial-scale manufacturing.

Keywords: omega-3 fatty acids; eicosapentaenoic acid; docosahexaenoic acid; reverse-phase medium-pressure liquid chromatography; AQ-C18; high-purity

1. Introduction

Omega-3 fatty acids are long-chain poly-unsaturated fatty acids that feature an initial conjugated double bond at the third carbon atom from the methyl end; principally the eicosapentaenoic acid (C20:5 ω-3; EPA) and docosahexaenoic acid (C22:6 ω-3; DHA) are of major interest [1], which are derived from α-linolenic acid (C18:3 ω-3) via a series of chain elongations and desaturations. However, due to insufficient Δ-12 and Δ-15 desaturases for conversion in the human body [2,3], omega-3 fatty acids have to be obtained via dietary means [4] and are considered essential to human health [5,6]. An increased consumption of EPA and DHA has been scientifically proven to be beneficial in the treatment and prevention of atherosclerosis [7], myocardial infarction [8], inflammation [9], arthritis [10], diabetes [11], infant brain development [12,13], and cancers [14]. Notably, many epidemiological, observational, and clinical studies have emphasized the efficacy of omega-3 fatty acids in reducing plasma triglyceride levels and preventing cardiovascular diseases [3]. As a result, the authorities that oversee the pharmaceutical industry have set standards for "Omega-3-acid ethyl ester", "Ethyl polyenoate", and correlative capsules and thus approved several prescription drugs for treating hypertriglyceridemia or preventing cardiovascular diseases, such as Lovaza® and Omacor® [46% EPA ethyl esters (EPA-EE) + 38% DHA ethyl esters (DHA-EE)], and Vascepa® (96% EPA-EE) [15]. It is recommended by heart associations worldwide that a prescription of omega-3 fatty acids (EPA+DHA or EPA only) at a dose of 4 g/day (>3 g/day total EPA + DHA) represents an effective therapeutic agent in reducing triglycerides. With this increasing attention, the demand for high-purity omega-3 fatty acid has surged significantly. However, prolonged overfishing has led to a sharp decline in the main fish sources, resulting in a rapid growth in the price of omega-3 fatty acid. Nevertheless, there are only a handful of companies worldwide with the capacity to manufacture pharmaceutical-grade omega-3 fatty acid [16]. Hence, it is necessary to develop a universally applicable and cost-effective technology that ensures the safe production of high-purity omega-3 fatty acid.

Omega-3 fatty acid ethyl esters have been concentrated using various techniques, including low-temperature crystallization [17,18], urea inclusion [19,20], molecular distillation [21,22], lipase catalysis [23,24], and supercritical fluid extraction [25]. However, the purity of the omega-3 fatty acids produced using these methods does not meet the standards of the Pharmacopoeia. Chromatography is considered the primary purification technology used in the industrial-scale manufacturing of commercially valuable compounds with high purity, including high-pressure liquid chromatography (HPLC) [26], high-speed counter-current chromatography (HSCCC) [27], supercritical fluid chromatography (SFC) [28], and simulated moving bed (SMB) chromatography [29]. These methods, which rely on high-pressure conditions, effectively improve the purity of omega-3 fatty acids [30,31]. However, several drawbacks to these methods hinder their broader application in the industry, such as the requirement for relatively high initial sample purity, high energy consumption, complex equipment requirements, and high costs during production. The HPLC method utilizes packings with a smaller particle size and has a lower sample-loading volume to ensure high resolution and sensitivity [30]. The SFC method leverages supercritical fluids of CO_2 as the mobile phase, necessitating precise control over the operating parameters (pressure and temperature), which often requires specialized expertise [31]. SMB chromatography is currently acknowledged for its continuous operation and enhanced efficiency in the purification process [32]; however, for SMB to be effective, a high concentration of raw materials is required, and it utilizes smaller-particle stationary phases to ensure effective purification, which consequently elevates the cost. Therefore, developing a cost-effective, high-throughput chromatographic method tailored to low-purity fish oil samples to produce omega-3 fatty acids with high purity is of significant importance.

Reversed-phase medium-pressure liquid chromatography (RP-MPLC) is a type of liquid chromatography that operates at pressures ranging between 0 and 200 psi; the main features of RP-MPLC include a fast separation speed and high efficiency [33]. With significantly reduced instrumentation and operational expenses relative to HPLC, SFC, and SMB, the cost-effective alternative of RP-MPLC operates under medium pressure and employs packing materials with larger particle sizes, which leads to larger sample preparation capacities, reduces the impact of impurities on the packing materials, and shortens the processing times. By accepting lower-purity materials and decreasing the frequency of preliminary purification stages, RP-MPLC plays an important role in the separation and purification of saponins [34], polysaccharides [35], and polypeptides [36] and has gradually become a popular method in recent years [35,37]. Additionally, Ishihara et al. [38] purified stearidonic acid (C18:4 ω-3) and hexadecatetraenoic acid (C16:4 ω-3) to more than 95% purity using RP-MPLC. These results indicate that RP-MPLC could potentially produce high-purity omega-3 fatty acid ethyl esters.

In this study, RP-MPLC was initially employed to deliver high-purity omega-3 fatty acid ethyl esters, targeting a total content of EPA and DHA of not less than (NLT) 84%, as stipulated by the Pharmacopoeia. Fundamental variables controlling separation were evaluated and optimized based on the purity and recovery rate, including the packing material, mobile phase, sample volume, sample concentration, flow rate, and mobile-phase composition. Furthermore, comparisons were made between RP-MPLC and RP-HPLC using the same packings, with the aim of assessing the effectiveness of the purification. Ultimately, this research provides a new approach and theoretical basis for the cost-effective and high-throughput production of high-purity omega-3 fatty acids on an industrial manufacturing scale.

2. Results and Discussion

2.1. Fatty Acid Composition of Fish Oil Ethyl Esters

Identifying fatty acids and esters with a similar equivalent carbon length from the EPA and DHA is a crucial stage in the development of the RP-MPLC method. The GC chromatogram for the fish oil ethyl esters is displayed in Figure 1, and the fatty acid composition is outlined in Table 1. We can see that 24 fatty acids were detected in the esters, including 6 saturated fatty acids (SFAs), 5 mono-unsaturated fatty acids (MUFAs), and 13 poly-unsaturated fatty acids (PUFAs), accounting for 4.45%, 7.84%, and 87.71% of the content, respectively. The major SFAs were C18:0 and C16:0, at 2.18% and 1.52%, respectively. The primary MUFAs were C18:1n9, C22:1n9, and C20:1n9, at 4.64%, 1.19%, and 1.17%, respectively. PUFAs made up the bulk of fatty acids, with ω-6 PUFAs at 9.12% (mainly C20:4n6, C18:4n6, and C22:5n6) and ω-3 PUFAs at a higher percentage, predominantly EPA (C20:5n3), DHA (C22:6n3), DPA (C22:5n3), C21:5n3, and C20:3n3. However, the sum of the content of EPA and DHA was 67.91%, falling short of the acceptance criteria of the Chinese Pharmacopoeia (2020 edition) for "Ethyl polyenoate", which require NLT 84% purity of the total content of EPA and DHA.

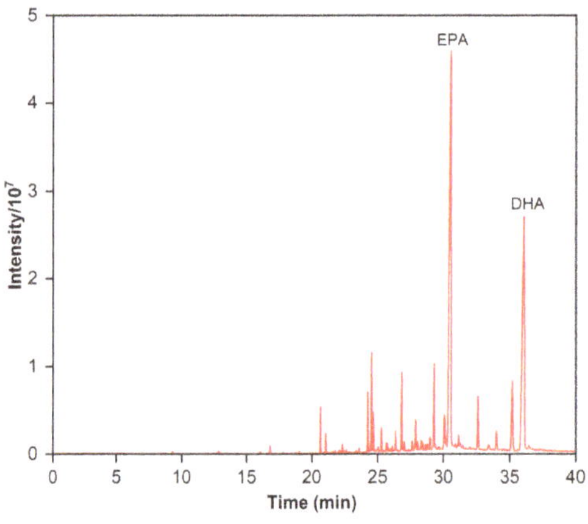

Figure 1. Gas chromatogram of the fish oil ethyl esters.

Table 1. Fatty acid composition of the fish oil ethyl esters.

Abbreviation	Compound Name	Content (%)
C14:0	Myristic acid	0.23 ± 0.06
C15:0	Pentadecanoic acid	0.04 ± 0.02
C16:0	Palmitic acid	1.52 ± 0.04
C18:0	Stearic acid	2.18 ± 0.08

Table 1. *Cont.*

Abbreviation	Compound Name	Content (%)
C19:0	Nonadecanoic acid	0.10 ± 0.04
C20:0	Arachidic acid	0.37 ± 0.02
C16:1n7	Palmitoleic acid	0.64 ± 0.06
C18:1n9	Octadecenoic acid	4.64 ± 0.05
C20:1n9	11-Eicosenoic acid	1.17 ± 0.03
C22:1n9	Erucic acid	1.19 ± 0.07
C24:1n9	Nervonic acid	0.21 ± 0.03
C18:3n3	α-Linolenic acid	0.54 ± 0.03
C20:3n3	11,14,17-Eicosatrienoic acid	2.43 ± 0.04
EPA C20:5n3	5,8,11,14,17-Eicosapentaenoic acid	40.83 ± 0.12
C21:5n3	Heneicosapentaenoic acid	2.58 ± 0.06
DPA C22:5n3	Docosapentaenoic acid	5.17 ± 0.07
DHA C22:6n3	4,7,10,13,16,19-Docosahexaenoic acid	27.08 ± 0.07
C18:2n6	Linolelaidic acid	0.29 ± 0.03
C18:3n6	Octadecatrienoic acid	0.80 ± 0.04
C18:4n6	Octadecatetraenoic acid	2.79 ± 0.06
C20:3n6	8,11,14-Eicosatrienoic acid	0.52 ± 0.04
C20:4n6	Arachidonic acid	3.46 ± 0.04
C22:4n6	7,10,13,16-Docosatetraenoic acid	0.19 ± 0.02
C22:5n6	4,7,10,13,16-Docosapentaenoate	1.06 ± 0.02
∑SFA	Saturated fatty acids	4.45 ± 0.1
∑MUFA	Mono-unsaturated fatty acids	7.84 ± 0.06
∑PUFA	Poly-unsaturated fatty acids	87.71 ± 0.06
∑ω-3 PUFA	ω-3 Poly-unsaturated fatty acids	78.59 ± 0.23
∑ω-6 PUFA	ω-6 Poly-unsaturated fatty acids	9.12 ± 0.19
EPA + DHA		67.91 ± 0.18

2.2. Fatty Acid Composition of Separate Fractions in RP-MPLC Chromatogram

An RP-MPLC chromatogram of fish oil ethyl esters is shown in Figure 2. Four major fractions (A–D), corresponding to the separated group, were collected, vacuum-evaporated, and analyzed using GC-MS. A GC chromatogram of the fish oil ethyl esters, with four fractions (A–D), is displayed in Figure 3. The varying fatty acid compositions across different fractions are demonstrated in Figure 4.

Fraction A comprised a total of 42.85% of EPA (29.76%) and DHA (13.09%), with the remainder predominantly made up of C18:4n6 (45.34%), C14:0 (3.76%), C18:0 (1.44%), and C20:4n6 (1.01%). Conversely, fraction B was mainly represented by EPA, at 90.66%, with the remainder mostly including DHA (2.63%), C18:0 (2.11%), and C18:3n3 (1.42%). Fraction C displayed the highest DHA content, at 75.01%, complemented by EPA (6.81%), C21:5n3 (6.48%), C18:0 (2.70%), C20:4n6 (1.90%), C18:3n6 (1.49%), and C20:1n9 (1.39%). Fraction D was notable for the highest DPA concentration, at 51.70%, with the rest primarily comprising DHA (13.17%), C20:4n6 (7.27%), C20:3n3 (5.67%), C20:5n3 (4.50%), C20:1n9 (4.10%), C22:5n6 (3.84%), and C18:1n9 (1.88%). In essence, fraction B and fraction C were identified as the target components of EPA and DHA, respectively, which enabled their precise collection for the further purification of omega-3 fatty acid ethyl esters.

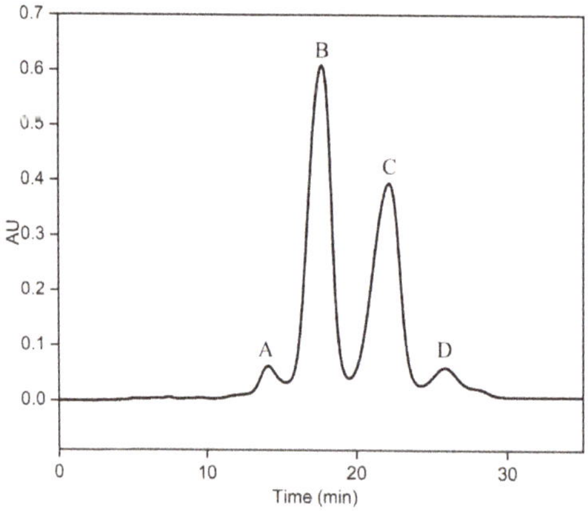

Figure 2. RP-MPLC chromatogram of fish oil ethyl esters with four fractions: A, B, C, and D. Conditions: stationary phase, C18; sample load, 0.4 mL fish oil ethyl esters; mobile phase, methanol–water (90:10, *v:v*); flow rate, 20 mL/min; detector, UV 210 nm; room temperature.

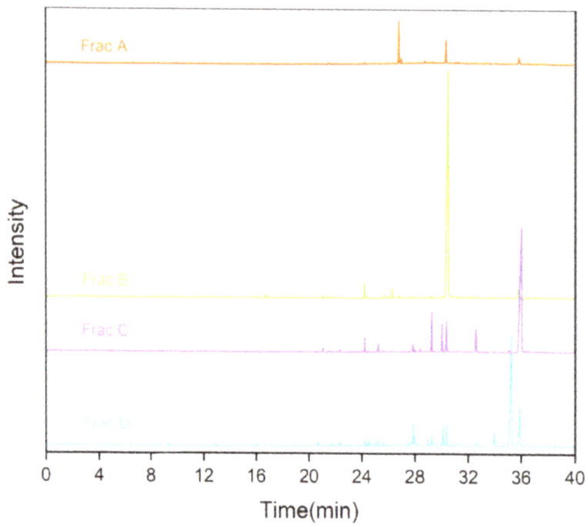

Figure 3. Gas chromatogram of fractions A, B, C, and D.

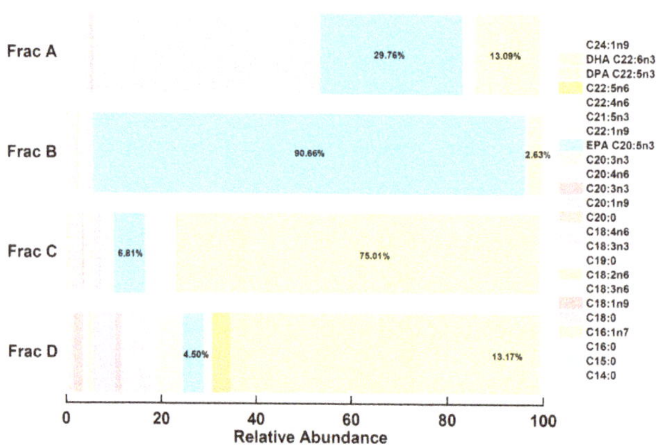

Figure 4. Stacked histograms of fatty acid compositions from different fractions.

2.3. Optimization of RP-MPLC

2.3.1. Effects of Packing Materials

The packing materials are the "heart" of the chromatographic system. The physicochemical properties of the packing materials, including the uniformity of the packing structure (monolithic, porous, or nonporous), the geometry (particle size, area of the bed, and pore size and shape), and the type of attached ligands, significantly influence the separation efficacy [5,39]. To identify column packings that provide high throughput, low back pressure, strong sensitivity, and high resolution for efficient separation, various bonded packing materials (CN, Diol, C4, C6, C8, C18, and AQ-C18) were evaluated in the purification of omega-3 fatty acid ethyl esters (Figure 5 and Table 2).

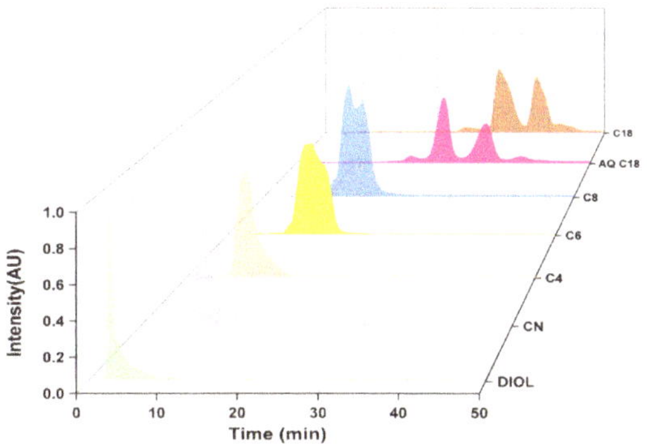

Figure 5. RP-MPLC chromatograms of omega-3 fatty acid ethyl esters using various packing materials.

Table 2. Effects of AQ-C18 and C18 on the esters of EPA and DHA purified via RP-MPLC.

Packing Materials	t_{R2} (min)	t_{R3} (min)	R_{S1}	R_{S2}
AQ-C18	17.09 ± 0.08 [b]	21.53 ± 0.07 [b]	1.43 ± 0.02 [a]	1.13 ± 0.03 [a]
C18	31.08 ± 0.14 [a]	37.90 ± 0.1 [a]	1.27 ± 0.03 [b]	1.02 ± 0.03 [b]

Note: t_{R2} represents the retention time of EPA; t_{R3} represents the retention time of DHA; R_{S1} represents the resolution of EPA and the preceding impurity (fraction A); R_{S2} represents the resolution of DHA and the following impurity (fraction D). Different letters within the group indicate significant differences ($p < 0.05$). The same notes apply to the following tables.

Analysis of the elution curves for CN, Diol, C4, C6, and C8 revealed that EPA and DHA did not acquire baseline separation from the preceding and subsequent impurities. CN, a normal-phase stationary phase with cyanide groups, was more suited to the reverse-phase separation of weakly polar compounds, showing a weak affinity for EPA and DHA [40]. Diol, with intermediate polarity, and C4, C6, and C8, with shorter carbon chains, exhibited weaker affinity, leading to shorter retention times and less effective purification of EPA and DHA. Thus, these materials were deemed unsuitable.

In contrast, the elution curves of C18 and AQ-C18 packing materials display clear target peaks for EPA and DHA, with good separation from adjacent impurity peaks. AQ-C18 demonstrates earlier peak emergence, shorter purification time, and lower solvent consumption compared to C18, showcasing superior separation efficiency. Both C18 and AQ-C18 are non-polar reversed-phase stationary phases with octadecyl carbon chains bonded to silica. However, AQ-C18 undergoes the polar end-capping of silica hydroxyl groups, minimizing the residual silanol on the surface [41] and thus enhancing the separation effects of omega-3 fatty acid ethyl esters (Figure 6). Based on these findings, AQ-C18 was chosen as the stationary phase for subsequent experiments.

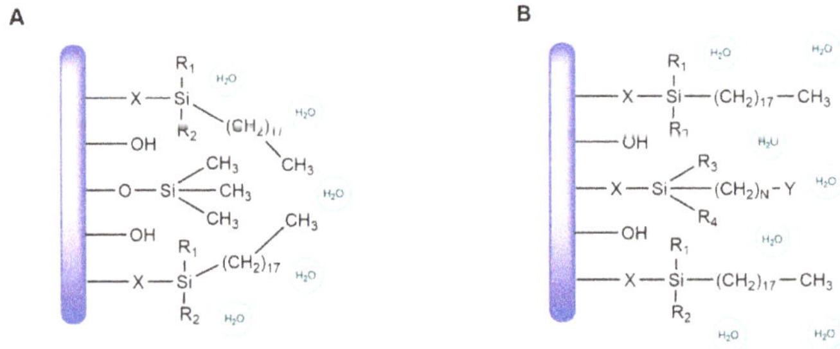

Figure 6. Structural differences between RP-MPLC stationary phases (**A**) C18 and (**B**) AQ-C18.

2.3.2. Effects of Mobile Phases

The appropriate mobile phase plays an important auxiliary role in improving separation efficiency. When choosing the mobile phase for industrial applications, emphasis has been placed on the separation efficacy, cost, column pressure drop, and ease of subsequent product purification [42]. As a result, solvents characterized by low viscosity, low boiling points, and low cost are preferred. In Figure 7 and Table 3, we can see that ethanol and acetonitrile were ineffective in achieving the baseline separation of impurities from omega-3 fatty acids, unlike methanol, which was successful. Despite its higher viscosity, methanol's lower boiling point facilitated easier separation from the product than acetonitrile and ethanol. More importantly, methanol is an effective solvent, with easy recovery from boiling point [43,44], which allows resource reuse and promotes the development of the circular economy [45]. Consequently, methanol was employed as the preferred mobile phase.

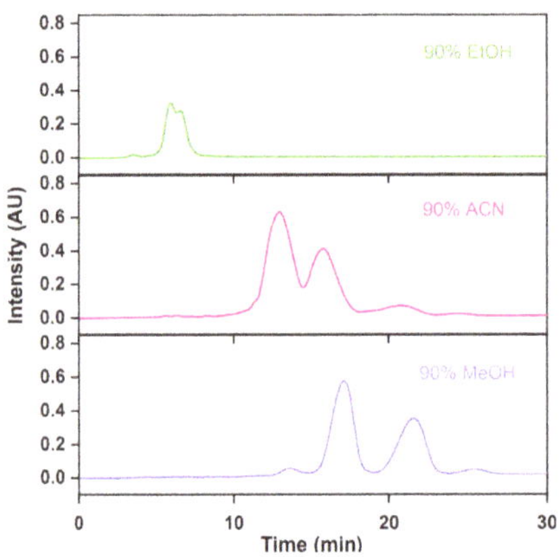

Figure 7. RP-MPLC chromatograms of omega-3 fatty acid ethyl esters via different mobile phases.

Table 3. Effects of different mobile phases on the esters of EPA and DHA purified via RP-MPLC.

Mobile Phases	t_{R2} (min)	t_{R3} (min)	R_{S1}	R_{S2}
Ethanol	6.29 ± 0.08 [c]	7.14 ± 0.04 [c]	0	0
Acetonitrile	13.95 ± 0.1 [b]	15.81 ± 0.08 [b]	0	1.32 ± 0.02 [a]
Methanol	17.08 ± 0.06 [a]	21.54 ± 0.08 [a]	1.42 ± 0.02 [a]	1.27 ± 0.03 [b]

Note: Different letters within the group indicate significant differences ($p < 0.05$).

2.3.3. Effects of Sample Load Volume

According to the nonlinear theory of chromatographic preparation, increasing the sample volume can improve the processing capacity of chromatography, boost the product recovery rate, and enhance production efficiency [46,47]. As shown with the growth loading volume in Table 4 and Figure 8, the retention time is delayed, peak shapes are widened, the resolution is reduced, and the purification time is increased. This is consistent with the experimental findings of Dillon et al. [48]. This could be due to the adsorption of more impurities on the AQ-C18 packings, which affects the separation of main and impurity peaks, thereby reducing the purity of the target substance [49]. With a sample volume of 0.6 mL, the recovery rates for the total ethyl esters for the EPA and DHA peak were the highest (83.57%). To maximize the loading volume while achieving better separation effects, the sample loading of 0.6 mL, equivalent to the 1.25% bed volume of the chromatographic column, was selected.

Table 4. Effects of different injection volumes on the esters of EPA and DHA purified via RP-MPLC.

Sample Volumes (mL)	Purity of EPA-EE/DHA-EE (%)	Recovery of EPA-EE/DHA-EE (%)	t_{R2} (min)	t_{R3} (min)	R_{S1}	R_{S2}
0.4	87.57 ± 0.30^{a}	58.44 ± 0.13^{c}	17.10 ± 0.04^{d}	21.47 ± 0.04^{d}	1.43 ± 0.02^{a}	1.07 ± 0.02^{a}
0.5	86.75 ± 0.08^{b}	65.43 ± 0.21^{b}	17.25 ± 0.05^{c}	21.80 ± 0.03^{c}	1.32 ± 0.03^{b}	1.02 ± 0.01^{b}
0.6	86.67 ± 0.24^{b}	83.57 ± 0.22^{a}	17.40 ± 0.05^{b}	22.07 ± 0.07^{b}	1.27 ± 0.02^{b}	1.02 ± 0.02^{b}
0.7	83.15 ± 0.30^{c}	63.59 ± 0.36^{bc}	17.51 ± 0.04^{a}	22.30 ± 0.06^{a}	1.06 ± 0.02^{c}	0.96 ± 0.02^{c}

Note: Different letters within the group indicate significant differences ($p < 0.05$).

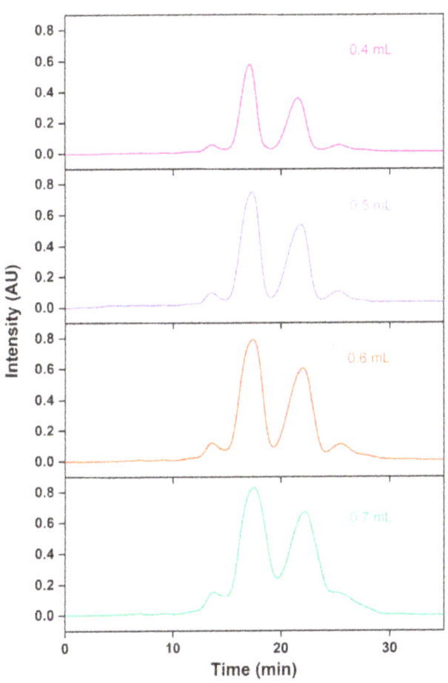

Figure 8. RP-MPLC chromatograms of omega-3 fatty acid ethyl esters with different injection volumes.

2.3.4. Effects of Sample Concentration

During industrial production, increasing the sample concentration can enhance the chromatographic processing capacity, while reducing the concentration helps to promote the partition and adsorption processes of analytes onto chromatographic packing material, thus enhancing the separation between target substances and impurities [50–53]. However, this improvement comes at the cost of a corresponding decrease in recovery rates. Chromatographic curves depicting various concentrations of fish oil ethyl esters diluted with methanol in RP-MPLC are presented in Figure 9, accompanied by Table 5. As the concentrations of fish oil ethyl esters increased, the purity of ethyl esters of EPA and DHA decreased, whereas the recovery rate, retention time, and resolution exhibited an increase. Conversely, the use of pure fish oil for injection reduced the resolution of ethyl esters of EPA and DHA, achieving a separation factor of 1.23 for front impurities and 1.10 for rear impurities, with a purity of 85.75%. The recovery rate of ethyl esters of EPA and DHA steadily increased with the sample concentration, peaking at 74.62% with pure fish oil. To maximize production efficiency, the pure fish oil ethyl esters were chosen.

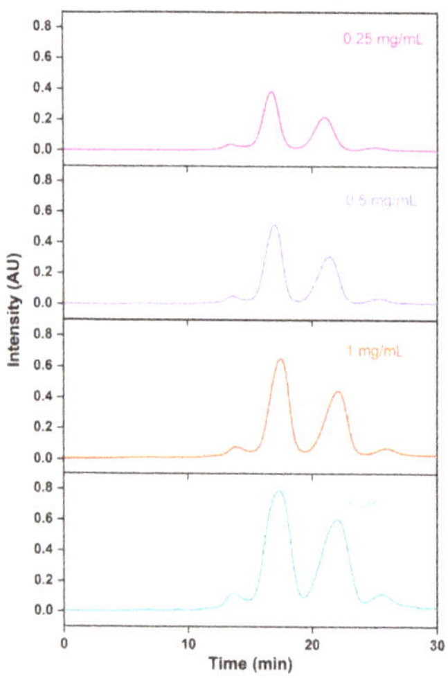

Figure 9. RP-MPLC chromatograms of omega-3 fatty acid ethyl esters of different concentrations.

Table 5. Effects of different concentrations on the esters of EPA and DHA purified via RP-MPLC.

Sample Concentrations (g/mL)	Purity of EPA-EE/DHA-EE (%)	Recovery Rate of EPA-EE/DHA-EE (%)	t_{R2} (min)	t_{R3} (min)	R_{S1}	R_{S2}
0.25	87.19 ± 0.19 [a]	50.47 ± 0.08 [d]	15.86 ± 0.03 [c]	18.07 ± 0.06 [c]	1.38 ± 0.03 [a]	1.31 ± 0.04 [a]
0.5	86.63 ± 0.28 [a]	58.65 ± 0.07 [c]	17.51 ± 0.04 [b]	20.69 ± 0.06 [b]	1.35 ± 0.02 [b]	1.27 ± 0.03 [a]
1	86.11 ± 0.11 [b]	62.21 ± 0.08 [b]	17.61 ± 0.03 [b]	21.47 ± 0.04 [a]	1.29 ± 0.02 [c]	1.13 ± 0.02 [b]
Pure	85.75 ± 0.15 [c]	74.62 ± 0.05 [a]	17.72 ± 0.02 [a]	21.92 ± 0.03 [a]	1.23 ± 0.04 [c]	1.10 ± 0.03 [b]

Note: Different letters within the group indicate significant differences ($p < 0.05$).

2.3.5. Effects of Flow Rate

In a liquid chromatography system, increasing the flow rate can shorten the elution time of analytes, albeit with the potential downside of compromising the robustness of chromatographic analysis [54,55]. As the flow rate accelerates, the peak elution time shifts earlier and the peak shape becomes more compact, resulting in closer proximity between the peaks of the ethyl esters EPA and DHA and the surrounding impurity peaks (Figure 10). As the flow rate speeds up, the elution time of the ethyl esters EPA and DHA consistently decreases, leading to a reduced retention time and resolution and thus causing a decline in purity (Table 6). Additionally, the recovery rates of the ethyl esters EPA and DHA reduce continuously with the growth in flow rate. This is similar to the trends observed by Oh et al. [30], which could be attributed to the fact that excessively high flow rates might impede the adsorption of target compounds to the stationary phase. To optimize the separation efficiency while minimizing time and solvent usage, the flow rate of 30 mL/min was selected.

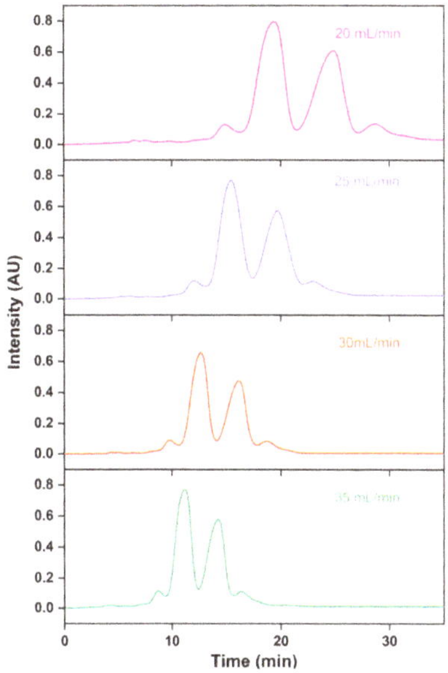

Figure 10. RP-MPLC chromatograms of omega-3 fatty acid ethyl esters with different flow rates.

Table 6. Effects of different flow rates on the esters EPA and DHA purified via RP-MPLC.

Flow Rate (mL/min)	Purity of EPA-EE/DHA-EE (%)	Recovery Rate of EPA-EE/DHA-EE (%)	t_{R2} (min)	t_{R3} (min)	R_{S1}	R_{S2}
20.00	86.17 ± 0.15 [a]	82.86 ± 0.18 [a]	17.33 ± 0.06 [a]	20.60 ± 0.07 [a]	1.27 ± 0.03 [a]	1.28 ± 0.04 [a]
25.00	86.01 ± 0.14 [a]	76.35 ± 0.01 [b]	14.08 ± 0.08 [b]	17.82 ± 0.08 [b]	1.26 ± 0.03 [a]	1.06 ± 0.02 [b]
30.00	85.27 ± 0.15 [b]	73.82 ± 0.16 [b]	11.87 ± 0.05 [c]	15.07 ± 0.06 [c]	1.23 ± 0.01 [a]	1.01 ± 0.02 [b]
35.00	84.16 ± 0.83 [c]	58.94 ± 0.14 [c]	9.62 ± 0.09 [d]	12.09 ± 0.08 [d]	1.14 ± 0.02 [b]	0.87 ± 0.03 [c]

Note: Different letters within the group indicate significant differences ($p < 0.05$).

2.3.6. Effects of Mobile-Phase Composition

The proportion of the organic solvent in the mobile phase modifies its polarity, thereby altering the distribution coefficient of the sample components in the stationary phase and affecting the separation efficiency [56]. Increasing the methanol proportions advances the peak emergence time, broadens the peak shape, and reduces the retention time, resolution, and purity of the ethyl esters EPA and DHA (Figure 11 and Table 7). This is attributed to the fact that increasing the polarity of the mobile phase has been found to improve the separation efficiency by delaying the retention time of the non-polar FAEE in the column [30,48]. With a methanol proportion of 86% to 90%, the purity of omega-3 fatty acid gradually declines; meanwhile, the recovery rate improves. Wei et al. [29] observed the same experimental trend. At a 92% methanol proportion, the purity of ethyl esters of EPA and DHA falls to 83.39%, which does not meet the national Pharmacopoeia standards. Methanol proportions over 90% are not conducive to the preparation of high-purity omega-

3 fatty acids. Therefore, a 90% methanol solution was chosen as the composition of the mobile phase.

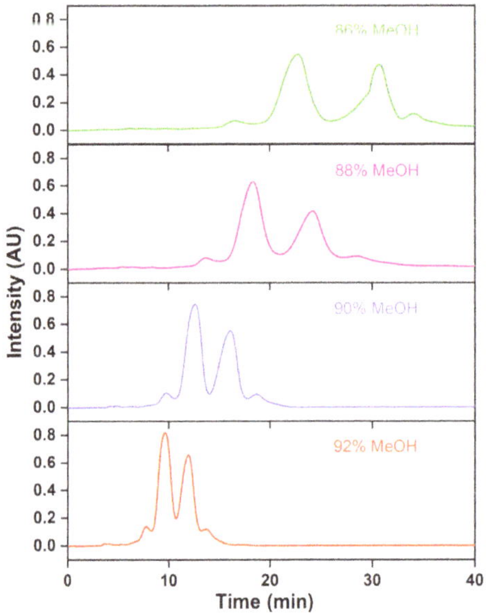

Figure 11. RP-MPLC chromatograms of omega-3 fatty acid ethyl esters with different mobile-phase compositions.

Table 7. Effects of different mobile-phase compositions on the esters of EPA and DHA purified via RP-MPLC.

Methanol–Water (v:v)	Purity of EPA-EE/DHA-EE (%)	Recovery Rate of EPA-EE/DHA-EE (%)	t_{R2} (min)	t_{R3} (min)	R_{S1}	R_{S2}
86:14	87.17 ± 0.15 [a]	54.51 ± 0.16 [c]	22.81 ± 0.05 [a]	30.48 ± 0.08 [a]	1.64 ± 0.04 [a]	1.41 ± 0.03 [a]
88:12	86.32 ± 0.10 [b]	65.24 ± 0.12 [b]	18.37 ± 0.07 [b]	24.26 ± 0.06 [b]	1.50 ± 0.02 [b]	1.26 ± 0.03 [b]
90:10	85.27 ± 0.15 [c]	74.30 ± 0.11 [a]	11.87 ± 0.05 [c]	15.07 ± 0.04 [c]	1.22 ± 0.04 [c]	1.02 ± 0.02 [c]
92:8	83.39 ± 0.14 [d]	53.28 ± 0.01 [c]	9.67 ± 0.1 [d]	12.02 ± 0.07 [d]	1.05 ± 0.03 [d]	0.84 ± 0.02 [d]

Note: Different letters within the group indicate significant differences ($p < 0.05$).

2.4. Fatty Acid Analysis of Purified Omega-3 Fatty Acid Ethyl Esters

Fish oil ethyl esters equivalent to 1.25% of the column volume were introduced into an RP-MPLC column equipped with AQ-C18 packings (20–40 μm particle size, 100 Å pore size, 320–340 m^2/g surface area) and eluted with a methanol–water (90:10, v:v) isocratic mobile phase at a flow rate of 30 mL/min under an operating pressure of 1–4 bar. The RP-MPLC chromatogram of fish oil ethyl esters under optimal purification conditions is shown in Figure 12. Fractions B and C were collected, concentrated under reduced pressure, and subsequently characterized via GC-MS. In total, 16 types of fatty acids were identified in the form of 2.39% SFAs, 3.14% MUFAs, and 94.47% PUFAs (Figure 13 and Table 8). Compared with Table 1, SFAs C14:0, C15:0, and C19:0; MUFA C24:1n9; and PUFAs C22:5n3, C20:3n6, and C22:5n6 were removed. The proportion of ω-3 PUFAs rose from 78.59% to 90.34%, the main components of which were EPA and DHA, accounting for 57.13% and 28.14%, respectively, and totaling 85.27%. Thus, the omega-3 fatty acid ethyl esters purified via AQ-C18 RP-MPLC meet the criteria for "Ethyl polyenoate" in the 2020 edition of the Pharmacopoeia, which states that the combined EPA and DHA content must be at least 84%.

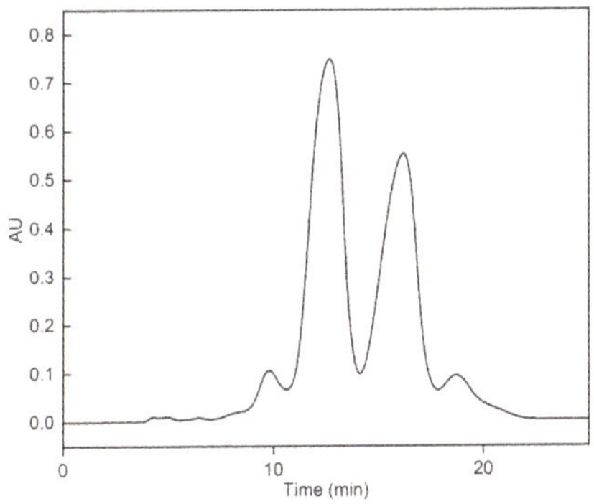

Figure 12. RP-MPLC chromatogram of fish oil ethyl ester under optimal purification conditions. Conditions: stationary phase, AQ-C18; sample load, pure fish oil ethyl esters with volume equivalent to 1.25% bed volume; mobile phase, methanol–water (90:10, *v:v*); flow rate, 30 mL/min; detector, UV 210 nm; room temperature.

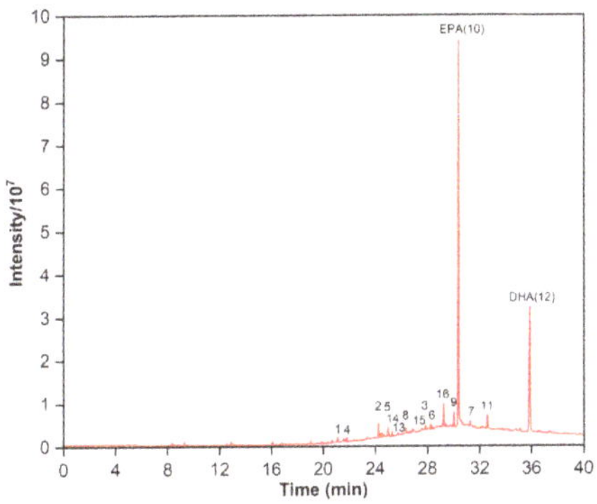

Figure 13. Gas chromatogram of the purified omega-3 fatty acid ethyl esters.

Table 8. Fatty acid composition of purified omega-3 fatty acid ethyl esters.

Number	Abbreviation	Compound Name	Content (%)
1	C16:0	Palmitic acid	0.19 ± 0.03
2	C18:0	Stearic acid	1.70 ± 0.07
3	C20:0	Arachidic acid	0.51 ± 0.04
4	C16:1n7	Palmitoleic acid	0.53 ± 0.04
5	C18:1n9	Octadecenoic acid	1.40 ± 0.07
6	C20:1n9	Eicosenoic acid	0.61 ± 0.04
7	C22:1n9	Erucic acid	0.57 ± 0.05
8	C18:3n3	α-Linolenic acid	0.58 ± 0.05
9	C20:3n3	11,14,17-Eicosatrienoic acid	2.20 ± 0.03
10	C20:5n3	5,8,11,14,17-Eicosapentaenoic acid	57.13 ± 0.08
11	C21:5n3	Heneicosapentaenoic acid	2.29 ± 0.04
12	C22:6n3	4,7,10,13,16,19-Docosahexaenoic acid	28.14 ± 0.08
13	C18:2n6	Linoleic acid	0.43 ± 0.19
14	C18:3n6	Octadecatrienoic acid	0.46 ± 0.04
15	C18:4n6	Octadecatetraenoic acid	0.18 ± 0.06
16	C20:4n6	Arachidonic acid	3.08 ± 0.07
17	∑SFA	Saturated fatty acids	2.40 ± 0.08
18	∑MUFA	Mono-unsaturated fatty acids	3.14 ± 0.07
19	∑PUFA	Poly-unsaturated fatty acids	94.47 ± 0.15
20	∑ω-3 PUFA	ω-3 Poly-unsaturated fatty acids	90.34 ± 0.17
21	∑ω-6 PUFA	ω-6 Poly-unsaturated fatty acids	4.13 ± 0.15
22	EPA + DHA		85.27 ± 0.15

2.5. Comparison of RP-MPLC and RP-HPLC

The purification of omega-3 fatty acid ethyl esters using RP-MPLC and RP-HPLC methods was evaluated using the same packing of AQ-C18, focusing on the purity, recovery rate, retention time, operation duration, and solvent consumption. The results are compared in Table 9; both the RP-MPLC and RP-HPLC methods achieved a purity $\geq$ 84% of the total content of the EPA and DHA ethyl esters, meeting the standards set by the Pharmacopoeia. This indicates that both methods are suitable for purification. Despite RP-MPLC exhibiting a slightly lower purity (85.27%) and recovery rate (74.30%) compared to RP-HPLC, it demonstrated notable advantages in shortening times for the ethyl esters EPA and DHA and the overall operation duration, with reductions of 58.35%, 63.47%, and 70.67%, respectively. Within the permissible operating pressure, the flow rate of RP-MPLC was ten-fold faster than that of RP-HPLC, which significantly improved the separation efficiency, leading to earlier peak emergence and reducing the purification time to one-third of that of RP-HPLC. However, this came at the cost of consuming 1.93 times more solvent. In industrial settings, solvents like methanol and ethanol can be recycled via dehydration with 3Å molecular sieves and rotary evaporation. This helps to cut costs and reduce environmental pollution [57]. In conclusion, when using AQ-C18 as the stationary phase, RP-MPLC not only meets the Pharmacopoeia purity standards for the ethyl esters EPA and DHA, but also enhances the unit time production efficiency by requiring only 29.33% of the purification time compared to RP-HPLC. Additionally, solvent recycling reduces costs, making RP-MPLC more cost-effective in terms of equipment, operation, and maintenance than RP-HPLC. Therefore, RP-MPLC emerges as the preferred method for the large-scale production of omega-3 fatty acids with high purity and yield.

Table 9. Comparison of purification effects of RP-MPLC and RP-HPLC methods on omega-3 fatty acid ethyl esters.

Methods	Purity of EPA-EE/DHA-EE (%)	Recovery Rate of EPA-EE/DHA-EE (%)	t_{R2} (min)	t_{R3} (min)	Duration (min)	Solvent Consumption (mL)
RP-MPLC	85.27 ± 0.15^{b}	74.30 ± 1.14^{b}	11.87 ± 0.07^{b}	15.08 ± 0.06^{b}	22.17 ± 0.76^{b}	665 ± 20.91^{a}
RP-HPLC	86.11 ± 0.96^{a}	78.56 ± 0.86^{a}	28.51 ± 0.06^{a}	41.27 ± 0.07^{a}	75 ± 1.05^{a}	225.50 ± 3.18^{b}

Note: Different letters within the group indicate significant differences ($p < 0.05$).

2.6. Summary and Prospect of RP-MPLC

Liquid chromatography is highly effective in separating structurally similar fatty acids and can be used to prepare high-purity omega-3 fatty acids [58–61], as demonstrated in this study. The RP-MPLC method ensures a safer and more reliable omega-3 fatty acid product. This process utilizes AQ-C18 as the stationary phase and methanol as the mobile phase, effectively preventing hazardous reagents such as silver mercury ions and acetone from contaminating the product and thereby eliminating potential risks to human health [62]. The RP-MPLC method requires packings with larger particle sizes and operation at lower pressures, thus allowing for increased sample loading. This enhances the chromatographic processing capacity and results in higher production efficiency. RP-MPLC is an ideal technology for enhancing the purity of substances with commercial value, especially those with low initial purity levels, such as the fish oil ethyl esters studied here, which initially had a combined EPA and DHA purity of only 67.91%. This method uses simpler purification equipment, and its cost-effectiveness is a notable advantage, making it an attractive option for the large-scale industrial purification of omega-3 fatty acid ethyl esters. Furthermore, the applications of RP-MPLC in conjunction with other chromatographic packing materials in separation and purification have been extensively studied [63,64]. It is essential to explore the purification mechanism for improving the efficiency of RP-MPLC.

3. Materials and Methods

3.1. Materials and Reagents

Fish oil ethyl esters, with an EPA purity of 40.83% and a DHA purity of 27.08%, were kindly provided by Fujian Coland Marine Bioengineering Co., Ltd (Fuzhou, China). Guaranteed-grade sodium chloride and anhydrous sodium sulfate were purchased from Xiya Chemical Technology Co., Ltd (Linyi, China). HPLC-grade methanol, ethanol, acetonitrile, and n-hexane; GC-grade boron trichloride–methanol solution (15% in methanol); standards of EPA ethyl and DHA ethyl; and the standard mixture of 40 fatty acid methyl esters were purchased from ANPLE Laboratory Technologies Inc. (Shanghai, China). Milli-Q water (Integral-3, Merck Millipore, Darmstadt, Germany) was used. All other chemicals were of chromatographic and analytical grade.

3.2. RP-MPLC Procedure

A Pure Chromatography Instrument C-815 Flash (Buchi, Switzerland) equipped with two binary gradient pumps and flashpure columns (26.2 mm × 152.3 mm) filled with various stationary phases (20–40 μm particle size, 100 Å pore size, 320–340 m^2/g surface area) was utilized for the purification of omega-3 fatty acid (Figure 14). Fish oil ethyl esters (equivalent to 0.8–1.4% of column volume) were injected manually through a sample loading loop and mixed with the mobile phase at an electric port valve. The eluent signal was recorded by UV detectors, and peak fractions with signals over 0.05 AU were collected, dried under reduced pressure for solvent removal using an R-100 rotary evaporator (Buchi, Switzerland), and subsequently characterized using a GCMS-QP2020NX gas chromatograph mass spectrometer (Shimadzu, Japan).

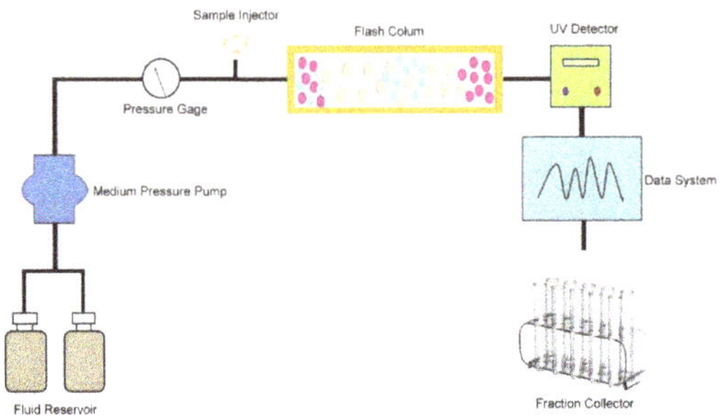

Figure 14. Schematic diagram of experimental apparatus.

3.3. Fatty Acid Analysis

The fatty acid composition of fish oil ethyl esters or RP-MPLC fraction aliquots was analyzed using GC-MS fitted with an SH-Wax capillary column (30 m × 0.25 mm × 0.25 μm). Helium (99.999%) was employed as the carrier gas at a constant flow rate of 3.0 mL/min; the injector temperature was set at 240 °C; the injection mode was splitless; and the pressure was maintained at 109.1 kPa, with a total flow rate of 30 mL/min and a column flow rate of 1.46 mL/min. The injector temperature was set at 230 °C. The temperature program was initiated at 100 °C and held for 3 min, followed by an increase at a rate of 5 °C/min to 230 °C, and then maintained for 15 min. MS detection was set to electron impact (EI) ionization mode, the ion source temperature was set at 230 °C, the interface temperature was 250 °C, the ionization energy was 70 eV, the detector voltage was 0.9 kV, the solvent delay time was 5 min, the mass scan range was set from 35 to 500 m/z, and the acquisition mode was set to scan.

The ethyl esters were identified by their retention times compared with a fatty acid methyl ester standard mixture. The relative purity (%) of various esters was calculated using integrated peak areas according to Equation (1):

$$P_i = \frac{A_{SI} \times F_{FAEEi\text{-}FAi}}{\sum A_{SI} \times F_{FAEEi\text{-}FAi}} \times 100\% \tag{1}$$

where P_i represents the percentage of a specific fatty acid in the sample relative to the total fatty acids as a percentage (%), A_{SI} is the peak area of each fatty acid ethyl ester in the sample, F_{FAEEi}-F_{Ai} refers to the coefficient for converting fatty acid ethyl esters to fatty acids, and $\sum A_{SI}$ denotes the sum of peak areas of all fatty acid ethyl esters in the sample.

3.4. Optimization of RP-MPLC

The column packing materials (CN, Diol, C4, C6, C8, C18, AQ-C18), sample load volume (0.4 mL, 0.5 mL, 0.6 mL, and 0.7 mL), sample concentration (pure fish oil, 1 g/mL, 0.5 g/mL, and 0.25 g/mL, with methanol as the dilution medium), mobile phase (methanol, ethanol, acetonitrile), flow rate (20 mL/min, 25 mL/min, 30 mL/min, and 35 mL/min), and the mobile-phase composition (92:8, 90:10, 88:12, 86:14, v:v) were varied for optimization. The performance of the RP-MPLC process was evaluated using the following indicators: the retention time (t_R), resolution from impurities (R_S), relative purity (P), and recovery rate of EPA/DHA (R_e).

The resolution, a measure of the degree of separation between two adjacent peaks in a chromatogram, reflects an important comprehensive indicator of column efficiency and selectivity. It was calculated according to Equation (2):

$$R_S = \frac{2\left|t_{R_\alpha} - t_{R_\beta}\right|}{W_1 + W_2} \times 100\% \quad (2)$$

where Rs (%) represents the resolution between omega-3 fatty acid ethyl esters and impurities, t_{R_α} (min) represents the retention time of the sample, t_{R_β} (min) represents the retention time of impurities, W_1 (min) is the peak width of omega-3 fatty acids, and W_2 (min) is the peak width of impurities.

The recovery rate is the ratio of the content of the target component to its theoretical content in the raw material, and it is an important indicator for measuring separation efficiency. The recovery calculation for the omega-3 fatty acid ethyl ester formula is as follows:

$$R_e = \frac{m_2 \times P_2}{m_1 \times P_1} \times 100\% \quad (3)$$

where R_e (%) represents the recovery rate of omega-3 fatty acid ethyl esters as a percentage; m_1 (g) is the mass of the fish oil ethyl esters raw sample, in grams; m_2 (g) is the mass of the purified fish oil ethyl esters, in grams; P_1 (%) is the purity of EPA + DHA in the fish oil ethyl esters raw sample; and P_2 (%) is the purity of EPA + DHA in the purified fish oil ethyl esters.

3.5. Comparison of RP-MPLC and RP-HPLC

By using AQ-C18 packing (20–40 μm, 100 Å pore size, 320–340 m^2/g surface area) as the stationary phase, the effects of RP-MPLC and RP-HPLC chromatography methods on the purification of omega-3 fatty acid ethyl esters were explored based on indicators such as the purity, recovery rate, retention time, operation duration, and solvent consumption.

RP-MPLC conditions: The Pure Chromatography Instrument C-815 Flash system equipped with an AQ-C18 RP-MPLC column (26.2 mm × 152.3 mm) was employed. Sample concentration: pure fish oil ethyl esters; mobile phase: methanol–water (90:10, *v:v*); isocratic elution; flow rate: 30 mL/min; UV (wavelength of 210 nm) detectors.

RP-HPLC conditions: The SMB-2–50 Liquid Chromatography System (Hanbon Sci & Tech, Jiangsu, China) fitted with an AQ-C18 RP-HPLC column (10 mm × 150 mm) was utilized. Sample concentration: pure fish oil ethyl esters; mobile phase: methanol–water (90:10, *v:v*); isocratic elution; flow rate: 3 mL/min; column temperature: 30 °C; UV detection wavelength: 210 nm.

3.6. Statistical Analysis

Samples were analyzed in triplicates, and the data are presented as mean ± standard deviation. One-way analysis of variance (ANOVA) with LSD and Turkey methods and a two-sample *t*-test were carried out using SPSS Statistics 25 (SPSS Inc., Chicago, IL, USA); a difference was considered significant at $p < 0.05$. Graphical analysis was carried out using Origin 2022b software (Origin Lab Corp., Northampton, MA, USA).

4. Conclusions

In this study, omega-3 fatty acid ethyl esters with a total content of EPA and DHA of 85.27% and the considerable recovery rate of 74.30% were obtained using AQ-C18 RP-MPLC, the optimal conditions of which were as follows: AQ-C18 packings (20–40 μm, 100 Å pore size, 320–340 m^2/g surface area), a fish oil sample injection that constituted 1.25% of the column volume; methanol–water (90:10, *v:v*) as the mobile phase, isocratic elution at a 30 mL/min flow rate; and an operating pressure of 1–4 bar. To our knowledge, this is the first study to highlight the greater suitability of AQ-C18 packing for purifying omega-3 fatty acids, as it enhanced the retention time via the polar end-capping of the silanol groups, resulting in better purification effects. A comparison between RP-MPLC and RP-HPLC revealed that, although the purity of EPA and DHA ethyl esters produced using both methods met national Pharmacopoeia standards, RP-MPLC allowed for greater sample loading, higher flow rates, and lower system pressure, thereby shortening the purification time, increasing the production efficiency, and reducing production costs. The preliminary RP-MPLC study has demonstrated great potential for industrial production, thereby laying the foundation for pilot-scale experiments. In the future, it will be necessary to conduct pilot-scale experiments to explore the RP-MPLC purification of high-purity omega-3 fatty acids and perform a techno-economic assessment of large-scale production costs.

Author Contributions: Writing—original draft, M.S.; writing—review and editing, N.P.; conceptualization, N.P. and Z.L.; methodology, N.P and J.W.; data curation, M.S. and H.F.; investigation, M.S., X.C. and S.C.; visualization, M.S. and N.P.; validation, X.J.; supervision, M.X. and Z.L.; funding acquisition, N.P. and Z.L. All authors have read and agreed to the published version of the manuscript.

Funding: This research was supported by the National Key Research and Development Program of China (Grant No. 2018YFD0901102), the Basic Special Project for Public Welfare Scientific Research Institutes in Fujian Province (Grant No. 2021R10130014), and the Special Program for High Quality Development of Marine Services and Fisheries in Fujian Province (Grant No. FJHY-YYKJ-2023-1-8).

Institutional Review Board Statement: Not applicable.

Data Availability Statement: The data presented in this study are available on request from the corresponding author.

Acknowledgments: We express our gratitude to Lan Lin and Xiaojun Chen from Fujian Coland Marine Bioengineering Co., Ltd. for generously providing the fish oil ethyl esters and their profound insights into fish oil production. We are grateful to Dawei Liu and Wei Yang from Hanbon Sci. & Tech. Co., Ltd. for providing the SMB-2-50 HPLC equipment and technical support.

Conflicts of Interest: The authors declare no conflicts of interest.

References

1. Xie, D.; Chen, Y.; Yu, J.; Yang, Z.; Wang, X.; Wang, X. Progress in enrichment of n-3 polyunsaturated fatty acid: A review. *Crit. Rev. Food Sci. Nutr.* **2023**, *63*, 11310–11326. [CrossRef] [PubMed]
2. Saini, R.K.; Keum, Y.S. Omega-3 and omega-6 polyunsaturated fatty acids: Dietary sources, metabolism, and significance—A review. *Life Sci.* **2018**, *203*, 255–267. [CrossRef] [PubMed]
3. Bhatt, D.L.; Steg, P.G.; Miller, M.; Brinton, E.A.; Jacobson, T.A.; Ketchum, S.B.; Doyle, R.T., Jr.; Juliano, R.A.; Jiao, L.; Granowitz, C.; et al. Cardiovascular risk reduction with icosapent ethyl for hypertriglyceridemia. *N. Engl. J. Med.* **2019**, *380*, 11–22. [CrossRef] [PubMed]
4. Laye, S.; Nadjar, A.; Joffre, C.; Bazinet, R.P. Anti-inflammatory effects of omega-3 fatty acids in the brain: Physiological mechanisms and relevance to pharmacology. *Pharmacol. Rev.* **2018**, *70*, 12–38. [CrossRef] [PubMed]
5. Mallik, A.K.; Qiu, H.; Takafuji, M.; Ihara, H. High molecular-shape-selective stationary phases for reversed-phase liquid chromatography: A review. *TrAC Trends Anal. Chem.* **2018**, *108*, 381–404. [CrossRef]
6. Jesionowska, M.; Ovadia, J.; Hockemeyer, K.; Clews, A.C.; Xu, Y. EPA and DHA in microalgae: Health benefits, biosynthesis, and metabolic engineering advances. *J. Am. Oil Chem. Soc.* **2023**, *100*, 831–842. [CrossRef]
7. Sheppard, J.P.; Lakshmanan, S.; Dahal, S.; Roy, S.K.; Bhatt, D.L.; Budoff, M.J.; Nelson, J.R. EPA versus mixed EPA/DHA plus statin for coronary atherosclerosis: Meta-analysis of prospective imaging trials. *JACC Cardiovasc. Imaging* **2022**, *15*, 1825–1828. [CrossRef] [PubMed]
8. Pertiwi, K.; Kupers, L.K.; de Goede, J.; Zock, P.L.; Kromhout, D.; Geleijnse, J.M. Dietary and circulating long-chain omega-3 polyunsaturated fatty acids and mortality risk after myocardial Infarction: A long-term follow-up of the alpha omega cohort. *J. Am. Heart Assoc.* **2021**, *10*, e022617. [CrossRef] [PubMed]
9. Crupi, R.; Cuzzocrea, S. Role of EPA in inflammation: Mechanisms, effects, and clinical relevance. *Biomolecules* **2022**, *12*, 242. [CrossRef]
10. Alfaddagh, A.; Elajami, T.K.; Saleh, M.; Elajami, M.; Bistrian, B.R.; Welty, F.K. The effect of eicosapentaenoic and docosahexaenoic acids on physical function, exercise, and joint replacement in patients with coronary artery disease: A secondary analysis of a randomized clinical trial. *J. Clin. Lipidol.* **2018**, *12*, 937–947.e932. [CrossRef]
11. Vitlov Uljevic, M.; Starcevic, K.; Masek, T.; Bocina, I.; Restovic, I.; Kevic, N.; Racetin, A.; Kretzschmar, G.; Grobe, M.; Vukojevic, K.; et al. Dietary DHA/EPA supplementation ameliorates diabetic nephropathy by protecting from distal tubular cell damage. *Cell Tissue Res.* **2019**, *378*, 301–317. [CrossRef] [PubMed]

12. Lauritzen, L.; Brambilla, P.; Mazzocchi, A.; Harsløf, L.B.; Ciappolino, V.; Agostoni, C. DHA effects in brain development and function. *Nutrients* **2016**, *8*, 6. [CrossRef]
13. Charles, C.N.; Swai, H.; Msagati, T.; Chacha, M. Development of a natural product rich in bioavailable omega-3 DHA from locally available ingredients for prevention of nutrition related mental illnesses. *J. Am. Coll. Nutr.* **2020**, *39*, 720–732. [CrossRef] [PubMed]
14. Rahman, M.; Ouhtit, A.; Saleh, H.A. Docosahexaenoic acid (DHA) as an adjunctive therapeutic agent for the treatment of cancer. *FASEB J.* **2019**, *33*, 652.17. [CrossRef]
15. Blair, H.A.; Dhillon, S. Omega-3 carboxylic acids (Epanova): A review of its use in patients with severe hypertriglyceridemia. *Am. J. Cardiovasc. Drugs* **2014**, *14*, 393–400. [CrossRef] [PubMed]
16. Lembke, P. Production techniques for omega-3 concentrates. In *Omega-6/3 Fatty Acids*; Meester, F., Watson, R., Zibadi, S., Eds.; Humana Press: Totowa, NJ, USA, 2013; pp. 353–364.
17. Zhang, Y.; Wang, X.; Xie, D.; Li, Y.; Wang, X. One-step concentration of highly unsaturated fatty acids from tuna oil by low-temperature crystallization. *J. Am. Oil Chem. Soc.* **2017**, *94*, 475–483. [CrossRef]
18. Mu, H.; Zhang, H.; Li, Y.; Zhang, Y.; Wang, X.; Jin, Q.; Wang, X. Enrichment of DPAn-6 and DHA from *Schizochytrium* sp. oil by low-temperature solvent crystallization. *Ind. Eng. Chem. Res.* **2016**, *55*, 737–746. [CrossRef]
19. González-Fernández, M.J.; Fabrikov, D.; Lyashenko, S.; Ferrón-Carrillo, F.; Guil-Guerrero, J.L. Highly concentrated very long-chain PUFA obtainment by urea complexation methodology. *Environ. Technol. Innov.* **2020**, *18*, 100736. [CrossRef]
20. Dovale-Rosabal, G.; Rodriguez, A.; Contreras, E.; Ortiz-Viedma, J.; Munoz, M.; Trigo, M.; Aubourg, S.P.; Espinosa, A. Concentration of EPA and DHA from refined salmon oil by optimizing the urea fatty acid adduction reaction conditions using response surface methodology. *Molecules* **2019**, *24*, 1642. [CrossRef]
21. Magallanes, L.M.; Tarditto, L.V.; Grosso, N.R.; Pramparo, M.C.; Gayol, M.F. Highly concentrated omega-3 fatty acid ethyl esters by urea complexation and molecular distillation. *J. Sci. Food Agric.* **2019**, *99*, 877–884. [CrossRef]
22. Rossi, P.; Gayol, M.F.; Renaudo, C.; Pramparo, M.C.; Nepote, V.; Grosso, N.R. The use of artificial neural network modeling to represent the process of concentration by molecular distillation of omega-3 from squid oil. *Grasas Y Aceites* **2014**, *65*, e052. [CrossRef]
23. Qiao, M.; Hua, X.; Yuan, Y.; Cao, X. Enrichment of EPA and DHA by enzymatic ethanolysis: Effects of lipases from *Thermomyces lanuginosus, Rhizomucor miehei, Rhizopus oryzae* and *Candida antarctica*. *J. Am. Oil Chem. Soc.* **2023**. [CrossRef]
24. Baloch, K.A.; Patil, U.; Pudtikajorn, K.; Khojah, E.; Fikry, M.; Benjakul, S. Lipase-catalyzed synthesis of structured fatty acids enriched with medium and long-chain n-3 fatty acids via solvent-free transesterification of skipjack tuna eyeball oil and commercial butterfat. *Foods* **2024**, *13*, 347. [CrossRef] [PubMed]
25. Molino, A.; Martino, M.; Larocca, V.; Di Sanzo, G.; Spagnoletta, A.; Marino, T.; Karatza, D.; Iovine, A.; Mehariya, S.; Musmarra, D. Eicosapentaenoic acid extraction from *Nannochloropsis gaditana* using carbon dioxide at supercritical conditions. *Mar. Drugs* **2019**, *17*, 132. [CrossRef]
26. Mu, H.; Jin, J.; Xie, D.; Zou, X.; Wang, X.; Wang, X.; Jin, Q. Combined urea complexation and argentated silica gel column chromatography for concentration and separation of PUFAs from tuna oil: Based on improved DPA level. *J. Am. Oil Chem. Soc.* **2016**, *93*, 1157–1167. [CrossRef]
27. Bárcenas-Pérez, D.; Lukeš, M.; Hrouzek, P.; Zápal, J.; Kuzma, M.; Kopecký, J.; Kubáč, D.; Arredondo-Vega, B.O.; Cheel, J. Bio-production of eicosapentaenoic acid from the diatom *Nanofrustulum shiloi* via two-step high performance countercurrent chromatography. *J. Appl. Phycol.* **2022**, *34*, 2995–3010. [CrossRef]
28. Zhao, Y.; Wang, N.; Gao, H.-L.; Guo, Z.-X.; Lu, A.-X.; Guo, X.-J.; Lu, J.-H.; Luan, Y.-X. Determination of aflatoxin B1 in lotus seed by high performance liquid chromatography with aptamer affinity column for purification and enrichment. *Chin. J. Anal. Chem.* **2020**, *48*, 662–669. [CrossRef]
29. Wei, B.; Wang, S. Separation of eicosapentaenoic acid and docosahexaenoic acid by three-zone simulated moving bed chromatography. *J. Chromatogr. A* **2020**, *1625*, 461326. [CrossRef]
30. Oh, C.-E.; Kim, G.-J.; Park, S.-J.; Choi, S.; Park, M.-J.; Lee, O.M.; Seo, J.-W.; Son, H.-J. Purification of high purity docosahexaenoic acid from *Schizochytrium* sp. SH103 using preparative-scale HPLC. *Appl. Biol. Chem.* **2020**, *63*, 56. [CrossRef]
31. Montañés, F.; Catchpole, O.J.; Tallon, S.; Mitchell, K.; Lagutin, K. Semi-preparative supercritical chromatography scale plant for polyunsaturated fatty acids purification. *J. Supercrit. Fluids* **2013**, *79*, 46–54. [CrossRef]
32. Li, M.; Bao, Z.; Xing, H.; Yang, Q.; Yang, Y.; Ren, Q. Simulated moving bed chromatography for the separation of ethyl esters of eicosapentaenoic acid and docosahexaenoic acid under nonlinear conditions. *J. Chromatogr. A* **2015**, *1425*, 189–197. [CrossRef] [PubMed]
33. Ji, D.; Wang, Q.; Wang, H.; Ma, Q.; Wang, M.; Lu, Y. Preparative separation of gallic acid from *Fallopia aubertii* using middle-pressure chromatogram isolated gel coupled with reversed-phase chromatography with hydrophilic groups. *RSC Adv.* **2021**, *11*, 27276–27282. [CrossRef] [PubMed]
34. Boshra, S.A.; El-Haddad, A.E. The protective effects of MPLC isolated glycyrrhizin and mangiferin against brain toxicity in rats. *Med. Chem. Res.* **2018**, *27*, 1449–1459. [CrossRef]
35. Lu, X.; Zheng, Z.; Miao, S.; Li, H.; Guo, Z.; Zhang, Y.; Zheng, Y.; Zheng, B.; Xiao, J. Separation of oligosaccharides from lotus seeds via medium-pressure liquid chromatography coupled with ELSD and DAD. *Sci. Rep.* **2017**, *7*, 44174. [CrossRef] [PubMed]

36. Sabana, I.; Naufal, M.; Wiani, I.; Zainuddin, A.; Hidayat, A.; Harneti, D.; Nurlelasari, N.; Al-Anshori, J.; Supratman, U.; Maharani, R. Synthesis of antioxidant peptide SCAP1 (Leu-Ala-Asn-Ala-Lys). *Egypt. J. Chem.* **2020**, *63*, 921–926. [CrossRef]
37. Wang, Y.; Liu, H.; Shen, L.; Yao, L.; Ma, Y.; Yu, D.; Chen, J.; Li, P.; Chen, Y.; Zhang, C. Isolation and purification of six iridoid glycosides from *Gardenia jasminoides* fruit by medium-pressure liquid chromatography combined with macroporous resin chromatography. *J. Sep. Sci.* **2015**, *38*, 4119–4126. [CrossRef] [PubMed]
38. Ishihara, K.; Murata, M.; Kaneniwa, M.; Saito, H.; Komatsu, W.; Shinohara, K. Purification of stearidonic acid (18: 4 (n-3)) and hexadecatetraenoic acid (16: 4 (n-3)) from algal fatty acid with lipase and medium pressure liquid chromatography. *Biosci. Biotechnol. Biochem.* **2000**, *64*, 2454–2457. [CrossRef] [PubMed]
39. Zuvela, P.; Skoczylas, M.; Jay Liu, J.; Ba Czek, T.; Kaliszan, R.; Wong, M.W.; Buszewski, B.; Heberger, K. Column characterization and selection systems in reversed-phase high-performance liquid chromatography. *Chem. Rev.* **2019**, *119*, 3674–3729. [CrossRef] [PubMed]
40. Ruiz-Angel, M.J.; Pous-Torres, S.; Carda-Broch, S.; Garcia-Alvarez-Coque, M.C. Performance of different C18 columns in reversed-phase liquid chromatography with hydro-organic and micellar-organic mobile phases. *J. Chromatogr. A* **2014**, *1344*, 76–82. [CrossRef]
41. Gritti, F.; Kazakevich, Y.V.; Guiochon, G. Effect of the surface coverage of endcapped C18-silica on the excess adsorption isotherms of commonly used organic solvents from water in reversed phase liquid chromatography. *J. Chromatogr. A* **2007**, *1169*, 111–124. [CrossRef]
42. Agrahari, V.; Bajpai, M.; Nanda, S. Essential concepts of mobile phase selection for reversed phase HPLC. *Res. J. Pharm. Technol.* **2013**, *6*, 459–464.
43. Stepnowski, P.; Blotevogel, K.-H.; Jastorff, B. Applied waste-free recovery of methanol: A sustainable solution for chromatography laboratories. *Environ. Sci. Pollut. Res.* **2002**, *9*, 34–38. [CrossRef] [PubMed]
44. Habaki, H.; Yoshimura, Y.; Egashira, R. Separation of aromatic components from light cycle oil by solvent extraction. *Sep. Sci. Technol.* **2019**, *54*, 1159–1166. [CrossRef]
45. Magda, R.; Toth, J. The connection of the methanol economy to the concept of the circular economy and its impact on sustainability. *Visegr. J. Bioecon. Sustain. Dev.* **2019**, *8*, 58–62. [CrossRef]
46. Schmidt-Traub, H. *Preparative Chromatography*; Wiley-VCH Verlag GmbH & Co. KGaA: Weinheim, Germany, 2012; pp. 154–190. [CrossRef]
47. Chen, Y.; Guo, Z.; Wang, X.; Qiu, C. Sample preparation. *J. Chromatogr. A* **2008**, *1184*, 191–219. [CrossRef] [PubMed]
48. Dillon, J.T.; Aponte, J.C.; Tarozo, R.; Huang, Y. Purification of omega-3 polyunsaturated fatty acids from fish oil using silver-thiolate chromatographic material and high performance liquid chromatography. *J. Chromatogr. A* **2013**, *1312*, 18–25. [CrossRef] [PubMed]
49. Ren, D.B.; Yang, Z.H.; Liang, Y.Z.; Fan, W.; Ding, Q. Effects of injection volume on chromatographic features and resolution in the process of counter-current chromatography. *J. Chromatogr. A* **2013**, *1277*, 7–14. [CrossRef] [PubMed]
50. McCalley, D.V. The challenges of the analysis of basic compounds by high performance liquid chromatography: Some possible approaches for improved separations. *J. Chromatogr. A* **2010**, *1217*, 858–880. [CrossRef] [PubMed]
51. Rafferty, J.L.; Siepmann, J.I.; Schure, M.R. Retention mechanism for polycyclic aromatic hydrocarbons in reversed-phase liquid chromatography with monomeric stationary phases. *J. Chromatogr. A* **2011**, *1218*, 9183–9193. [CrossRef]
52. Rafferty, J.L.; Siepmann, J.I.; Schure, M.R. The effects of chain length, embedded polar groups, pressure, and pore shape on structure and retention in reversed-phase liquid chromatography: Molecular-level insights from Monte Carlo simulations. *J. Chromatogr. A* **2009**, *1216*, 2320–2331. [CrossRef]
53. Rafferty, J.L.; Siepmann, J.I.; Schure, M.R. Molecular-level comparison of alkylsilane and polar-embedded reversed-phase liquid chromatography systems. *Anal. Chem.* **2008**, *80*, 6214–6221. [CrossRef]
54. Terol, A.; Ardini, F.; Grotti, M.; Todoli, J.L. High temperature liquid chromatography-inductively coupled plasma mass spectrometry for the determination of arsenosugars in biological samples. *J. Chromatogr. A* **2012**, *1262*, 70–76. [CrossRef] [PubMed]
55. Duan, Y.; Liu, H.; Li, J.; Ma, J.; Gu, Y.; Yan, C.; Yang, G. Preparation and evaluation of a porous P(NIPAAm-MAA-EDMA) monolithic column for HPLC. *Chromatographia* **2011**, *75*, 87–93. [CrossRef]
56. Zheng, Z.; Dai, Z.; Cao, Y. Isolation, purification of DPAn-3 from the seal oil ethyl ester. *Eur. J. Lipid Sci. Technol.* **2018**, *120*, 1800225. [CrossRef]
57. Kamarei, F.; Vajda, P.; Guiochon, G. Comparison of large scale purification processes of naproxen enantiomers by chromatography using methanol–water and methanol–supercritical carbon dioxide mobile phases. *J. Chromatogr. A* **2013**, *1308*, 132–138. [CrossRef]
58. Namal Senanayake, S.P.J. Methods of concentration and purification of omega-3 fatty acids. In *Separation, Extraction and Concentration Processes in the Food, Beverage and Nutraceutical Industries*; Woodhead Publishing: Cambridge, UK, 2013; pp. 483–505.
59. Dołowy, M.; Pyka, A. Chromatographic methods in the separation of long-chain mono- and polyunsaturated fatty acids. *J. Chem.* **2015**, *2015*, 120830. [CrossRef]
60. Rubio-Rodríguez, N.; Beltrán, S.; Jaime, I.; de Diego, S.M.; Sanz, M.T.; Carballido, J.R. Production of omega-3 polyunsaturated fatty acid concentrates: A review. *Innov. Food Sci. Emerg. Technol.* **2010**, *11*, 1–12. [CrossRef]
61. Mansour, M.P. Reversed-phase high-performance liquid chromatography purification of methyl esters of C(16)-C(28) polyunsaturated fatty acids in microalgae, including octacosaoctaenoic acid [28:8(n-3)]. *J. Chromatogr. A* **2005**, *1097*, 54–58. [CrossRef]

62. Rezvani, E.; Rafferty, A.; McGuinness, C.; Kennedy, J. Adverse effects of nanosilver on human health and the environment. *Acta Biomater.* **2019**, *94*, 145–159. [CrossRef]
63. Jansen, R.; Sood, S.; Mohr, K.I.; Kunze, B.; Irschik, H.; Stadler, M.; Müller, R. Nannozinones and sorazinones, unprecedented pyrazinones from myxobacteria. *J. Nat. Prod.* **2014**, *77*, 2545–2552. [CrossRef]
64. Feng, J.; He, X.; Zhou, S.; Peng, F.; Liu, J.; Hao, L.; Li, H.; Ao, G.; Yang, S. Preparative separation of crocins and geniposide simultaneously from gardenia fruits using macroporous resin and reversed-phase chromatography. *J. Sep. Sci.* **2014**, *37*, 314–322. [CrossRef] [PubMed]

Disclaimer/Publisher's Note: The statements, opinions and data contained in all publications are solely those of the individual author(s) and contributor(s) and not of MDPI and/or the editor(s). MDPI and/or the editor(s) disclaim responsibility for any injury to people or property resulting from any ideas, methods, instructions or products referred to in the content.

marine drugs

MDPI

Article

Integrated Process for *Schizochytrium* Oil Extraction, Enzymatic Modification of Lipids and Concentration of DHA Fatty Acid Esters Using Alternative Methodologies

Gonzalo Berzal, Paz García-García and Francisco Javier Señoráns *

Healthy-Lipids Group, Food Science Department, Faculty of Sciences, Universidad Autónoma de Madrid, Francisco Tomás y Valiente, 7, 28049 Madrid, Spain; gonzalo.berzal@inv.uam.es (G.B.); mariap.garcia@uam.es (P.G.-G.)

* Correspondence: javier.senorans@uam.es; Tel.: +34-91-4973-775

Abstract: Marine microalgae *Schizochytrium* sp. have a high content of docosahexaenoic acid (DHA), an omega-3 fatty acid that is attracting interest since it prevents certain neurodegenerative diseases. The obtention of a bioactive and purified DHA fatty acid ester using a whole-integrated process in which renewable sources and alternative methodologies are employed is the aim of this study. For this reason, lyophilized *Schizochytrium* biomass was used as an alternative to fish oil, and advanced extraction techniques as well as enzymatic modification were studied. Microalgal oil extraction was optimized via a surface-response method using pressurized liquid extraction (PLE) obtaining high oil yields (29.06 ± 0.12%) with a high concentration of DHA (51.15 ± 0.72%). Then, the enzymatic modification of *Schizochytrium* oil was developed by ethanolysis using immobilized *Candida antarctica* B lipase (Novozym® 435) at two reaction temperatures and different enzymatic loads. The best condition (40 °C and 200 mg of lipase) produced the highest yield of fatty acid ethyl ester (FAEE) (100%) after 8 h of a reaction attaining a cost-effective and alternative process. Finally, an enriched and purified fraction containing DHA-FAEE was obtained using open-column chromatography with a remarkably high concentration of 93.2 ± 1.3% DHA. The purified and bioactive molecules obtained in this study can be used as nutraceutical and active pharmaceutical intermediates of marine origin.

Keywords: microalgae; *Schizochytrium* sp.; DHA; Novozym®435; pressurized liquid extraction; enzymatic ethanolysis

1. Introduction

In recent years, several studies have focused on polyunsaturated fatty acids (PUFA), specially omega-3, and their beneficial effects on health [1,2]. In particular, docosahexaenoic acid (DHA, C22:6 *n*-3), a long-chained PUFA (LC-PUFA) [3], is being investigated due to its capability to modulate many inflammatory processes and its key role in brain development [4]. Thus, DHA is proposed to prevent neurodegenerative diseases, such as Alzheimer's or Parkinson's, as well as inflammatory diseases [5–7].

Nevertheless, to efficiently increase the amount of DHA present in neuronal tissues, DHA may be incorporated in the diet via nutritional supplements and nutraceuticals, since conversion from α-linolenic acid to DHA in adults is limited [8,9]. In this way, the preferred sources of DHA are fish and krill, but microalgae have emerged as an alternative and ecological source of DHA [8,10] that is in agreement with the sustainability of marine resources. Therefore, bioactive compounds from microalgal oils are gaining importance as a renewable and sustainable source of DHA [8,11,12].

Among different microalgae that are harvested worldwide and accepted as novel food, *Schizochytrium* sp. is emphasized due to its high lipid content, especially regarding the high amounts of DHA (around 40% DHA), whose accumulation depends on culture conditions [13,14]. This microalgal oil is advantageous in the industry because of its heterotrophic culture conditions, which enable elevated DHA production in the form of triacylglycerols (TAG) [15]. Moreover, *Schizochytrium* oil is also composed of three other principal fatty acids, myristic (C14:0), palmitic (C16:0), and docosapentaenoic acid (C22:5 n-6, DPA), the main acids found in its composition [13,14]. For their extraction, modern technologies such as pressurized liquid extraction (PLE) arise as an alternative to traditional extraction procedures [16,17]. PLE uses high temperatures and pressures that can extract bioactive compounds extremely quickly, avoiding their oxidation and deterioration in a process that follows the principles of green chemistry [18–22]. Thus, the possibility of modifying these parameters enables the optimization of lipid extraction for its consequent hydrolysis by lipases.

Lipases (EC 3.1.1.3) are used in a wide range of industrial applications [23–25] in the pharmaceutical and food industries. Over the last decade, oil hydrolysis and transesterification by lipases have become more important [26–28]. However, only a few reports focus on the ethanolysis of microalgae oil using lipases [29–32]. The esterification of LC-PUFA, such as DHA, is developed by immobilized lipases under mild conditions to maintain the labile structure of the fatty acid. The use of immobilized lipases as industrial biocatalysts is the most suitable method for developing more selective, controlled, and rapid procedures in the industry [33–36]. Furthermore, immobilization allows for the industrial reuse of the biocatalyst for several cycles as a result of an increase in its stability, and the easy separation of the desired product, obtaining cost-efficient procedures [37–40].

Candida antarctica lipase B (CALB) is an immobilized lipase, commercially known as Novozym®435, characterized by its versatility, high activity, and stability [41]. CALB is used for oil modification, mainly in biodiesel production, as well as food industry [42]. It is able to cause oil ethanolysis in solvent-free systems according to green chemistry [26]. The molecules produced in the reaction (fatty acid ethyl esters, FAEE) [29] could serve as a scaffold to develop structured lipids enriched in DHA with improved properties and composition [1,43–47] using regioselective lipases [48–50].

In this study, a novel strategy to obtain an enriched DHA oil from *Schizochytrium* sp. was proposed. For this aim, different extraction conditions using PLE technology were compared and optimized, and the extracted oil was characterized by GC-MS. The enzymatic ethanolysis of microalgal oil was developed using CALB Novozym®435 to produce FAEE with the highest concentration of DHA, which can serve as a food supplement or nutraceutical for structured phospholipid synthesis to prevent neurodegenerative diseases. Thus, the hypothesis of this paper was to determine the possibility of producing DHA concentrates from sustainable raw materials as microalgae using environmentally friendly technologies.

2. Results and Discussion

2.1. Lipid Extraction from Schizochytrium sp. by Pressurized Liquids Compared to the Soxhlet Method

Lipid extraction of *Schizochytrium* lipids was compared using two different methods. On the one hand, Soxhlet was used as the traditional procedure for obtaining a reference lipidic yield. On the other hand, PLE was proposed as an alternative method for the advanced and fast extraction of bioactive compounds preserving the bioactivity.

The result obtained with Soxhlet was an extraction yield of 24.04 ± 0.25%. Lipid extraction using pressurized liquids was optimized using surface-response methodology. Parameters such as temperature and solvent polarity (hexane and ethanol) were evaluated in a previous study on *Nannochloropsis* lipids [51]. In this study, hexane was used as a non-polar solvent, ethanol as a polar solvent and a mixture of both with mild polarity (1:1) for lipid extraction, as reported for other microalgae species [52]. Moreover, different extraction times (5, 10 and 15 min) and different temperatures (from 80 to 120 °C) were also evaluated.

Using the Statgraphics 19 program, the surface-response plots shown in Figure 1 were drawn.

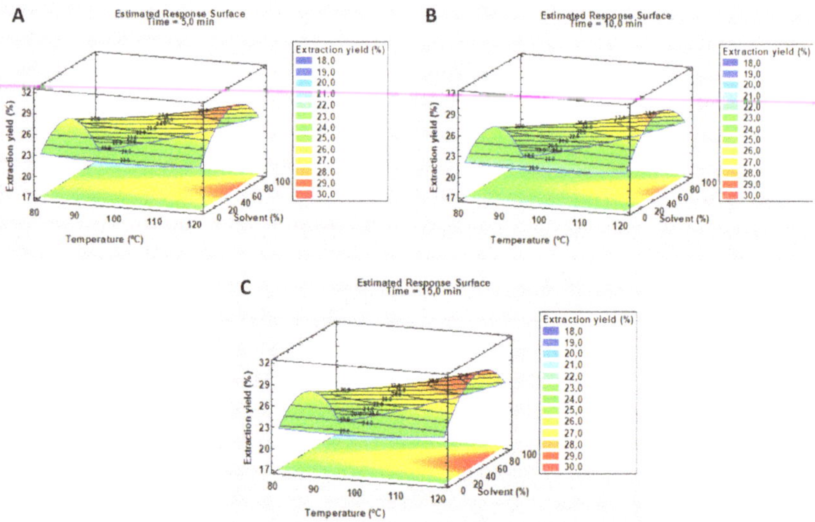

Figure 1. Surface-response plots obtained with the Statgraphics 19 program with different extraction times. (**A**) Corresponds to a static extraction time of 5 min. (**B**) Corresponds to a static extraction time of 10 min. (**C**) Corresponds to a static extraction time of 15 min.

As can be seen, the influence of different solvent mixtures on the extraction was revealed. Thus, using hexane and ethanol in equal parts (1:1), the yield improved compared to extraction only with hexane or ethanol for all the times studied. With respect to the static extraction time, there was no difference in the oil yield when this parameter was modified (Appendix A). In addition, an increase in the extraction yield was obtained, as previously found in the literature, with oil yields ranging from 17.13 $\pm$ 1.17% to a maximum yield of 29.06 $\pm$ 0.12% at 120 °C for 15 min, and with a hexane–ethanol 1:1 solvent mixture.

Thus, the extraction with a mixture of hexane–ethanol in a 1:1 proportion was used to scale up the method, with a final volume of 70 mL of solvent. In this case, a 26.15 $\pm$ 1.13% extraction yield was achieved, notably similar to that obtained by the standard protocol with less volume. In the scale-up, a solvent saving system was configured with up to 10 times the amount of microalgae. In both cases, the results found were also comparable to the yield found using the Soxhlet method, representing the valuable alternative of PLE against traditional methods, decreasing both the amount of non-friendly solvents, such as hexane, and the necessary time and energy. Therefore, the scale-up of the extraction method was useful as it allowed a higher amount of microalgal oil to be produced, leading to the development of further experiments of enzymatic modification.

2.2. *Characterization of Schizochytrium sp. Oils*

Starting with the HPLC-ELSD characterization (see Section 3.7), all oils used for the experimental development were analyzed, and the results show that they contained only TAG in their composition (see Appendix B), in accordance with the consulted literature [53].

For characterization of fatty acids of the TAG by GC-MS, an analysis of *Schizochytrium* sp. commercial oil and microalgae oils obtained using PLE in our laboratory was carried out. To compare their fatty acid profiles, as oils contained only TAG in their composition (no free fatty acids), their characterization in a basic medium is an effective method (see Section 3.5). GC-MS was employed to obtain different fractions of decreasing polarity and separate the fatty acid methyl esters. Fatty acids such as myristic (14:0), palmitic (16:0), stearic (18:0), oleic (18:1), linoleic (18:2), arachidonic (20:4), eicosapentaenoic (EPA), docosapentaenoic (DPA *n*-3 and *n*-6), and docosahexaenoic (DHA) were found.

According to results shown in Table 1, there were some differences between commercial oil and oils extracted using PLE with different solvents. In commercial oil a lack of myristic acid was found in the fatty acid profile, despite the fact that myristic acid is commonly found in other oils of *Schizochytrium* sp. described in the literature [13]. In our case, a ratio of 14:0 fatty acids was present in all microalgae oils obtained using PLE in around 10–13% of the total fatty acid composition.

Table 1. Fatty acid profiles of different *Schizochytrium* sp. oils (commercial, extracted from biomass by PLE with different solvents and extracted from biomass using the Soxhlet method) analyzed by GC-MS. Microalgal oil 1 referred to *Schizochytrium* oil extracted with PLE using hexane. Microalgal oil 2 referred to *Schizochytrium* oil extracted with PLE using hexane–ethanol (1:1). Data were calculated as a percentage of fatty acid composition related to total ± standard deviation (SD).

	Commercial Oil	Microalgal Oil 1	Microalgal Oil 2	Soxhlet Extraction
14:0	-	10.41 ± 0.19	9.96 ± 0.45	12.72 ± 0.18
16:0	20.67 ± 0.04	20.30 ± 3.89	19.34 ± 0.57	23.02 ± 0.07
18:0	0.99 ± 0.02	1.15 ± 0.55	0.53 ± 0.09	0.41 ± 0.05
18:1 n-9	0.33 ± 0.09	0.80 ± 0.05	0.63 ± 0.03	0.64 ± 0.05
18:2 n-6	0.41 ± 0.03	0.27 ± 0.02	0.20 ± 0.07	0.14 ± 0.01
20:4	0.46 ± 0.08	0.81 ± 0.06	0.77 ± 0.04	0.83 ± 0.25
20:5 n-3 (EPA)	0.38 ± 0.02	0.98 ± 0.08	0.75 ± 0.04	0.83 ± 0.01
22:5 (DPA n-6)	15.42 ± 0.12	14.61 ± 0.58	15.49 ± 0.09	14.12 ± 0.30
22:5 (DPA n-3)	0.49± 0.04	0.36 ± 0.06	0.45 ± 0.03	0.33 ± 0.05
22:6 n-3 (DHA)	60.85 ± 0.32	49.25 ± 4.72	51.15 ± 0.72	45.58 ± 0.34

Moreover, the composition of palmitic acid was similar in all studied oils (around 20% of total composition) as well as DPA *n*-6 (around 15% in all cases). Even though the DHA percentage in the composition was higher in the commercial oil, there was an impact on the whole fatty acid profile, as it did not have myristic oil. Therefore, DHA representativeness was diverse, and the results cannot be directly compared. Accordingly, the oil with the highest amount of DHA from all extracted oils tested was that produced by PLE with 1:1 hexane–ethanol. More than 50% of the total fatty acid composition corresponded to DHA, exceeding that extracted by Soxhlet by 5%. This result may be due to the methodology applied, since the Soxhlet method uses high temperatures for several hours. In the case of PLE, the sample is in an environment free of oxygen and light, unlike the Soxhlet method, so there is no oxidation of PUFA during short time extraction, nor when high temperatures such as 120 °C are used [54–56].

Regarding the fatty acid composition of *Schizochytrium* sp. oils extracted with different solvent mixtures, the extraction by PLE influenced extraction yield (see Section 2.1), but it did not have much impact on the fatty acid composition of the extracted oil.

2.3. Enzymatic Ethanolysis of Schizochytrium Oil

To optimize the enzymatic ethanolysis reactions of *Schizochytrium* oil and produce ethyl esters of the fatty acids, reaction kinetics at different temperatures (30 °C and 40 °C) and different loads of commercial lipase Novozym® 435 (CALB) were studied.

The results represented in Figure 2 show that the conversion of TAG into FAEE was similar at both reaction temperatures of 30 °C or 40 °C. However, remarkable results were obtained for the first two hours of reaction, where large differences between temperatures in the conversion of the initial TAG were observed. In this case, at 30 °C (Figure 2A) there was a 64.35% TAG compared to 20.68% FAEE. Comparing these results with those obtained at 40 °C (Figure 2B), at the same time as the reaction, there was 54.24% of TAG versus 30.18% of FAEE.

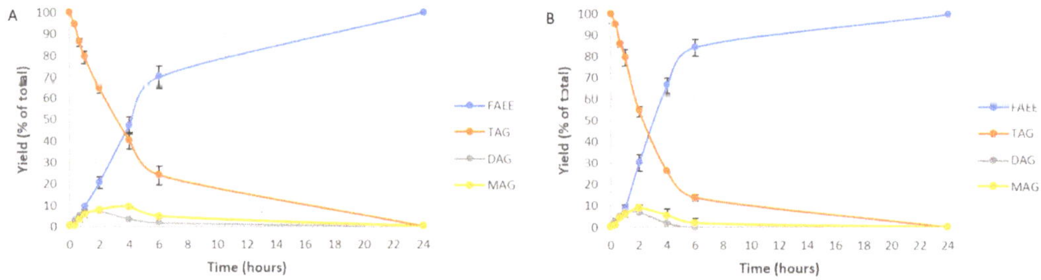

Figure 2. Kinetics of enzymatic ethanolysis, using CALB at 30 °C (**A**) and 40 °C (**B**), both at 200 rpm.

From this point, there was a higher conversion in FAEE at 40 °C than at 30 °C, while after 6 h of ethanolysis, at 40 °C, the yield of FAEE reached 84.47%; the yield at 30 °C was 69.84%. When considering the 24-h aliquot, in both cases, a yield of 100% was achieved in the conversion of TAG into FAEE. In all cases, no relevant differences in reaction intermediates (DAG and MAG) were observed, either at 30 °C or 40 °C.

Therefore, the increase in temperature favors the course of the enzymatic reaction, as was already discussed in the literature on Novozym® 435 lipase [30]. Subsequently, enzymatic ethanolysis was also performed at 40 °C by adding twice the amount of the enzyme to study the influence of this parameter.

As seen in Figure 3, the reaction with a double load of CALB evolved more rapidly than the reaction with a normal load at the same temperature.

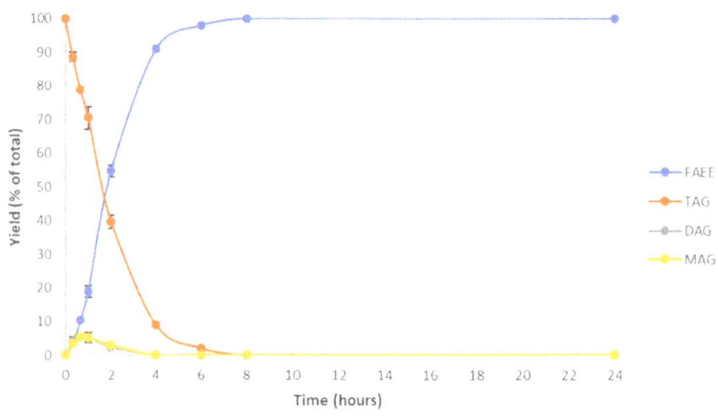

Figure 3. Kinetics of enzymatic ethanolysis with a double load of CALB at 40 °C and 200 rpm.

After 4 h, a yield of FAEE of 66.58% was achieved by adding the normal enzymatic load (Figure 2B). In comparison, when adding a double amount of enzyme, the yield obtained was 91.03%. Moreover, 100% FAEE was reached after only 8 h of reaction with a double enzymatic load, while it took 24 h to obtain this yield with a simple load. Therefore, by adding twice the enzymatic load at 40 °C and 200 rpm, 100% FAEE was produced after 8 h, so the time was considerably reduced, but more enzyme was used.

With this process, *Schizochytrium* oil composed by TAG was enzymatically modified in a mild process at 40 °C to produce FAEE that can be separated to produce concentrates of DHA for different purposes.

2.4. Open-Column Chromatography

Therefore, open-column chromatography was performed to separate and purify DHA from the rest of FAEE produced in the enzymatic ethanolysis described before. The sample was divided into different fractions that were eluted with different mixtures of solvents.

The first four fractions were discarded as FAEE was kept inside the column because of its affinity. A second round of hexane–ethyl acetate (95:5), which correspond to fraction 5 (F5) was needed for FAEE elution. F5 was analyzed by GC-MS and the corresponding chromatogram is represented in Figure 4 (blue). It can be seen that a large amount of saturated fatty acids (myristic acid and palmitic acid), as well as squalene and a low amount of DPA, were eluted.

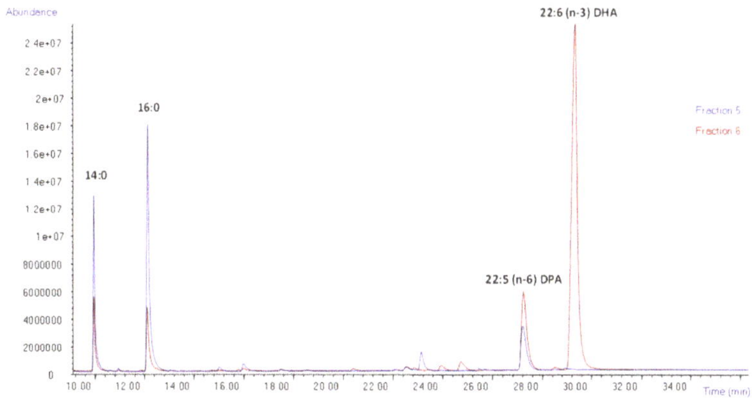

Figure 4. Analysis of FAEE profiles in fraction 5 (blue) and fraction 6 (red) obtained by open-column chromatography using GC-MS.

Then, fraction 6 (F6) was eluted with hexane–ethyl acetate in proportion 90:10 and analyzed by GC-MS (Figure 4, red). In this case, the following values of FAEE were obtained: 5.0 ± 0.8% myristic acid, 6.3 ± 1.3% palmitic acid, 1.7 ± 0.1% EPA, 14.9 ± 0.1% DPA, and 72.0 ± 2.2% DHA. The initial proportion of fatty acids is explained in the characterization of microalgal oil 2 (see Section 2.2). Overall, a fraction with almost 90% of LC-PUFA was isolated, constituting mostly DHA. Moreover, in the next fraction of open-column chromatography (F7), there was not any peak corresponding to FAEE. Therefore, all the DHA extracted from *Schizochytrium* sp. was eluted in fraction 6, with a maximum recovery.

In order to attain a DHA-enriched fraction, the protocol of open-column chromatography was slightly modified as expressed in Section 3.9. The first four fractions were discarded. Moreover, either P5 or P6 of the purification protocol (duplicates) were composed of myristic and palmitic acid, squalene, a large part of DPA, and a low fraction of DHA. Subsequently, modifying the relation between hexane and ethyl acetate from 95:5 to 92:8 and using two cycles of elution under the same conditions enabled the total elution of certain FAEE. However, part of DHA was also eluted in this fraction, avoiding the full recovery of DHA.

Furthermore, in the last step of purification, hexane–ethyl acetate in proportion 90:10 was used. This protocol was also carried out in duplicate, obtaining a fraction with the following values of FAEE: 2.0 ± 0.4% EPA, 4.8 ± 0.9% DPA, and 93.2 ± 1.3% DHA (Figure 5). In this case, a purity higher than 90% DHA was achieved.

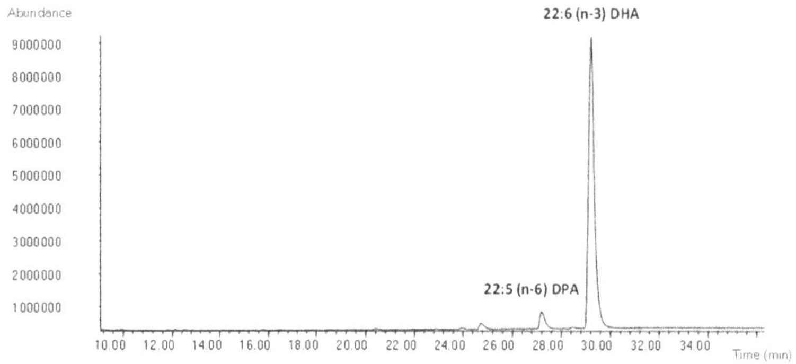

Figure 5. Analysis of FAEE profile in an enriched DHA fraction isolated by purification in open column chromatography using GC-MS.

Hence, a fraction with 75.0 ± 14.9 mg with a purity of 93.2% of DHA was obtained from 0.3 g of FAEE that was introduced in the open column and produced by enzymatic ethanolysis. If the recovery of each of the duplicates was calculated in terms of average and standard deviation, a yield of 44.72 ± 6.99% DHA was obtained. Taking into account that the percentage of DHA extracted with PLE using hexane–ethanol (1:1) was 51.2% of the total fatty acid composition, almost 90% of DHA was recovered from *Schizochytrium* sp. with a high purity.

3. Materials and Methods

3.1. Materials

Lyophilized *Schizochytrium* sp. biomass was provided by Cianoalgae S.L. (Madrid, Spain). *Schizochytrium* sp. commercial oil was acquired from Nutilab S.L. (Valencia, Spain). The oil and microalgae were kept at cool temperatures and protected from light to avoid oxidation. Commercial immobilized lipase Novozym® 435 (Candida Antarctica B lipase, CALB) was kindly donated from Novozymes A/S (Bagsvaerd, Denmark). Hexane, ethyl acetate, and methyl tert-butyl ether (MTBE) were purchased from Avantor Performance Materials (Gliwice, Poland). Molecular sieves (3Å) were purchased from Scharlab S.L. (Sentmenat, Spain). Methanol, acetone, and isopropanol (HPLC-grade) were provided by Lab-Scan Analytical Sciences (Gliwice, Poland). Ethanol and reagents of common use were provided by Panreac Quimica S.A.U. (Barcelona, Spain).

3.2. Traditional Lipid Extraction by Soxhlet Method

For Soxhlet extraction [57], 4.0 g of lyophilized *Schizochytrium* sp. was used, and 150 mL of hexane was added in a Soxhlet extractor with continuous boiling and condensation cycles of the solvent for 6 h (matrix/solvent ratio 1:37.5). Subsequently, the extracts were evaporated with a rotary evaporator (Heidolph Hei-Vap Value HB/G3, Germany, Berlin) at 35 °C followed by nitrogen stream to constant weight. Lipid content was determined gravimetrically and calculated as percentage by weight of dry biomass (Equation (1)).

$$\text{Lipid content}(\%) = \frac{\text{Evaporated extract weight}(\text{mg})}{\text{Dry biomass weight}(\text{mg})} \times 100 \quad (1)$$

The extracts obtained were stored in a nitrogen atmosphere in the dark at 4 °C until analysis. In all cases, the experiments were carried out at least in triplicate.

3.3. Pressurized Liquid Extraction of Microalgal Biomass

PLE was carried out with a DIONEX ASE 350 extractor (Sunnyvale, California) equipped with stainless steel extraction cells (10 mL volume). Lyophilized *Schizochytrium* sp. was weighed (1 g) and loaded into the extraction cell, adding 2 g of sand at both ends of the cell.

The extraction cell was then filled with the different solvents used: hexane, ethanol, and a mixture of both (1:1), different temperatures (80, 100, and 120 °C) and different static extraction times (5, 10, and 15 min) were also tested. For this purpose, a surface-response experimental design was created by introducing all of the variables mentioned above, using the Statgraphics 19 statistical program. The volume of solvent used was 20–25 mL, depending on cell temperature and pressure. Finally, the extract was recovered under a stream of nitrogen in 50 mL vials and kept at 4 °C until analysis. Each of the extraction conditions was performed in duplicate [51].

3.4. Scale-up of Pressurized Liquid Extraction

The extraction procedure was scaled up using 100 mL stainless steel cells. In this case, 20.00 g of lyophilized *Schizochytrium* sp. was weighed and loaded into the extraction cell, adding sand at both ends of the cell. A solvent saving mode with a flow rate of 1.5 mL/min was used. The static extraction time was 15 min and the volume of the solvent used was 69–75 mL. The scale-up was performed in triplicate.

3.5. Characterization of Schizochytrium sp. Oils by GC-MS

The fatty acid profile of *Schizochytrium* sp. oil was determined in duplicate, both from commercial oil and from different extractions carried out by PLE. Derivatization in a basic medium was developed following ISO TC34/SC 5 standard method (See Appendix C for explanation) [58]. The obtained fatty acid methyl esters (FAME) were analyzed by gas chromatography coupled to a mass spectrometer (GC-MS) using 100 µL of FAMEs and 400 µL of hexane for GC-MS analysis.

The fatty acid analysis was performed on an Agilent Technologies (Palo Alto, Cal., USA) 5975 MSD Series gas-mass chromatograph with an automatic injector and He as the carrier gas. An Agilent Technologies HP-88 capillary column was used, with dimensions of 100 m x 0.25 mm x 0.20 µm. The injection temperature was 250 °C. The oven was kept at 175 °C for 8 min. The temperature was then raised at a rate of 3 °C/min to 230 °C, which was maintained for an additional 10 min. The temperature of the detector was 230 °C. The amount of sample injected was 1 µL with a 1:20 split. The mass spectrometer used an ionization potential of 70 eV and an atomic mass range from 30 to 400 µ (atomic mass units).

The fatty acids were identified by comparing their retention times and the mass spectra (NIST Mass Spectral Library Version 2.0) with those obtained from the standards, expressing the amounts as percentages of the total FA content. The findings were compared with updated bibliographic references.

3.6. Enzymatic Ethanolysis of Schizochytrium sp. Oil

First, 200 mg of 3Å molecular sieves, which remove water to prevent hydrolysis reactions, and 100 mg of commercial CALB were weighed and placed in a 30 mL capacity glass vial. Then, 2.05 mL of hexane, 150 µL of absolute ethanol, and 300 µL of *Schizochytrium* sp. oil extracted by PLE were incorporated. Both solvents were dehydrated when in contact with molecular sieves. The reaction was carried out in a Heidolph incubator equipped with a platform shaker (Unimax 1010) and a heating unit, with constant agitation (200 rpm) at different temperatures (30 °C and 40 °C). The reaction was also tested at 40 °C with a double enzymatic load, according to the previously described method.

To study the reaction kinetics, 25 µL of the suspension (reaction medium) was taken at different times (0, 20 min, 40 min, 1 h, 2 h, 4 h, 6 h and 24 h), diluted 25 times in hexane and analyzed by high-performance liquid chromatography coupled with a light-scattering detector (HPLC-ELSD). All reactions were conducted in duplicates.

3.7. Analysis by HPLC-ELSD

HPLC-ELSD analysis was performed using an Agilent 1260 Infinity Auto-injector Chromatograph (G1329B) with a quaternary pump (G1311B/C), equipped with an Agilent 385 Evaporative Scattered Light Detector (Palo Alto, CA, USA). The chromatographic separation of the different reaction products obtained by enzymatic ethanolysis was carried out with a silica normal phase column (250 mm x 4.6 mm i.d., 5 µm) maintained at 30 °C [59] using a ternary gradient as follows: 0–2 min, 99.5% A and 0.5% B; at t = 6.5 min, 70% A and 30% B; at t = 11 min, 63% A, 27% B and 10% C; at t = 18 min, 99.5% A and 0.5% B; and at t = 20 min, and 99.5% A and 0.5% B. Eluent A consisted of 2,2,4- trimethylpentane, eluent B consisted of methyl tert-butyl ether, and eluent C consisted of 2-propanol.

The optimal signal and resolution of the ELSD detector were achieved with the following conditions: evaporator and nebulizer temperature of 30 °C, and evaporator gas (N_2) at 1.6 SLM.

To identify different lipids represented in the sample such as TAG, diacylglycerols (DAG), monoacylglycerols (MAG), and FAEE, standards were individually injected and compared with standards already analyzed in previous studies.

3.8. Fractionation of Fatty Acid Ethyl Esters by Open-Column Chromatography

First, the filler material (6 g of silica) was added to the burette, as it has a glass frit in the lower part so that it remain fixed, adding the necessary hexane to dilute the silica to the burette. Then, more hexane was added to prevent the formation of bubbles in the column, and the sample was loaded into the column (0.3 g of the FAEE obtained from the scale-up of the enzymatic ethanolysis). To start separating the sample components in fractions (F), different mixtures of hexane and ethyl acetate were added to the column: 10 mL of hexane (F1); 10 mL of hexane–ethyl acetate (99:1) (F2); 10 mL of hexane–ethyl acetate (98:2) (F3); 15 mL of hexane–ethyl acetate (95:5) (F4); 5 mL of hexane–ethyl acetate (95:5) (F5); 5 mL of hexane–ethyl acetate (90:10) (F6) and another 5 mL of hexane–ethyl acetate (90:10) (F7). Experiments were made in duplicates. Fractions were kept refrigerated for GC-MS analysis.

3.9. Purification of an Enriched Dha Fraction by Open-Column Chromatography

The burette was prepared as in the previous section (see Section 3.8). The column was conditioned, and the sample was loaded (0.3 g of FAEE). Solvent mixtures to extract the first three fractions (P1, P2 and P3) were the same as in the previous protocol (see Section 3.8). Moreover, different mixtures were added: 10 mL of hexane–ethyl acetate (95:5) (P4); 5 mL of hexane–ethyl acetate (92:8) (P5); 5 mL of hexane–ethyl acetate (92:8) (P6); 5 mL of hexane–ethyl acetate (90:10) (P7); and 5 mL of hexane–ethyl acetate (90:10) (P8). Experiments were made in duplicate. Fractions were kept refrigerated for GC-MS analysis.

3.10. Analysis of Fatty Acid Composition by GC-MS

The fatty acid profile of the different fractions obtained by open-column chromatography was analyzed by GC-MS. Fractions were derivatized as reported in Section 3.5, and FAME were analyzed using GC-MS according to the described previously method. In this case, 200 µL of each fraction was collected on 400 µL of hexane for GC-MS analysis.

4. Conclusions

In conclusion, it should be noted that the use of alternative techniques of extraction from microalgal biomass, such as pressurized liquids, with subsequent enzymatic ethanolysis of the produced *Schizochytrium* oil, enables a sustainable and environmentally friendly procedure to effectively generate FAEE of DHA. Ethyl esters separated by column chromatography were fractionated and two objectives were achieved: On the one hand, an enriched fraction of DHA was obtained with complete recovery. On the other hand, it was possible to produce a fraction with high purity and the recovery of DHA, 93.2% and 86%, respectively, which may be used as a food supplement and to develop nutraceuticals and active pharmaceutical intermediates of marine origin in an integrated process.

Author Contributions: Conceptualization, G.B., P.G.-G. and F.J.S.; methodology, G.B.; investigation, G.B., P.G.-G.; resources, F.J.S.; writing—original draft preparation, G.B., P.G.-G. and F.J.S.; writing—reviewing and editing, G.B., P.G.-G. and F.J.S.; supervision, P.G.-G. and F.J.S.; project administration: F.J.S.; funding acquisition, F.J.S. All authors have read and agreed to the published version of the manuscript.

Funding: This research was financially supported by the Comunidad de Madrid (Spain) through project ALGATEC-CM (P2018/BAA-4532), co-financed by the European Social Fund, and from the Ministry of Science and Innovation, Spain (Projects No. RTI2018–093583-B-I00 and PID2022–143229NB-I00).

Data Availability Statement: The original data presented in the study are included in the article; further inquiries can be directed to the corresponding author.

Acknowledgments: The authors would like to thank Novozymes and Ramiro Martinez for the generous gift of commercial lipases.

Conflicts of Interest: The authors declare no conflicts of interest.

Appendix A

Table A1. Results obtained with the experimental design of PLE extraction. [1] 0 = 100% hexane; 50 = 50% hexane: 50% ethanol; 100 = 100% ethanol.

Temperature (°C)	Time (min)	Solvent [1]	Extraction Yield (%)
80	10	0	21.85 ± 0.06
100	5	0	22.38 ± 0.34
100	15	0	22.62 ± 1.12
120	10	0	22.86 ± 0.45
80	5	50	26.11 ± 0.71
100	10	50	26.08 ± 0.13
100	10	50	25.87 ± 1.01
100	10	50	25.97 ± 1.23
80	15	50	25.71 ± 0.51
120	5	50	28.66 ± 0.78
120	15	50	29.06 ± 0.12
80	10	100	17.13 ± 1.17
100	5	100	21.41 ± 0.75
100	15	100	22.03 ± 1.85
120	10	100	24.81 ± 0.35

Appendix B

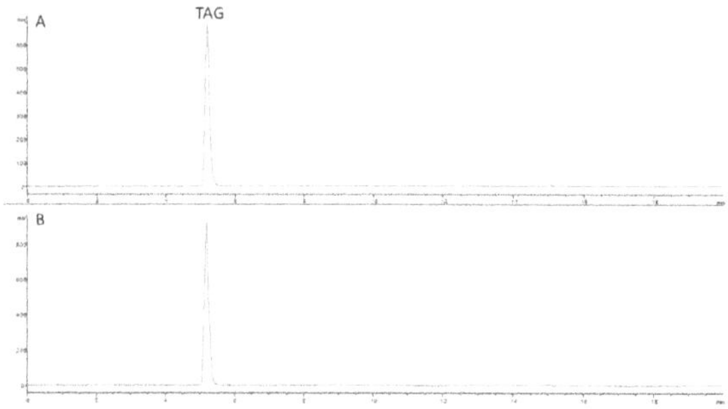

Figure A1. HPLC-ELSD analysis. Chromatograms correspond to the following: (**A**) microalgal oil 2; (**B**) commercial oil.

Appendix C

To obtain the FAMEs in a basic medium, ISO TC34/SC 5 was followed:

- First, 25 mg of oil was mixed with 200 µL of hexane.
- Next, 50 µL of 2N KOH in methanol (prepared daily) was added.
- The mixture was shaken for 1 min in a vortex (Velp Scientifica ZX3).
- The mixture is rested for 5 min before the reaction proceeds.
- Then, 125 mg of sodium hydrogen sulfate monohydrate ($NaHSO_4$ H_2O) was added to stop the reaction and vortexed.
- Finally, the mixture was centrifuged for 5 min at 50 rpm (Hettich zentrifugen Mikro 120): Supernatant = FAMEs

For further analysis, 100 µL of FAMEs was collected from 400 µL of hexane.

References

1. Jin, J.; Jin, Q.; Wang, X.; Akoh, C.C. High Sn-2 Docosahexaenoic Acid Lipids for Brain Benefits, and Their Enzymatic Syntheses: A Review. *Engineering* **2020**, *6*, 424–431. [CrossRef]
2. García, J.L.; de Vicente, M.; Galán, B. Microalgae, Old Sustainable Food and Fashion Nutraceuticals. *Microb. Biotechnol.* **2017**, *10*, 1017–1024. [CrossRef] [PubMed]
3. Tocher, D.R.; Betancor, M.B.; Sprague, M.; Olsen, R.E.; Napier, J.A. Omega-3 Long-Chain Polyunsaturated Fatty Acids, EPA and DHA: Bridging the Gap between Supply and Demand. *Nutrients* **2019**, *11*, 89. [CrossRef] [PubMed]
4. Echeverría, F.; Valenzuela, R.; Catalina Hernandez-Rodas, M.; Valenzuela, A. Docosahexaenoic Acid (DHA), a Fundamental Fatty Acid for the Brain: New Dietary Sources. *Prostaglandins Leukot. Essent. Fat. Acids* **2017**, *124*, 1–10. [CrossRef]
5. Jayatunga, D.P.W.; Hone, E.; Fernando, W.M.A.D.B.; Garg, M.L.; Verdile, G.; Martins, R.N. A Synergistic Combination of DHA, Luteolin, and Urolithin A Against Alzheimer's Disease. *Front. Aging Neurosci.* **2022**, *14*, 780602. [CrossRef] [PubMed]
6. Sun, G.Y.; Simonyi, A.; Fritsche, K.L.; Chuang, D.Y.; Hannink, M.; Gu, Z.; Greenlief, C.M.; Yao, J.K.; Lee, J.C.; Beversdorf, D.Q. Docosahexaenoic Acid (DHA): An Essential Nutrient and a Nutraceutical for Brain Health and Diseases. *Prostaglandins Leukot. Essent. Fat. Acids* **2018**, *136*, 3–13. [CrossRef]
7. Remize, M.; Brunel, Y.; Silva, J.L.; Berthon, J.-Y.; Filaire, E. Microalgae N-3 PUFAs Production and Use in Food and Feed Industries. *Mar. Drugs* **2021**, *19*, 113. [CrossRef]
8. Harwood, J.L. Algae: Critical Sources of Very Long-Chain Polyunsaturated Fatty Acids. *Biomolecules* **2019**, *9*, 708. [CrossRef] [PubMed]
9. Souza, C.M.M.; de Lima, D.C.; Bastos, T.S.; de Oliveira, S.G.; Beirão, B.C.B.; Félix, A.P. Microalgae *Schizochytrium* Sp. as a Source of Docosahexaenoic Acid (DHA): Effects on Diet Digestibility, Oxidation and Palatability and on Immunity and Inflammatory Indices in Dogs. *Anim. Sci. J.* **2019**, *90*, 1567–1574. [CrossRef]
10. Yalagala, P.C.R.; Sugasini, D.; Zaldua, S.B.; Tai, L.M.; Subbaiah, P.V. Lipase Treatment of Dietary Krill Oil, but Not Fish Oil, Enables Enrichment of Brain Eicosapentaenoic Acid and Docosahexaenoic Acid. *Mol. Nutr. Food Res.* **2020**, *64*, 2000059. [CrossRef]
11. Lupette, J.; Benning, C. Human Health Benefits of Very-Long-Chain Polyunsaturated Fatty Acids from Microalgae. *Biochimie* **2020**, *178*, 15–25. [CrossRef]
12. Medipally, S.R.; Yusoff, F.M.; Banerjee, S.; Shariff, M. Microalgae as Sustainable Renewable Energy Feedstock for Biofuel Production. *BioMed Res. Int.* **2015**, *2015*, 519513. [CrossRef] [PubMed]

13. Kujawska, N.; Talbierz, S.; Dębowski, M.; Kazimierowicz, J.; Zieliński, M. Cultivation Method Effect on *Schizochytrium* Sp. Biomass Growth and Docosahexaenoic Acid (DHA) Production with the Use of Waste Glycerol as a Source of Organic Carbon. *Energies* **2021**, *14*, 2952. [CrossRef]
14. Rodríguez-España, M.; Mendoza-Sánchez, L.G.; Magallón-Servín, P.; Salgado-Cervantes, M.A.; Acosta-Osorio, A.A.; García, H.S. Supercritical Fluid Extraction of Lipids Rich in DHA from *Schizochytrium* Sp. *J. Supercrit. Fluids* **2022**, *179*, 105391. [CrossRef]
15. Sahin, D.; Tas, E.; Altindag, U.H. Enhancement of Docosahexaenoic Acid (DHA) Production from *Schizochytrium* Sp. S31 Using Different Growth Medium Conditions. *AMB Express* **2018**, *8*, 7. [CrossRef] [PubMed]
16. Wang, Y.; Peng, D.; Zhu, Y.; Xie, S.; Pan, Y.; Chen, D.; Tao, Y.; Yuan, Z. Establishment of Pressurized Liquid Extraction Followed by HPLC–MS/MS Method for the Screening of Adrenergic Drugs, Steroids, Sedatives, Colorants and Antioxidants in Swine Feed. *J. Sep. Sci.* **2019**, *42*, 1915–1929. [CrossRef]
17. Chemat, F.; Vian, M.A.; Cravotto, G. Green Extraction of Natural Products: Concept and Principles. *Int. J. Mol. Sci.* **2012**, *13*, 8615. [CrossRef]
18. Blanco-Llamero, C.; Señoráns, F.J. Biobased Solvents for Pressurized Liquid Extraction of *Nannochloropsis Gaditana* Omega-3 Lipids. *Mar. Drugs* **2021**, *19*, 107. [CrossRef] [PubMed]
19. Señoráns, M.; Castejón, N.; Señoráns, F.J. Advanced Extraction of Lipids with DHA from *Isochrysis galbana* with Enzymatic Pre-Treatment Combined with Pressurized Liquids and Ultrasound Assisted Extractions. *Molecules* **2020**, *25*, 3310. [CrossRef]
20. He, Y.; Huang, Z.; Zhong, C.; Guo, Z.; Chen, B. Pressurized Liquid Extraction with Ethanol as a Green and Efficient Technology to Lipid Extraction of *Isochrysis* Biomass. *Bioresour. Technol.* **2019**, *293*, 122049. [CrossRef]
21. Waseem, R.; Low, K.H. Advanced Analytical Techniques for the Extraction and Characterization of Plant-Derived Essential Oils by Gas Chromatography with Mass Spectrometry. *J. Sparation Sci.* **2015**, *38*, 483–501. [CrossRef]
22. Sousa, S.C.; Freitas, A.C.; Gomes, A.M.; Carvalho, A.P. Extraction of Nannochloropsis Fatty Acids Using Different Green Technologies: The Current Path. *Mar. Drugs* **2023**, *21*, 365. [CrossRef] [PubMed]
23. Bornscheuer, U.T. Enzymes in Lipid Modification: An Overview. In *Lipid Modification by Enzymes and Engineered Microbes*; AOCS Press: Champaign, IL, USA, 2018; pp. 1–9. [CrossRef]
24. Rodrigues, R.C.; Fernandez-Lafuente, R. Lipase from *Rhizomucor Miehei* as a Biocatalyst in Fats and Oils Modification. *J. Mol. Catal. B Enzym.* **2010**, *66*, 15–32. [CrossRef]
25. Contesini, F.J.; Davanço, M.G.; Borin, G.P.; Vanegas, K.G.; Cirino, J.P.G.; de Melo, R.R.; Mortensen, U.H.; Hildén, K.; Campos, D.R.; Carvalho, P. de O. Advances in Recombinant Lipases: Production, Engineering, Immobilization and Application in the Pharmaceutical Industry. *Catalysts* **2020**, *10*, 1032. [CrossRef]
26. Moreno-Perez, S.; Turati, D.F.M.; Borges, J.P.; Luna, P.; Señorans, F.J.; Guisan, J.M.; Fernandez-Lorente, G. Critical Role of Different Immobilized Biocatalysts of a given Lipase in the Selective Ethanolysis of Sardine Oil. *J. Agric. Food Chem.* **2017**, *65*, 117–122. [CrossRef] [PubMed]
27. Kuan, D.; Dai, L.; Liu, D.; Liu, H.; Du, W. Efficient Biodiesel Conversion from Microalgae Oil of *Schizochytrium* Sp. *Catalysts* **2019**, *9*, 341. [CrossRef]
28. Moreno-Perez, S.; Luna, P.; Señorans, F.J.; Guisan, J.M.; Fernandez-Lorente, G. Enzymatic Synthesis of Triacylglycerols of Docosahexaenoic Acid: Transesterification of Its Ethyl Esters with Glycerol. *Food Chem.* **2015**, *187*, 225–229. [CrossRef]
29. Tian, X.; Dai, L.; Liu, M.; Liu, D.; Du, W.; Wu, H. Lipase-Catalyzed Methanolysis of Microalgae Oil for Biodiesel Production and PUFAs Concentration. *Catal. Commun.* **2016**, *84*, 44–47. [CrossRef]
30. Kuan, D.; Dai, L.; Liu, D.; Du, W.; Liu, H. A Novel Clean Process for the Combined Production of Fatty Acid Ethyl Esters (FAEEs) and the Ethyl Ester of Polyunsaturated Fatty Acids (PUFAs) from Microalgae Oils. *Renew. Energy* **2019**, *143*, 772–778. [CrossRef]
31. He, Y.; Wang, X.; Zhang, Y.; Guo, Z.; Jiang, Y.; Chen, F. Enzymatic Ethanolysis Subjected to *Schizochytrium* Biomass: Sequential Processing for DHA Enrichment and Biodiesel Production. *Energy Convers. Manag.* **2019**, *184*, 159–171. [CrossRef]
32. Wang, X.; Wang, X.; Wang, W.; Jin, Q.; Wang, X. Synthesis of Docosapentaenoic Acid-Enriched Diacylglycerols by Enzymatic Glycerolysis of *Schizochytrium* Sp. Oil. *Bioresour. Technol.* **2018**, *262*, 278–283. [CrossRef]
33. Prasad, S.; Roy, I. Converting Enzymes into Tools of Industrial Importance. *Recent Pat. Biotechnol.* **2017**, *12*, 33–56. [CrossRef] [PubMed]
34. Fernandez-Lorente, G.; Filice, M.; Lopez-Vela, D.; Pizarro, C.; Wilson, L.; Betancor, L.; Avila, Y.; Guisan, J.M. Cross-Linking of Lipases Adsorbed on Hydrophobic Supports: Highly Selective Hydrolysis of Fish Oil Catalyzed by RML. *JAOCS J. Am. Oil Chem. Soc.* **2011**, *88*, 801–807. [CrossRef]
35. Orrego, A.H.; Ghobadi, R.; Moreno-Perez, S.; Mendoza, A.; Fernandez-Lorente, G.; Guisan, J.M.; Rocha-Martin, J. Stabilization of Immobilized Lipases by Intense Intramolecular Cross-Linking of Their Surfaces by Using Aldehyde-Dextran Polymers. *Int. J. Mol. Sci.* **2018**, *19*, 553. [CrossRef] [PubMed]
36. Wang, X.; Li, D.; Qu, M.; Durrani, R.; Yang, B.; Wang, Y. Immobilized MAS1 Lipase Showed High Esterification Activity in the Production of Triacylglycerols with N-3 Polyunsaturated Fatty Acids. *Food Chem.* **2017**, *216*, 260–267. [CrossRef] [PubMed]
37. Mateo, C.; Bolivar, J.M.; Godoy, C.A.; Rocha-Martin, J.; Pessela, B.C.; Curiel, J.A.; Muñoz, R.; Guisan, J.M.; Fernández-Lorente, G. Improvement of Enzyme Properties with a Two-Step Immobilizaton Process on Novel Heterofunctional Supports. *Biomacromolecules* **2010**, *11*, 3112–3117. [CrossRef]
38. Di Cosimo, R.; Mc Auliffe, J.; Poulose, A.J.; Bohlmann, G. Industrial Use of Immobilized Enzymes. *Chem. Soc. Rev.* **2013**, *42*, 6437–6474. [CrossRef] [PubMed]

39. Mateo, C.; Palomo, J.M.; Fernandez-Lorente, G.; Guisan, J.M.; Fernandez-Lafuente, R. Improvement of Enzyme Activity, Stability and Selectivity via Immobilization Techniques. *Enzym. Microb. Technol.* **2007**, *40*, 1451–1463. [CrossRef]
40. Guisan, J.M. *Immobilization of Enzymes and Cells IN Series Editor*; Springer Science+Business Media: New York, NY, USA, 2013; Volume 1051, ISBN 9781627035491. [CrossRef]
41. Kuo, C.H.; Huang, C.Y.; Lee, C.L.; Kuo, W.C.; Hsieh, S.L.; Shieh, C.J. Synthesis of Dha/Epa Ethyl Esters via Lipase-Catalyzed Acidolysis Using Novozym® 435: A Kinetic Study. *Catalysts* **2020**, *10*, 565. [CrossRef]
42. Urrutia, P.; Arrieta, R.; Alvarez, L.; Cardenas, C.; Mesa, M.; Wilson, L. Immobilization of Lipases in Hydrophobic Chitosan for Selective Hydrolysis of Fish Oil: The Impact of Support Functionalization on Lipase Activity, Selectivity and Stability. *Int. J. Biol. Macromol.* **2018**, *108*, 674–686. [CrossRef]
43. Lagarde, M.; Hachem, M.; Bernoud-Hubac, N.; Picq, M.; Véricel, E.; Guichardant, M. Biological Properties of a DHA-Containing Structured Phospholipid (AceDoPC) to Target the Brain. *Prostaglandins Leukot. Essent. Fat. Acids* **2015**, *92*, 63–65. [CrossRef] [PubMed]
44. Hachem, M.; Géloën, A.; Lo Van, A.; Foumaux, B.; Fenart, L.; Gosselet, F.; Da Silva, P.; Breton, G.; Lagarde, M.; Picq, M.; et al. Efficient Docosahexaenoic Acid Uptake by the Brain from a Structured Phospholipid. *Mol. Neurobiol.* **2016**, *53*, 3205–3215. [CrossRef]
45. Castejón, N.; Señoráns, F.J. Enzymatic Modification to Produce Health-Promoting Lipids from Fish Oil, Algae and Other New Omega-3 Sources: A Review. *New Biotechnol.* **2020**, *57*, 45–54. [CrossRef] [PubMed]
46. Yalagala, P.C.R.; Sugasini, D.; Dasarathi, S.; Pahan, K.; Subbaiah, P.V. Dietary Lysophosphatidylcholine-EPA Enriches Both EPA and DHA in the Brain: Potential Treatment for Depression. *J. Lipid Res.* **2019**, *60*, 566–578. [CrossRef] [PubMed]
47. Chojnacka, A.; Gładkowski, W.; Grudniewska, A. Lipase-Catalyzed Transesterification of Egg-Yolk Phophatidylcholine with Concentrate of n-3 Polyunsaturated Fatty Acids from Cod Liver Oil. *Molecules* **2017**, *22*, 1771. [CrossRef] [PubMed]
48. Ang, X.; Chen, H.; Xiang, J.Q.; Wei, F.; Quek, S.Y. Preparation and Functionality of Lipase-Catalysed Structured Phospholipid—A Review. *Trends Food Sci. Technol.* **2019**, *88*, 373–383. [CrossRef]
49. Xu, X. Production of Specific-Structured Triacylglycerols by Lipase-Catalyzed Reactions: A Review. *Eur. J. Lipid Sci. Technol.* **2000**, *102*, 287–303. [CrossRef]
50. Abed, S.M.; Wei, W.; Ali, A.H.; Korma, S.A.; Mousa, A.H.; Hassan, H.M.; Jin, Q.; Wang, X. Synthesis of Structured Lipids Enriched with Medium-Chain Fatty Acids via Solvent-Free Acidolysis of Microbial Oil Catalyzed by *Rhizomucor Miehei* Lipase. *LWT* **2018**, *93*, 306–315. [CrossRef]
51. Castejón, N.; Señoráns, F.J. Simultaneous Extraction and Fractionation of Omega-3 Acylglycerols and Glycolipids from Wet Microalgal Biomass of *Nannochloropsis Gaditana* Using Pressurized Liquids. *Algal Res.* **2019**, *37*, 74–82. [CrossRef]
52. Li, Y.; Ghasemi Naghdi, F.; Garg, S.; Adarme-Vega, T.C.; Thurecht, K.J.; Ghafor, W.A.; Tannock, S.; Schenk, P.M. A Comparative Study: The Impact of Different Lipid Extraction Methods on Current Microalgal Lipid Research. *Microb. Cell Factories* **2014**, *13*, 14. [CrossRef]
53. Martins, D.; Custódio, L.; Barreira, L.; Pereira, H.; Ben-Hamadou, R.; Varela, J.; Abu-Salah, K. Alternative Sources of N-3 Long-Chain Polyunsaturated Fatty Acids in Marine Microalgae. *Mar. Drugs* **2013**, *11*, 2259–2281. [CrossRef]
54. Ferreira de Mello, B.T.; Stevanato, N.; Filho, L.C.; da Silva, C. Pressurized Liquid Extraction of Radish Seed Oil Using Ethanol as Solvent: Effect of Pretreatment on Seeds and Process Variables. *J. Supercrit. Fluids* **2021**, *176*, 105307. [CrossRef]
55. Perez-Vazquez, A.; Carpena, M.; Barciela, P.; Cassani, L.; Simal-Gandara, J.; Prieto, M.A. Pressurized Liquid Extraction for the Recovery of Bioactive Compounds from Seaweeds for Food Industry Application: A Review. *Antioxidants* **2023**, *12*, 612. [CrossRef] [PubMed]
56. Jaime, L.; Rodríguez-Meizoso, I.; Cifuentes, A.; Santoyo, S.; Suarez, S.; Ibáñez, E.; Señorans, F.J. Pressurized Liquids as an Alternative Process to Antioxidant Carotenoids' Extraction from *Haematococcus Pluvialis* Microalgae. *LWT—Food Sci. Technol.* **2010**, *43*, 105–112. [CrossRef]
57. Soxhlet, F. Die gewichtsanalytische Bestimmung des Milchfettes. *Dinglers Polytech. J* **1879**, *232*, 461.
58. Castejón, N.; Luna, P.; Señoráns, F.J. Alternative Oil Extraction Methods from *Echium Plantagineum* L. Seeds Using Advanced Techniques and Green Solvents. *Food Chem.* **2018**, *244*, 75–82. [CrossRef]
59. Castejón, N.; Moreno-Pérez, S.; Abreu Silveira, E.; Fernández Lorente, G.; Guisán, J.M.; Señoráns, F.J. Synthesis of Omega-3 Ethyl Esters from Chia Oil Catalyzed by Polyethylene Glycol-Modified Lipases with Improved Stability. *Food Chem.* **2019**, *271*, 433–439. [CrossRef]

Disclaimer/Publisher's Note: The statements, opinions and data contained in all publications are solely those of the individual author(s) and contributor(s) and not of MDPI and/or the editor(s). MDPI and/or the editor(s) disclaim responsibility for any injury to people or property resulting from any ideas, methods, instructions or products referred to in the content.

marine drugs

Article

The Presence of Bioactive Compounds in European Eel (*Anguilla anguilla*) Skin: A Comparative Study with Edible Tissue

Antía Bote [1], Marcos Trigo [1], Sidonia Martínez [2] and Santiago P. Aubourg [1,*]

[1] Marine Research Institute (CSIC), c/E. Cabello, 6, 36208 Vigo, Spain; antia.bote.chamorro@alumnos.uvigo.es (A.B.); mtrigo@iim.csic.es (M.T.)
[2] Food Technology, Faculty of Science, University of Vigo, 32004 Ourense, Spain; sidonia@uvigo.gal
* Correspondence: saubourg@iim.csic.es

Abstract: The presence of bioactive compounds in European eel (*Anguilla anguilla*) skin was studied. Proximate and lipid class compositions and analysis of the fatty acid (FA) profile (individual FAs; FA groups, i.e., saturated, monounsaturated, and polyunsaturated; FA ratios, i.e., polyunsaturated/saturated, ω3/ω6) were determined and compared to the composition of the eel muscle. As a result, higher ($p < 0.05$) levels of proteins (271.6 g·kg^{-1}), lipids (38.0 g·kg^{-1}), ash (27.7 g·kg^{-1}), and ω6 FAs were observed in the skin tissue. Contrary, the muscle tissue showed higher ($p < 0.05$) moisture, ω3 FA, and ω3/ω6 ratio values. Regarding lipid classes, a higher ($p < 0.05$) proportion of phospholipids (111.1 g·kg^{-1} lipids), free sterols (104.7 g·kg^{-1} lipids), α-tocopherol (274.0 mg·kg^{-1} lipids), and free FAs (43.6 g·kg^{-1} lipids) was observed in the skin tissue. No differences ($p > 0.05$) between both tissues could be detected for triacylglycerol and FA group (saturated, monounsaturated, and polyunsaturated) values and for the polyunsaturated/saturated FA ratio. It is concluded that European eel skin, a by-product resulting from commercial processing, can be considered a valuable source for the food and pharmaceutical industries by providing value-added constituents such as proteins, lipids, ω3 FAs, phospholipids, and α-tocopherol.

Keywords: European eel (*Anguilla anguilla*); skin; muscle; proteins; phospholipids; free sterols; α-tocopherol; ω3 fatty acids; ω3/ω6 ratio; source

1. Introduction

A wide range of studies have recognised the fish fatty acid (FA) profile and the lipid class composition as being responsible for the health benefits resulting from the employment of fish-enriched diets [1,2]. Among polyunsaturated FAs (PUFAs), eicosapentaenoic (EPA) and docosahexaenoic (DHA) acid consumption has been associated with a low prevalence of several human diseases such as cardiovascular and neurodegenerative concerns [3,4]. According to their amphiphilic character, phospholipid (PL) compounds have shown to be valuable drug delivery systems for their high bioavailability and protecting effects on different kinds of diseases [5,6]. Regarding tocopherol compounds, fishery products have been described as an important source of this effective lipid-soluble antioxidant system [7,8].

As a result of fish processing, a large volume of undesired by-products is obtained that constitute an important environmental contamination source, unless efforts for their recovery are attained and their commercial value can be enhanced [9,10]. Remarkably, fish by-products have been reported to include bioactive and profitable components such as amino acids, enzymes, collagen, pigments, chitin, vitamins, and minerals [11,12]. Among fish by-products, skin tissue has attracted a great attention and has been studied in different kinds of fish species. Thus, Spanish mackerel (*Scomberomorous niphonius*) [13] and great hammerhead shark (*Sphyrna lewini*) [14] skin showed to be a remarkable source of collagen. Additionally, skin tissue from bigeye snapper (*Priacanthus tayenus* and *Priacanthus macracanthus*) [15], snakehead, and shark [16] showed to be a valuable substrate for gelatine extraction. Notably, fish skin-derived peptides have shown antioxidant [17–19], antimicrobial [20], and antifreezing [21] properties.

European eel (*Anguilla anguilla*), a teleost fish belonging to the family *Anguillidae*, is a commercially valuable species in Europe and Asia. As a result of recent overfishing on coasts and several biological concerns [22], great attention has been accorded to the development of this fish species as a farmed product [23,24]. With the extension of farming, the total annual production is assumed to be over 10,500 tonnes, The Netherlands consuming ca. 50% of this [25]. Previous studies account for the analysis of the proximate composition and FA profile [26,27] and the presence of essential and toxic elements in the muscle [28]. Additionally, the evolution of the eel muscle quality has been studied during different processing conditions such as refrigeration [29,30], cooking [31], and canning [32].

However, previous research on European eel by-products can be considered very scarce. Thus, Sila et al. [33] carried out the extraction and characterisation of sulphated glycosaminoglycans. Taktak et al. [34] developed novel eco-friendly, gelatine-based microfibers from eel skin for fish encapsulation. Teng et al. [35] prepared peptide-chelated calcium from European eel bones. Regarding eel skin, it is considered a thick substrate that is commonly treated as a waste material during the commercial processing of eel and is normally converted into low-value products or discarded. The unemployment of this by-product not only results in the loss of a large amount of bioactive constituents, but also leads to environmental concerns.

The current study focused on the presence of bioactive compounds in European eel (*A. anguilla*) skin. Determination of proximate and lipid class compositions and analysis of the FA profile, i.e., individual FAs, FA groups (saturated, STFAs; monounsaturated, MUFAs; PUFAs), and FA ratios (PUFAs/STFAs and ω3 FAs/ω6 FAs) was carried out. On the basis of the wide consumer acceptance of eel muscle as a valuable food for the human diet, the composition of the edible tissue was also analysed in this study and was compared to the composition of the skin.

2. Results and Discussion

2.1. Determination of the Proximate Composition

Values obtained for the proximate composition are included in Table 1. Water was shown to be the most abundant constituent in both eel tissues, with a higher ($p < 0.05$) value being detected in the muscle. Crude protein levels higher than 160 $g \cdot kg^{-1}$ wet tissue were observed in both tissues; notably, values obtained in the skin tissue (ca. 272 $g \cdot kg^{-1}$ wet tissue) were higher ($p < 0.05$) than in the muscle. The crude lipid content of the present eel samples depicted values included in the 28–38 $g \cdot kg^{-1}$ wet tissue range. As for crude protein content, crude lipid values were found to be higher ($p < 0.05$) in the skin tissue. Regarding the ash content, skin samples (27.7 $g \cdot kg^{-1}$ wet tissue) showed higher values ($p < 0.05$) than their counterparts, corresponding to the muscle tissue (9.9 $g \cdot kg^{-1}$ wet tissue).

Table 1. Proximate composition ($g \cdot kg^{-1}$ wet tissue) of eel skin and muscle [1].

Chemical Constituent	Tissue	
	Skin	Muscle
Water	677.0 ± 7.0	783.3 ± 3.5 *
Crude protein	271.6 ± 7.2	166.4 ± 1.8 *
Crude lipid	38.0 ± 0.9	28.6 ± 1.6 *
Ash	27.7 ± 2.1	9.9 ± 1.0 *

[1] Average values ± standard deviations of four (n = 4) replicates. Muscle values followed by an asterisk denote significant differences ($p < 0.05$) with skin values, according to the LSD test.

The crude protein content obtained in the present study is higher than the one found in the muscle of most commercial fish species [36–38]. Therefore, this by-product can be considered a protein-rich substrate. Regarding the current crude lipid content of European eel skin, this substrate maybe ranked as a medium-fat substrate [36] and could be considered a valuable source of lipid components.

To the best of our knowledge, no previous research has focused on the proximate composition of European eel (*A. anguilla*) skin. However, previous research provides information regarding the muscle tissue of this fish species. Thus, higher lipid contents (5.0%) than in the present study were obtained by Özoğul et al. [26] in individuals caught in the northeastern Mediterranean. Additionally, higher protein (19.2–19.6%), lipid (5.0–10.21%), and ash (1.23–1.50%) levels were detected in European eel (*A. anguilla*) muscle when studying freshwater individuals corresponding to several sizes [27].

Previous studies have also addressed the proximate composition of the edible tissue of other eel species. Thus, Oku et al. [39] carried out a comparative study on wild and cultured Japanese eel (*Anguilla japonica*) muscle; as a result, higher protein (19.0 and 18.9%, respectively) and lipid (11.6 and 13.1%, respectively) values than in the present study were obtained, although moisture values were lower (69.1 and 67.4%, respectively). A higher protein content (ca. 18.1%) than in the current study was also detected in farmed and freshwater eel (*Monopterus albus*) muscle [40]. A varying lipid content (3.6–20.4%) resulting from the catching season and location was proved for freshwater eel (*A. japonica*) muscle [41], as well as from comparing *A. japonica* individuals in the initial and terminal stages of spawning migration (0.3–20.6%) [42].

Previous research accounts for the proximate composition of skin in different kinds of fish species. Thus, Njinkoué et al. [43] studied the lipid content of different fish species from the Senegalese coast; according to the present results, higher values were detected in the skin than in the white muscle for *Sardinella maderensis* (26 vs. 5.0%), *Sardinella aurita* (24 vs. 3.5%), and *Cepahlopholis taeniops* (2.4 vs. 1.3%). According to the present results, Pateiro et al. [44] found higher protein (ca. 25%) and lipid (ca. 27%) values in gilthead seabream (*Sparus aurata*) skin than in the counterpart muscle (ca. 21 and 8%, respectively). A similar protein content (ca. 28%) was detected by Ahmmed et al. [45] in blue mackerel (*Scomber australasicus*) skin; however, lower lipid (ca. 21%) and higher moisture (ca. 50%) values were obtained. The same authors (Ahmmed et al. [46]) obtained similar moisture values (ca. 65%) than in the present study in king salmon (*Oncorhynchus tshawytscha*) skin; however, protein and lipid contents were notably lower (20 and 13%, respectively). Recently, Park et al. [47] obtained a protein content included in the 11.0–40.9% range for *Conger myriaster* skin, by employing green extracting technologies.

2.2. Analysis of the FA Composition

A similar FA profile was detected in both eel tissues (Table 2). Thus, the two major FAs were C16:0 and C18:1ω9. Additionally, relatively abundant FAs were C18:0, C16:1ω7, C18:1ω7, C20:4ω6, C20:5ω3, C22:5ω3, and C22:6ω3. However, the comparative analysis of both tissues revealed remarkable quantitative differences. Thus, a higher content ($p < 0.05$) of C17:0, C20:1ω9, C22:1ω9, and C20:2ω6 was detected in the skin tissue. Contrary, C14:0, C16:1ω7, C20:4ω6, C22:4ω6, C20:5ω3, C22:5ω3, and C22:6ω3 revealed a higher presence ($p < 0.05$) in the muscle samples.

Comparison of both kinds of tissues did not show significant differences ($p > 0.05$) in the contents of STFA, MUFA, and PUFA groups (Figure 1). According to the individual FA profile, the MUFA group was shown to be the most abundant ($p < 0.05$) in both tissues, while the PUFA group depicted the lowest ($p < 0.05$) presence. In agreement with this similar composition for the FA groups, no differences ($p > 0.05$) between both tissues could be outlined for the PUFA/STFA ratio (Figure 2).

Table 2. Fatty acid (FA) profile (g·100 g^{-1} total FAs) of eel skin and muscle [1].

FA	Tissue	
	Skin	**Muscle**
14:0	2.70 ± 0.03	3.09 ± 0.04 *
15:0	0.70 ± 0.06	0.65 ± 0.01
16:0	22.37 ± 0.15	22.39 ± 0.09
17:0	1.48 ± 0.05	1.05 ± 0.02 *
18:0	5.97 ± 0.13	5.91 ± 0.20
16:1 ω7	6.63 ± 0.12	6.92 ± 0.15 *
18:1 ω7	6.93 ± 0.06	6.84 ± 0.16
18:1 ω9	25.34 ± 0.40	25.09 ± 0.47
20:1 ω9	1.25 ± 002	1.12 ± 0.01 *
22:1 ω9	0.16 ± 0.01	0.13 ± 0.01 *
24:1 ω9	0.45 ± 0.03	0.15 ± 0.02 *
18:2 ω6	2.35 ± 0.05	2.25 ± 0.14
20:2 ω6	1.25 ± 0.05	1.05 ± 0.08 *
20:4 ω6	5.45 ± 0.09	4.25 ± 0.13 *
22:4 ω6	3.23 ± 0.14	2.44 ± 0.10 *
20:5 ω3	5.83 ± 0.14	6.74 ± 0.07 *
22:5 ω3	4.82 ± 0.18	5.14 ± 0.08 *
22:6 ω3	3.06 ± 0.08	4.23 ± 0.16 *

[1] Average values ± standard deviations of four (n = 4) replicates. Muscle values followed by an asterisk denote significant differences ($p < 0.05$) with skin values, according to the LSD test.

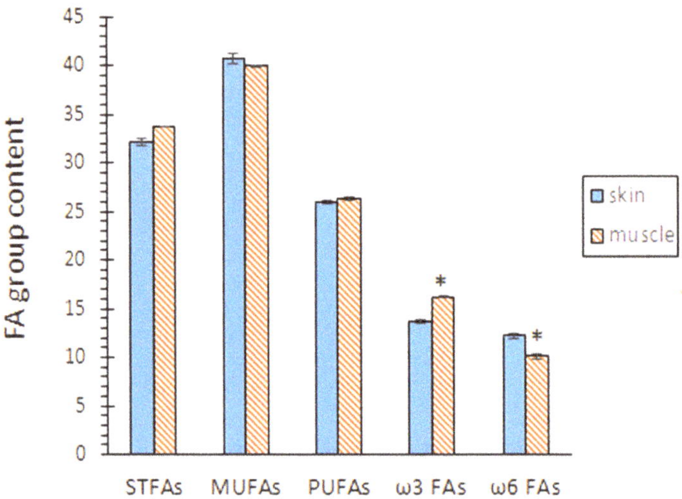

Figure 1. Fatty acid (FA) groups of eel skin and muscle (g·100 g^{-1} total FAs). Average values of four (n = 4) replicates; standard deviations are indicated by bars. Muscle values accompanied by an asterisk denote significant differences ($p < 0.05$) with skin values, according to the LSD test. Abbreviations: STFAs (saturated FAs), MUFAs (monounsaturated FAs), and PUFAs (polyunsaturated FAs).

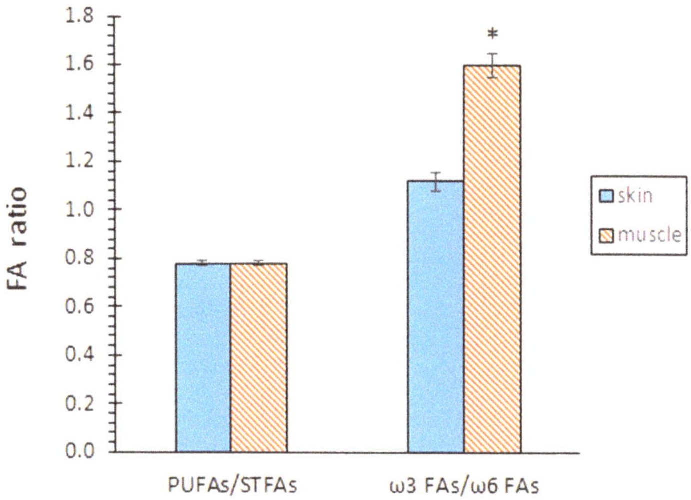

Figure 2. Fatty acid (FA) ratios of eel skin and muscle. Average values of four (n = 4) replicates; standard deviations are indicated by bars. Muscle values accompanied by an asterisk denote significant differences ($p < 0.05$) with skin values, according to the LSD test. Abbreviations as expressed in Figure 1.

The contents of total ω3 FAs and total ω6 FAs revealed remarkable differences between both tissues (Figure 1). Thus, samples corresponding to the skin tissue showed a lower content ($p < 0.05$) of ω3 FAs, but a higher content ($p < 0.05$) of ω6 FAs. As a result, a higher ω3/ω6 ratio ($p < 0.05$) was proved in samples corresponding to the muscle tissue (Figure 2).

A great interest has been paid to the presence of ω3 PUFAs, according to their beneficial health effects [48,49]. Based on epidemiological and clinical studies, EPA consumption has been related to circulatory, inflammatory, and coronary diseases [50], while DHA has been associated with the prevention of neurodegenerative diseases, foetal development, and the correct functioning of the nervous system and visual organs in the foetus [51]. Meanwhile, a relevant interest has also been given to the ω3/ω6 FA ratio [52,53]. Remarkably, recent studies have proved that Western populations do not include appropriate levels of ω3 FAs in their diet through natural dietary sources. In an attempt to avoid cardiovascular, neurological, and inflammatory concerns, the World Health Organization (WHO) recommends a higher ratio than 1:10 in the human diet [54]. Current results on both tissues have shown lower levels for EPA, DHA, and ω3 FAs than those present in the muscle of marine fish and invertebrate species [36–38]. However, results can be considered notably higher than in non-aquatic food such as poultry and egg [55], milk [56], and meat [57]. Remarkably, the ω3/ω6 ratio in both eel tissues was shown to be higher than 1 and was notably higher than 1/10, as recommended by the WHO [54].

No previous research is available regarding the FA composition of European eel (*A. anguilla*) skin. However, previous studies have focused on the FA composition of European eel muscle. According to the present results, a decreasing sequence for FA groups in muscle samples was described, i.e., MUFAs > STFAs > PUFAs, in individuals obtained in the northeastern Mediterranean Sea [26], in both wild and cultivated fish from Tunisian Mediterranean coasts [27], and in freshwater individuals from the River Ulla (Galicia, NW Spain) [58]. In such studies, and also in agreement with the current results, C18:1ω9 and C16:0 were the most abundant FAs.

Where FA ratios are concerned, lower PUFA/STFA ratios than in the present case were obtained in European eel (*A. anguilla*) muscle from the Mediterranean Sea (0.37) [26] and from both wild (0.46) and cultivated (0.52) individuals caught in the Tunisian Mediterranean coasts [58]. Regarding the ω3/ω6 ratio, higher values were detected in the muscle of freshwater individuals from the Ulla River (1.66–2.07) [27]. Additionally, Achouri et al. [58] found higher (3.28) and lower (1.31) ω3/ω6 ratio values in the muscle of both cultivated and wild individuals, respectively, obtained from Tunisian Mediterranean coasts.

Previous studies account for the FA composition of the skin and muscle tissues of related eel species. Thus, the same FA group distribution (MUFAs > STFAs > PUFAs) as in the current study was detected in the skin from *C. myriaster* eel from South Korea [47], in freshwater eel *A. japonica* muscle [41], in wild and cultivated Japanese eel (*A. japonica*) muscle [39], and in Japanese freshwater eel (*A. japonica*) muscle [42]. Contrary, the FA analysis of marbled eel (*Anguilla marmorata*) skin showed that C18:0 (ca. 50%) and C16:0 (ca. 22%) were the most abundant, with values for EPA and DHA being under 0.05% [59]; for the muscle tissue, this study proved that C18:1ω9 (ca. 45%) and C16:0 (ca. 19%) were the most abundant FAs, and EPA and DHA were present at a value below 0.03%. Regarding FA ratios, a higher ω3/ω6 ratio (4.48–5.41) than in the present work was detected by Park et al. [47] in *C. myriaster* eel skin from South Korea. Contrary, Lee et al. [41] obtained a similar ω3/ω6 ratio (0.90–1.67) in *A. japonica* muscle than in the current study.

Previous studies have also been focused on the FA composition of skin obtained from other fish species. During the analysis of the FA composition of skin obtained from three species from the Senegalese coast, Njinkoué et al. [43] proved that C16:0 and EPA were the most abundant (20.5% for both FAs) in *S. maderensis*, C16:0, C18:1ω9, and EPA (20.5, 15.5, and 10.4%, respectively) in *S. aurita*, and C16:0 and C18:1ω9 (28.4 and 12.5%, respectively) in *C. taeniops*; regarding DHA, values of 4.2, 2.5, and 6.9% were detected in such species, respectively. C18:1ω9 was the most abundant (ca. 36%), followed by C18:2ω6 (ca. 19%) and C16:0 (ca. 14%) in gilthead seabream (*S. aurata*) skin [44]; the presence of ω3 FAs was low (ca. 4 and 2% for DHA and EPA, respectively), so that a lower ω3/ω6 ratio (0.64) than in the present case was detected. The following FA distribution was detected by Ahmmed et al. [45] in blue mackerel (*S. australasicus*) skin: 19.5% (DHA), 17.3% (C16:0), 14.7% (C18:1ω9), and 2.30% (EPA); as a result, higher values than in the present study were obtained for PUFA/STFA (1.23) and ω3/ω6 (11.11) ratios. A very different FA distribution was found by the same authors [46] in king salmon (*O. tshawytscha*) skin; thus, C18:1ω9 was shown to be the most abundant (ca. 40%), followed by C16:0 (ca. 16%) and very low EPA and DHA values (<3%).

2.3. Determination of Lipid Classes

Results obtained for the composition of lipid classes is described in Table 3. Triacylglycerols (TAGs) were shown to be the most abundant lipid class in both tissues; values were included in the 400–412 $g \cdot kg^{-1}$ lipid range. A higher average content was observed in the muscle tissue; however, differences were not found significant ($p > 0.05$).

Free FAs (FFAs), compounds resulting from the hydrolysis of higher molecular weight compounds (i.e., TAGs and PLs) [60–62], provided values included in the 30–44 $g \cdot kg^{-1}$ lipids range; notably, higher values ($p < 0.05$) were detected in the skin tissue than in the counterpart edible substrate.

A remarkable presence of structured lipid classes (PLs and sterols, STs) were detected in the skin tissue (ca. 111 and 105 $g \cdot kg^{-1}$ lipids, respectively). Values were higher ($p < 0.05$) than those obtained in counterparts corresponding to the muscle tissue, especially for the ST compounds.

The analysis of the tocopherol composition of the current substrates indicated that the only tocopherol compound present was α-tocopherol. The content of this compound was found to be notably higher ($p < 0.05$) in the skin (274 $mg \cdot kg^{-1}$ lipids) than in the muscle (178 $mg \cdot kg^{-1}$ lipids).

Table 3 also indicates the lipid class content, expressed on a tissue basis. As for the previously mentioned results, on a lipid basis, higher ($p < 0.05$) FFA, PL, ST, and α-tocopherol values were detected in the skin tissue than in the counterpart muscle substrate.

Notably, a higher average value of TAGs was observed in skin samples, although differences with muscle tissues were not significant ($p > 0.05$).

Table 3. Composition of lipid classes of eel skin and muscle [1].

Lipid Class	Tissue	
	Skin	Muscle
Triacylglycerols	400.6 ± 28.0 (15.2 ± 3.0)	411.6 ± 7.3 (11.8 ± 0.6)
Free fatty acids	43.6 ± 1.4 (1.7 ± 0.4)	30.9 ± 0.8 * (0.9 ± 0.1 *)
Phospholipids	111.1 ± 5.5 (4.2 ± 0.6)	93.4 ± 5.5 * (2.7 ± 0.6 *)
Free sterols	104.7 ± 5.8 (4.0 ± 0.6)	24.2 ± 1.1 * (0.7 ± 0.2 *)
Alpha-tocopherol	274.0 ± 14.7 (10.4 ± 1.8)	178.0 ± 38.8 * (5.1 ± 1.9 *)

[1] Average values ± standard deviations of four ($n = 4$) replicates. Data expressed as $g \cdot kg^{-1}$ lipids, except for alpha-tocopherol ($mg \cdot kg^{-1}$ lipids). Data in brackets indicate the content, on a tissue basis, expressed as $g \cdot kg^{-1}$ tissue, except for α-tocopherol ($mg \cdot kg^{-1}$ tissue). Muscle values followed by an asterisk denote significant differences ($p < 0.05$) with skin values, according to the LSD test.

Based on their amphiphilic character, remarkable attention has been accorded to the PL compounds present in fish [5,6]. Thus, remarkable functions of PL compounds have been related to food production and pharmaceutical industries [63,64], these include antioxidant properties during food processing [65,66]. Based on their important role as lipid-soluble, chain-breaking antioxidants, tocopherol compounds have received great attention from marine technologists for their important role as lipophilic antioxidants [7,8]. Among them, α-tocopherol has been shown to be the most abundant in fish species [36]. The PL and α-tocopherol contents observed in the current eel by-product can be considered remarkable and correspond to the edible muscle of a medium-fat fish substrate [36,67], but lower than in a lean fish species [37,38].

Previous research regarding the analysis of lipid classes of European eel (*A. anguilla*) samples (skin or muscle) can be considered very scarce. According to the present results, TAGs were shown to be the most abundant lipid class in the muscle tissue from pre-migrant and migrant eel individuals [68]; in this study, phosphatidylcholine (PC) was shown to be the most abundant PL class. Regarding a related eel species, Park et al. [47] detected α-, (β + γ)-, and δ-tocopherol in *C. myriaster* skin from South Korea; as in the present case, α-tocopherol was shown to be the most abundant, with a content included in the 31–100 mg/100 g skin range. The presence of PL compounds was found to be relatively similar to the present concentrations in previous studies related to the skin substrate of other kinds of fish species. Thus, blue mackerel (*S. australasicus*) skin showed a 13.4 $g \cdot 100\ g^{-1}$ lipids concentration [45] and king salmon (*O. tshawytscha*) skin showed a 9.31 $g \cdot 100\ g^{-1}$ lipids value [46].

Previous studies also account for the lipid class analysis of the edible tissues of other eel species. Thus, Saito et al. [42] compared the lipid class composition of the initial and terminal stages of spawning migration of wild Japanese freshwater eel (*A. japonica*) muscle; as a result, TAGs were the major component in the initial-phase eels, but presented a remarkable content decrease in individuals corresponding to the terminal phase. A comparative study of the lipid class profile in both wild and cultivated individuals of the Japanese eel (*A. japonica*) was carried out by Oku et al. [39]; both in wild and cultivated individuals, TAGs were shown to be the most abundant lipid class of muscle (67.9–68.2%); other lipid classes detected were sterylesters (9.5–10.2%), FFAs (9.9–11.2%), STs (4.5%), PC (2.3–2.4%), and phosphatidylethanolamine (1.2–1.5%).

3. Materials and Methods

3.1. Solvents, Chemicals, and Standards

Solvents and chemical reagents used were of reagent grade and purchased from Merck (Darmstadt, Germany). The following solvents were employed: chloroform, methanol, hexane, toluene, ethyl ether, and isopropanol. In the case of tocopherol analysis, solvents used were liquid chromatographic grade. The following reagents were used: ammonium molybdate, hydroxylamine, cupric acetate, pyridine, acetic acid, acetic anhydride, acetyl chloride, and ferric trichloride.

Quantitative standards (1,2-dipalmitoyl-rac-glycero-3-phosphocholine, oleic acid, cholesterol, methyl stearate, nonadecanoic acid, and α-, β-, γ-, and δ-tocopherol) were purchased from Sigma-Aldrich (St. Louis, MO, USA). The following FAME standards were employed: Qualmix Fish (Larodan, Malmo, Sweden) and the Supelco 37 Component FAME Mix (Sigma-Aldrich, Laramie, WY, USA).

3.2. Fish Material and Sampling

European eels (*A. anguila*) (weight: 53–83 g; length: 32–39 cm) were captured during the authorised season (i.e., winter) and were slaughtered by ice water immersion at a local market (Mariscos Vivos del Grove, Quintela de Canedo, Ourense, Spain). Once at the post-slaughtered stage, eels were transported to the laboratory in insulated boxes on ice (0–1 °C). The fish (60 individuals) were distributed into four groups (fifteen individuals per group), which were considered independently for the statistical analysis (n = 4). As a first processing step, the fish were eviscerated and washed with running water. In each individual fish, the skin and muscle were excised and considered separately. Inside each of the four groups, portions corresponding to the same tissue (skin or muscle) were pooled together and subjected to the different chemical analyses.

3.3. Proximate Composition Analysis

Moisture content was determined as the weight difference in the homogenised tissue (1–2 g) before and after 4 h at 105 °C [69]. Results were calculated as g water·kg^{-1} tissue.

Crude protein content was measured using the Kjeldahl method [69] with a conversion factor of 6.25. Results were calculated as g protein·kg^{-1} wet tissue.

The crude lipid fraction was extracted using the Bligh and Dyer [70] method, which employs a single-phase solubilisation of the lipids using a chloroform–methanol (1:1) mixture. Results were calculated as g lipids·kg^{-1} wet tissue.

Ash content was measured according to the AOAC [69] method by heating the fish tissue at 550 °C. Results were calculated as g ash·kg^{-1} wet tissue.

3.4. FA Analysis

Lipid extracts were converted into fatty acid methyl esters (FAMEs) using acetyl chloride in methanol and then analysed using gas chromatography (Perkin-Elmer 8700 chromatograph, Madrid, Spain) according to an established procedure [71]. For it, a fused silica capillary column SP-2330 (0.25 mm i.d. × 30 m, Supelco, Inc., Bellefonte, PA, USA) was employed and the temperature program was as follows: increased from 145 to 190 °C at 1.0 °C min^{-1} and from 190 °C to 210 °C at 5.0 °C min^{-1}; held for 13.5 min at 210 °C. The carrier gas was nitrogen at 10 psig and detection was performed with a flame ionisation detector at 250 °C. A programmed temperature vaporiser injector was employed in the split mode (150:1) and was heated from 45 to 275 °C at 15 °C min^{-1}.

Peaks corresponding to FAMEs were identified by comparing their retention times with those of standard mixtures (Qualmix Fish and Supelco 37 Component FAME Mix). Peak areas were automatically integrated. C19:0 was used as an internal standard for quantitative purposes; for it, 100 µL (i.e., 40 µg C19:0) of a 0.4 mg·mL^{-1} solution in toluene were added to each sample before the methylation reaction with acetyl chloride. Limits of detection and quantification were 500 and 1500 area units, respectively. Quantitative calibration was carried out by means of the above-mentioned Supelco FAME Mix. Content of each FA was expressed as g·100 g^{-1} total FAs.

Results concerning FA groups (STFAs, MUFAs, and PUFAs; ω3 and ω6 FAs) and FA ratios (total ω3 FAs/total ω6 FAs and total PUFAs/total STFAs) were calculated taking into account the results obtained in individual FAs.

3.5. Analysis of Lipid Classes

To measure the TAG content, the total lipid extracts were first purified on 20 × 20 cm thin-layer chromatography plates coated with a 0.5 mm layer of silica gel G (Merck, Darmstadt, Germany) using a mixture of hexane-ethyl ether-acetic acid (90/10/1, *v*/*v*/*v*; two developments) as eluent [72]. Once the TAG fraction was purified, the method of Vioque and Holman [73] was used to measure the ester linkage content, according to the conversion of the esters into hydroxamic acids and their subsequent complexion with Fe (III). For quantification purposes, different quantities (0, 2, 5, 10, 20, and 40 µL) of a methyl stearate solution in toluene (41.0 mg/5 mL) were employed. The validity range was 16.4–328.0 µg methyl stearate and the R^2 value of the analytical procedure was 0.9995. Results were calculated as g tristearine·kg^{-1} lipids and g tristearine·kg^{-1} tissue.

FFA content of the total lipid extracts was determined following the Lowry and Tinsley [74] method, which is based on the formation of a complex with cupric acetate-pyridine. In this study, benzene was replaced by toluene as organic solvent. For quantification purposes, different quantities (0, 5, 10, 20, 40, 60, 80, 100, 130, and 150 µL) of an oleic acid solution in toluene (705.3 mg/25 mL) were employed. The validity range was 0.5–15.0 µmol oleic acid and the R^2 value of the analytical procedure was 0.9998. Results were calculated as g oleic acid·kg^{-1} lipids and g oleic acid·kg^{-1} tissue.

PLs were quantified by measuring the organic phosphorus in the total lipid extracts according to the Raheja et al. [75] method, which is based on a complex formation with ammonium molybdate. For quantification purposes, different quantities (0, 5, 10, 20, 40, 60, 80, 100, 130, and 150 µL) of a 1,2-dipalmitoyl-rac-glycero-3-phosphocholine (DPPC) solution in chloroform (15.3 mg/5 mL) were employed. The validity range was 16.1–483.0 µg DPPC and the R^2 value of the analytical procedure was 0.9995. Results were calculated as g DPPC·kg^{-1} lipids and g DPPC·kg^{-1} tissue.

Free STs were determined on total lipid extracts using the method of Huang et al. [76] based on the reaction with acetic anhydride in acetic acid (Liebermann–Buchardt reaction). For quantification purposes, different quantities (0, 5, 10, 20, 40, 60, 80, 100, 130, and 150 µL) of a cholesterol solution in acetic acid (12.2 mg/5 mL) were employed. The validity range was 11.5–345.0 µg cholesterol and the R^2 value of the analytical procedure was 0.9998. Results were calculated as g cholesterol·kg^{-1} lipids and g cholesterol·kg^{-1} tissue.

The content of tocopherol compounds was determined in both tissues according to the method of Cabrini et al. [77], with some modifications. For this purpose, a lipid fraction was carried out to dryness under nitrogen flux, dissolved in isopropanol, and analysed using HPLC (C18 5µm, 4.6 × 250 mm column; XBridge, Waters, Milfoird, MA, USA). The column was fluxed with methanol for 2 min; then, a gradient from 0 to 50% of isopropanol in 10 min was applied. A 1.5 mL·min^{-1} flow rate was employed and detection was carried out at 280 nm. The possible presence of α-, β-, γ-, and δ-tocopherol molecules was checked. For quantitative purposes, the content of each tocopherol compound present in the lipid extract was calculated with calibration curves prepared with the corresponding commercial tocopherol molecule (10–100 µL of a 1000 ppm tocopherol/methanol solution) and calculated as mg·kg^{-1} lipids and mg·kg^{-1} tissue.

3.6. Statistical Analysis

Data (n = 4) obtained from the different chemical analyses (proximate composition, individual FAs, FA groups and ratios, and lipid classes) were subjected to the ANOVA method to investigate differences between both tissues, i.e., skin and muscle (Statistica version 6.0, 200; Statsoft Inc., Chicago, Il, USA). A comparison of means was performed using the least-squares difference (LSD) test. The 95% confidence intervals of each chemical parameter were calculated; for it, the standard deviation of each sample and the number of replicates were considered.

4. Conclusions

The present study provides a first approach focused on the chemical composition of European eel (*A. anguilla*) skin. The presence of bioactive compounds in this waste substrate was comparatively analysed with the edible tissue, a seafood widely accepted by the consumer. Thus, higher ($p < 0.05$) levels of proteins (271.6 $g \cdot kg^{-1}$ wet tissue), lipids (38.0 $g \cdot kg^{-1}$ wet tissue), ash (27.7 $g \cdot kg^{-1}$ wet tissue), and ω6 FAs were observed in the skin tissues. Contrary, the muscle tissue showed higher ($p < 0.05$) moisture, ω3 FA, and ω3/ω6 values. Regarding lipid classes, higher proportions of PLs (111.1 $g \cdot kg^{-1}$ lipids), STs (104.7 $g \cdot kg^{-1}$ lipids), α-tocopherol (274.0 $mg \cdot kg^{-1}$ lipids), and FFAs (43.6 $g \cdot kg^{-1}$ lipids) were observed in the skin tissue. No differences ($p > 0.05$) between both tissues could be detected in TAG and FA group (STFAs, MUFAs, and PUFAs) values and in the total PUFA/total STFA ratio.

It is concluded that eel skin, a by-product resulting from commercial processing, can be considered a valuable source for the food and pharmaceutical industries by providing value-added constituents such as proteins, lipids, ω3 FAs, PLs, and α-tocopherol. Further research taking into account variations of the chemical composition resulting from internal and external factors ought to be addressed. The study agrees with the current search for alternative sources of healthy and nutritive compounds from waste substrates. As for the edible parts of fish, convenient handling and storage during the skin processing ought to be carried out to avoid the development of damage mechanisms such as autolysis, microbial activity, and lipid oxidation.

Author Contributions: Conceptualization, S.M. and S.P.A.; methodology, A.B. and M.T.; data curation, A.B. and M.T.; writing—original draft preparation, S.P.A.; writing—review and editing, S.M. and S.P.A. All authors have read and agreed to the published version of the manuscript.

Funding: This research received no external funding.

Institutional Review Board Statement: Not applicable.

Data Availability Statement: Data are contained within the article.

Conflicts of Interest: The authors declare no conflicts of interest.

References

1. Tilami, S.K.; Sampels, S. Nutritional Value of Fish: Lipids, proteins, vitamins, and minerals. *Rev. Fish. Sci.* **2018**, *26*, 242–253.
2. Xu, H.; Turchini, G.M.; Francis, D.S.; Liang, M.; Mock, T.S.; Rombenso, A.; Ai, Q. Are fish what they eat? A fatty acid's perspective. *Prog. Lipid Res.* **2020**, *80*, 101064. [CrossRef] [PubMed]
3. Minihane, A.; Armah, C.; Miles, E.; Madden, J.; Clark, A.; Caslake, M.; Calder, P. Consumption of fish oil providing amounts of eicosapentaenoic acid and docosahexaenoic acid that can be obtained from the diet reduces blood pressure in adults with systolic hypertension: A retrospective analysis. *J. Nutr.* **2016**, *146*, 516–523. [CrossRef] [PubMed]
4. Schunck, W.; Konkel, A.; Fischer, R.; Weylandt, K. Therapeutic potential of omega-3 fatty acid-derived epoxy eicosanoids in cardiovascular and inflammatory diseases. *Pharmacol. Ther.* **2018**, *183*, 177–204. [CrossRef]
5. Küllenberg, D.; Taylor, L.A.; Schneider, M.; Massing, U. Health effects of dietary phospholipids. *Lipids Health Dis.* **2012**, *11*, 3. [CrossRef]
6. Li, J.; Wang, X.; Zhang, T.; Huang, Z.; Luo, X.; Deng, Y. A review on phospholipids and their main applications in drug delivery systems. *Asian J. Pharm. Sci.* **2015**, *10*, 81–98. [CrossRef]
7. Kalogeropoulos, N.; Chiou, A. Antioxidants. In *Handbook of Seafood and Seafood Products Analysis*; Nollet, L.M., Toldrá, F., Eds.; CRC Press; Taylor and Francis Group: Boca Raton, FL, USA, 2010; pp. 309–326.
8. Kim, Y.N. Vitamins. In *Handbook of Seafood and Seafood Products Analysis*; Nollet, L.M., Toldrá, F., Eds.; CRC Press; Francis and Taylor Group: Boca Raton, FL, USA, 2010; pp. 327–350.
9. Özyurt, G.; Özkütük, S. Advances in discard and by-product processing. In *Innovative Technologies in Seafood Processing*; Özoğul, Y., Ed.; CRC Press; Taylor and Francis Group: Boca Raton, FL, USA, 2020; pp. 323–350.
10. Olsen, R.L.; Toppe, J.; Karunasagar, I. Challenges and realistic opportunities in the use of by-products from processing of fish and shellfish. *Trends Food Sci. Technol.* **2014**, *36*, 144–152. [CrossRef]

11. Atef, M.; Ojagh, M. Health benefits and food applications of bioactive compounds from fish byproducts: A review. *J. Funct. Foods* **2017**, *35*, 673–681. [CrossRef]
12. Yuan, Z.; Ye, X.; Hou, Z.; Chen, S. Sustainable utilization of proteins from fish processing by-products: Extraction, biological activities and applications. *Trend Food Sci. Technol.* **2024**, *143*, 104276. [CrossRef]
13. Li, Z.R.; Wang, B.; Chi, C.F.; Zhang, Q.H.; Gong, Y.D.; Tang, J.J.; Luo, H.Y.; Ding, G.F. Isolation and characterization of acid soluble collagens and pepsin soluble collagens from the skin and bone of Spanish mackerel (*Scomberomorous niphonius*). *Food Hydrocoll.* **2013**, *31*, 103–113. [CrossRef]
14. Chi, C.F.; Wang, B.; Li, Z.R.; Luo, H.Y.; Ding, G.F.; Wu, C.W. Characterization of acid-soluble collagen from the skin of hammerhead shark (*Sphyrna lewini*). *J. Food Biochem.* **2014**, *38*, 236–247. [CrossRef]
15. Benjakul, S.; Oungbho, K.; Visessanguan, W.; Thiansilakul, Y.; Roytrakul, S. Characteristics of gelatin from the skins of bigeye snapper, *Priacanthus tayenus* and *Priacanthus macracanthus*. *Food Chem.* **2009**, *116*, 445–451. [CrossRef]
16. Yu, E.; Pan, C.; Chen, W.; Ruan, Q.; Luo, X.; Lv, M.; Fang, Y.; Jiang, L.; Ma, H. Gelatin from specific freshwater and saltwater fish extracted using six different methods: Component interactions, structural characteristics, and functional properties. *LWT-Food Sci. Technol.* **2024**, *191*, 115656. [CrossRef]
17. Chansuwan, C.; Chinachoti, P. Antioxidative properties and hydrolysis profile of Skipjack tuna dark muscle and skin. *Int. Food Res. J.* **2015**, *22*, 1968–1976.
18. Nurilmala, M.; Hizbullah, H.H.; Karnia, E.; Kusumaningtyas, E.; Ochiai, Y. Characterization and antioxidant activity of collagen, gelatin, and the derived peptides from yellowfin tuna (*Thunnus albacares*) skin. *Mar. Drugs* **2020**, *18*, 98. [CrossRef]
19. Zhang, S.Y.; Zhao, Y.Q.; Wang, Y.M.; Yang, X.R.; Chi, C.F.; Wang, B. Gelatins and antioxidant peptides from Skipjack tuna (*Katsuwonus pelamis*) skins: Purification, characterization, and cytoprotection on ultraviolet-A injured human skin fibroblasts. *Food Biosci.* **2022**, *50*, 102138. [CrossRef]
20. Zhang, A.; Fan, X.; Zeng, X.; Xu, J.; Zhou, C.; Xia, Q.; Sun, Y.; Wu, Z.; Pan, D. Enhancing physicochemical, antimicrobial, and release properties of fish skin gelatin films using dual-layer nanoparticles loaded with tea polyphenols/kojic acid for air-dried chicken preservation. *Food Hydrocoll.* **2024**, *149*, 109580. [CrossRef]
21. Fu, W.; Wang, P.; Chen, Y.; Lin, J.; Zheng, B.; Zeng, H.; Zhang, Y. Preparation, primary structure and antifreeze activity of antifreeze peptides from *Scomberomorus niphonius* skin. *LWT-Food Sci. Technol.* **2019**, *101*, 670–677. [CrossRef]
22. Goymer, A.; Steele, K.; Jenkins, F.; Burgess, G.; Andrews, L.; Baumgartner, N.; Gubili, C.; Griffiths, A.M. For R-eel?! Investigating international sales of critically endangered species in freshwater eel products with DNA barcoding. *Food Control* **2023**, *150*, 109752. [CrossRef]
23. Heinsbroek, L.T.N.; Støttrup, J.G.; Jacobsen, C.; Corraze, G.; Kraiem, M.M.; Holst, L.K.; Tomkiewicz, J.; Kaushik, S.J. A review on broodstock nutrition of marine pelagic spawners: The curious case of the freshwater eels (*Anguilla* spp.). *Aquac. Nutr.* **2013**, *19*, 1–24. [CrossRef]
24. Støttrup, J.G.; Tomkiewicz, J.; Jacobsen, C.; Butts, I.A.E.; Holst, L.K.; Krüger-Johnsen, M.; Graver, C.; Lauesen, P.; Fontagne-Dicharry, S.; Heinsbroek, L.T.N.; et al. Development of a broodstock diet to improve developmental competence of embryos in European eel, *Anguilla anguilla*. *Aquac. Nutr.* **2016**, *22*, 725–737. [CrossRef]
25. FAO/WHO/UNU. Food and Agriculture Organization of the United Nations/World Health Organization *Anguilla anguilla*. Cultured Aquatic Species Information Programme. Text by The Danish Aquaculture Development Group (DANAQ). Fisheries and Aquaculture Division. 2024. Available online: https://www.fao.org/fishery/en/culturedspecies/anguilla_anguilla/en (accessed on 1 January 2024).
26. Özoğul, Y.; Ucar, Y.; Takadaş, F.; Durmus, M.; Köşker, A.R.; Polat, A. Comparison of green and conventional extraction methods on lipid yield and fatty acid profiles of fish species. *Eur. J. Lipid Sci. Technol.* **2018**, *120*, 1800107. [CrossRef]
27. Gómez-Limia, L.; Cobas, N.; Martínez, S. Proximate composition, fatty acid profile and total amino acid contents in samples of the European eel (*Anguilla anguilla*) of different weights. *Int. J. Gastron. Food Sci.* **2021**, *25*, 100364. [CrossRef]
28. Rudovica, V.; Bartkevics, V. Chemical elements in the muscle tissues of European eel (*Anguilla anguilla*) from selected lakes in Latvia. *Environ. Monit. Assess.* **2015**, *87*, 608. [CrossRef] [PubMed]
29. Kucukgulmez, A.; Yanar, Y.; Gerçek, G.; Gülnaz, O.; Celik, M. Effects of chitosan on color, sensory and microbiological properties of European eel (*Anguilla anguilla*) fillets during refrigerated storage. *J. Food Proc. Preserv.* **2013**, *37*, 766–771. [CrossRef]
30. Özoğul, I.; Polat, A.; Özogul, Y.; Boga, E.K.; Özogul, F.; Ayas, D. Effects of laurel and myrtle extracts on the sensory, chemical and microbiological properties of vacuum-packed and refrigerated European eel (*Anguilla anguilla*) fillets. *Int. J. Food Sci. Technol.* **2014**, *49*, 847–853. [CrossRef]
31. Ersoy, B. Effects of cooking methods on the proximate, mineral and fatty acid composition of European eel (*Anguilla anguilla*). *Int. J. Food Sci. Technol.* **2011**, *46*, 522–527. [CrossRef]
32. Gómez-Limia, L.; Carballo, J.; Rodríguez-González, M.; Martínez, S. Impact of the filling medium on the colour and sensory characteristics of canned European eels (*Anguilla anguilla* L.). *Foods* **2022**, *11*, 1115. [CrossRef] [PubMed]
33. Sila, A.; Bougatef, H.; Capitani, F.; Krichen, F.; Mantovani, V.; Ben Amor, I.; Galeotti, F.; Maccari, F.; Nedjar, N.; Volpi, N.; et al. Studies on European eel skin sulfated glycosaminoglycans: Recovery, structural characterization and anticoagulant activity. *Int. J. Biol. Macromol.* **2018**, *115*, 891–899. [CrossRef]

34. Taktak, W.; Nasri, R.; López-Rubio, A.; Chentir, I.; Gómez-Mascaraque, L.G.; Boughriba, S.; Nasri, M.; Karra-Chaabouni, M. Design and characterization of novel ecofriendly European fish eel gelatin-based electrospun microfibers applied for fish oil encapsulation. *Process Biochem.* **2021**, *106*, 10–19. [CrossRef]
35. Teng, H.; Qian, Y.; Fan, X.; Cao, H.; Tian, Y.; Chen, L. Nutritional properties of Europen eel (*Anguilla anguilla*) bone peptide-calcium and its apoptosis effect on Caco-2 cells. *Food Sci Hum. Wellness* **2022**, *11*, 1482–1490. [CrossRef]
36. Piclet, G. Le poisson aliment. Composition-Intérêt nutritionnel. *Cah. Nutr. Diététique* **1987**, *XXII*, 317–335.
37. Prego, R.; Pazos, M.; Medina, I.; Aubourg, S.P. Comparative chemical composition of different muscle zones in angler (*Lophius piscatorius*). *J. Food Compos. Anal.* **2012**, *28*, 81–87. [CrossRef]
38. Barbosa, R.G.; Trigo, M.; Prego, R.; Fett, R.; Aubourg, S.P. The chemical composition of different edible locations (central and edge muscles) of flat fish (*Lepidorhombus whiffiagonis*). *Int. J. Food Sci. Technol.* **2018**, *53*, 271–281. [CrossRef]
39. Oku, T.; Sugawara, A.; Choudhury, M.; Komatsu, M.; Yamada, S.; Ando, S. Lipid and fatty acid compositions differentiate between wild and cultured Japanese eel (*Anguilla japonica*). *Food Chem.* **2009**, *115*, 436–440. [CrossRef]
40. Nurfaidah; Metusalach, M.; Sukarno; Mahendradatta, M. Protein and albumin contents in several freshwater fish species of Makassar, South Sulawesi, Indonesia. *Int. Food Res. J.* **2021**, *28*, 745–751. [CrossRef]
41. Lee, K.; Kim, Y.J.; Hong, Y.K.; Song, M.Y.; Lee, W.O.; Hwang, K.T. Lipid content and fatty acid composition of freshwater eels *Anguilla japonica* caught in different seasons and locations in South Korea. *Fish. Sci.* **2020**, *86*, 573–580. [CrossRef]
42. Saito, H.; Kurogi, H.; Chow, S.; Mochioka, N. Variation of lipids and fatty acids of the Japanese eel, *Anguilla japonica*, during spawning migration. *J. Oleo Sci.* **2015**, *64*, 603–616. [CrossRef] [PubMed]
43. Njinkoué, J.M.; Barnathan, G.; Miralles, J.; Gaydou, E.M.; Samb, A. Lipids and fatty acids in muscle, liver and skin of three edible fish from the Senegalese coast: *Sardinella maderensis*, *Sardinella aurita* and *Cephalopholis taeniops*. *Comp. Biochem. Physiol. B Biochem. Mol. Biol.* **2002**, *131*, 395–402. [CrossRef] [PubMed]
44. Pateiro, M.; Munekata, P.E.; Domínguez, R.; Wang, M.; Barba, F.J.; Bermúdez, R.; Lorenzo, J.M. Nutritional profiling and the value of processing by-products from gilthead sea bream (*Sparus aurata*). *Mar. Drugs* **2020**, *18*, 101. [CrossRef]
45. Ahmmed, M.K.; Ahmmed, F.; Stewart, I.; Carne, A.; Tian, H.; Ahmed Bekhit, A.E. Omega-3 phospholipids in Pacific blue mackerel (*Scomber australasicus*) processing by-products. *Food Chem.* **2021**, *353*, 129451. [CrossRef]
46. Ahmmed, M.K.; Carne, A.; Ahmmed, F.; Stewart, I.; Tian, H.S.; Ahmed Bekhit, A.E. Positional distribution of fatty acids and phospholipid composition in King salmon (*Oncorhynchus tshawytscha*) head, roe and skin using nuclear magnetic resonance spectroscopy. *Food Chem.* **2021**, *363*, 130302. [CrossRef] [PubMed]
47. Park, J.S.; Chandra Roy, V.; Kim, S.Y.; Lee, S.C.; Chun, B.S. Extraction of edible oils and amino acids from eel by-products using clean compressed solvents: An approach of complete valorization. *Food Chem.* **2022**, *38*, 132949. [CrossRef]
48. Magalhães, J.P.; Müller, M.; Rainger, G.; Steegenga, W. Fish oil supplements, longevity and aging. *Aging* **2016**, *8*, 1578–1582. [CrossRef]
49. Devassy, J.G.; Leng, S.; Gabbs, M.; Monirujjaman, M.; Aukema, H.M. Omega-3 polyunsaturated fatty acids and oxylipins in neuroinflammation and management of Alzheimer disease. *Adv. Nutr.* **2016**, *7*, 905–916. [CrossRef]
50. Swanson, S.; Block, R.; Mousa, S. Omega-3 fatty acids EPA and DHA: Health benefits throughout life. *Adv. Nutr.* **2012**, *3*, 1–7. [CrossRef]
51. Ofosu, F.K.; Daliri, E.B.M.; Lee, B.H.; Yu, X. Current trends and future perspectives on omega-3 fatty acids. *Res. J. Biol.* **2017**, *5*, 11–20.
52. Uauy, R.; Valenzuela, A. Marine oils: The health benefits of n-3 fatty acids. *Nutrition* **2000**, *16*, 680–684. [CrossRef] [PubMed]
53. Šimat, V.; Vlahović, J.; Soldo, B.; Mekinić, I.G.; Čagalj, M.; Hamed, I.; Skroza, D. Production and characterization of crude oils from seafood processing by-products. *Food Biosci.* **2020**, *33*, 100484. [CrossRef]
54. Kumari, P.; Kumar, M.; Reddy, C.R.; Jha, B. Algal lipids, fatty acids and sterols. In *Functional Ingredients from Algae for Foods and Nutraceuticals*; Domínguez, H., Ed.; Woodhead Publishing: Cambridge, UK, 2013; pp. 87–134.
55. Cantor, A.H.; Decker, E.A.; Collins, V.P. Fatty acids in poultry and egg products. In *Fatty Acids in Foods and Their Health Implications*, 3rd ed.; Chow, C.K., Ed.; CRC Press: Boca Raton, FL, USA, 2008; pp. 127–154.
56. Palmquist, D.L.; Jensen, R.G. Fatty acids in milk fat. In *Fatty Acids in Foods and Their Health Implications*, 3rd ed.; Chow, C.K., Ed.; CRC Press: Boca Raton, FL, USA, 2008; pp. 109–125.
57. Wood, J.D.; Enser, M.; Richardson, R.I.; Whittington, F.M. Fatty acids in meat and meat products. In *Fatty Acids in Foods and Their Health Implications*, 3rd ed.; Chow, C.K., Ed.; CRC Press: Boca Raton, FL, USA, 2008; pp. 87–107.
58. Achouri, A.; Kharrat, N.; Smichi, N.; Miled, N.; Gargouri, Y.; Fendri, A. Nutritional properties, oxidative stability, and in vitro digestibility of oils extracted from muscles of wild and breeding eels (*Anguilla anguilla*). *J. Food Process. Preserv.* **2018**, *42*, e13519. [CrossRef]
59. Huyen, K.T.; Long, D.T.; Linh, N.Q. Some biological components in natural marbled eel in Thua Thien Hue, Vietnam. *J. Appl. Anim. Res.* **2023**, *51*, 397–405. [CrossRef]
60. Medina, I.; Sacchi, R.; Aubourg, S.P. A ^{13}C-NMR study of lipid alterations during fish canning: Effect of filling medium. *J. Sci. Food Agric.* **1995**, *69*, 445–450. [CrossRef]
61. Aubourg, S.P.; Quitral, V.; Larraín, M.A.; Rodríguez, A.; Gómez, J.; Maier, L.; Vinagre, J. Autolytic degradation and microbiological activity in farmed Coho salmon (*Oncorhynchus kisutch*) during chilled storage. *Food Chem.* **2007**, *104*, 369–375. [CrossRef]

62. Vázquez, M.; Torres, J.A.; Gallardo, J.M.; Saraiva, J.A.; Aubourg, S.P. Lipid hydrolysis and oxidation development in frozen mackerel (*Scomber scombrus*): Effect of a high hydrostatic pressure pre-treatment. *Innov. Food Sci. Emerg. Technol.* **2013**, *18*, 24–30. [CrossRef]
63. Takahashi, K.; Inoue, Y. Marine by-product phospholipids as booster of medicinal compounds. *Adv. Food Nutr. Res.* **2012**, *65*, 31–46.
64. Köhler, A.; Sarkinnen, E.; Tapola, N.; Niskanen, T.; Bruheim, I. Bioavailability of fatty acids from krill oil, krill meal and fish oil in healthy subjects—A randomized, single-dose, cross-over trial. *Lipids Health Dis.* **2015**, *14*, 19. [CrossRef]
65. Bandarra, N.M.; Campos, R.M.; Batista, I.; Nunes, M.L.; Empis, J.M. Antioxidant synergy of alpha-tocopherol and phospholipids. *J. Am. Oil Chem. Soc.* **1999**, *76*, 905–913. [CrossRef]
66. Hidalgo, F.J.; Leon, M.M.; Zamora, R. Antioxidative activity of amino phospholipids and phospholipid/amino acid mixtures in edible oils as determined by the rancimat method. *J. Agric. Food Chem.* **2006**, *54*, 5461–5467. [CrossRef]
67. Medina, I.; Aubourg, S.P.; Pérez-Martín, R. Composition of phospholipids of white muscle of six tuna species. *Lipids* **1995**, *30*, 1127–1135. [CrossRef] [PubMed]
68. Van Ginneken, V.; Hekman, M.; Verheij, E. The lipid composition and biochemistry of the migrating European eel (*Anguilla anguilla* L.): A LCMS-study following a lipidomics based systems biology approach. *Adv. Biochem. Biotechnol.* **2018**, *3*, 165.
69. AOAC. *Official Methods for Analysis of the Association of Analytical Chemistry*, 15th ed.; Association of Official Chemists, Inc.: Arlington, VA, USA, 1990; pp. 931–937.
70. Bligh, E.; Dyer, W. A rapid method of total extraction and purification. *Can. J. Biochem. Physiol.* **1959**, *37*, 911–917. [CrossRef]
71. Aubourg, S.P.; Medina, I.; Pérez-Martín, R. Polyunsaturated fatty acids in tuna phospholipids: Distribution in the sn-2 location and changes during cooking. *J. Agric. Food Chem.* **1996**, *44*, 585–589. [CrossRef]
72. Álvarez, V.; Medina, I.; Prego, R.; Aubourg, S.P. Lipid and mineral distribution in different zones of farmed and wild blackspot seabream (*Pagellus bogaraveo*). *Eur. J. Lipid Sci. Technol.* **2009**, *111*, 957–966. [CrossRef]
73. Vioque, E.; Holman, R. Quantitative estimation of esters by thin-layer chromatography. *J. Am. Oil Chem. Soc.* **1962**, *39*, 63–66. [CrossRef]
74. Lowry, R.; Tinsley, I. Rapid colorimetric determination of free fatty acids. *J. Am. Oil Chem. Soc.* **1976**, *53*, 470–472. [CrossRef] [PubMed]
75. Raheja, R.; Kaur, C.; Singh, A.; Bhatia, A. New colorimetric method for the quantitative determination of phospholipids without acid digestion. *J. Lipid Res.* **1973**, *14*, 695–697. [CrossRef] [PubMed]
76. Huang, T.; Chen, C.; Wefler, V.; Raftery, A. A stable reagent for the Liebermann-Buchardt reaction. *Anal. Chem.* **1961**, *33*, 1405–1407. [CrossRef]
77. Cabrini, L.; Landi, L.; Stefanelli, C.; Barzanti, V.; Sechi, A. Extraction of lipid and lipophilic antioxidants from fish tissues: A comparison among different methods. *Comp. Biochem. Physiol. Biochem. Mol. Biol.* **1992**, *101*, 383–386. [CrossRef]

Disclaimer/Publisher's Note: The statements, opinions and data contained in all publications are solely those of the individual author(s) and contributor(s) and not of MDPI and/or the editor(s). MDPI and/or the editor(s) disclaim responsibility for any injury to people or property resulting from any ideas, methods, instructions or products referred to in the content.

marine drugs

MDPI

Review

A Review on Marine Microbial Docosahexaenoic Acid Production Through Circular Economy, Fermentation Engineering, and Antioxidant Technology

Fengwei Yin, Xiaolong Sun, Xi Luo, Weilong Zheng, Longfei Yin, Yingying Zhang and Yongqian Fu *

Taizhou Key Laboratory of Biomass Functional Materials Development and Application, School of Life Sciences, Taizhou University, Taizhou 318000, China; yinfengwei@126.com (F.Y.); sunxiaolong198901@163.com (X.S.); tzcluoxi@163.com (X.L.); z3037713@163.com (W.Z.); yinlongfei@163.com (L.Y.); zyying@tzc.edu.cn (Y.Z.)
* Correspondence: bioengineer@163.com

Abstract: Marine microbial-derived docosahexaenoic acid (DHA) has garnered significant attention as a sustainable and health-promoting alternative to fish oil-derived DHA. However, its industrial production from marine heterotrophic microorganisms faces challenges related to high costs and suboptimal oil quality, which hinder its broader application. This review focuses on recent strategies aimed at achieving low-cost and high-quality marine microbial DHA production, emphasizing heterotrophic systems that dominate commercial supply. Key aspects include: Fermentation optimization using waste-derived feedstocks and bioprocess engineering to enhance DHA yields; Critical refining techniques—including degumming, neutralization, decolorization, and deodorization—are analyzed for improving DHA oil purity and quality, with emphasis on process optimization to adapt to the unique biochemical properties of microbial-derived oils. Additionally, strategies for oxidative stabilization, such as antioxidant protection, are discussed to extend the shelf life and preserve the nutritional value of marine microbial DHA oil. By integrating techno-economic and biochemical perspectives, this work outlines a holistic framework to guide the industrial optimization of marine microbial-sourced DHA oil production, addressing cost and quality challenges to facilitate its large-scale application as functional foods and nutraceuticals, thereby reducing reliance on marine resources and advancing sustainable omega-3 production.

Keywords: marine microbial DHA; circular economy; heterotrophic fermentation; oil quality enhancement; antioxidant technology

1. Introduction

The increasing demand for healthier diets, fueled by rising living standards, is supported by scientific and technological advancements that enable material progress and meet evolving dietary requirements. Lipids, essential for human growth and development, encompass fatty acids that are classified as either saturated (SFAs) or unsaturated (UFAs). UFAs are further divided into monounsaturated fatty acids (MUFAs) and polyunsaturated fatty acids (PUFAs). PUFAs, owing to their unique chemical structure, possess a range of medicinal and health-promoting properties, such as anti-cardiovascular disease effects, cholesterol reduction, and the prevention of hypertension and diabetes. These properties have established PUFAs as integral to health-conscious dietary practices. Among PUFAs, ω-3 fatty acids (e.g., docosahexaenoic acid (DHA, $C_{22}H_{32}O_2$) and eicosapentaenoic acid (EPA, $C_{20}H_{30}O_2$)) and ω-6 fatty acids (e.g., gamma-linolenic acid (GLA, $C_{18}H_{30}O_2$) and

 https://doi.org/10.3390/md23060256

arachidonic acid (ARA, $C_{20}H_{32}O_2$)) [1,2] are highly valued for their health benefits, including regulating anti-inflammatory pathways, supporting neurovascular health, and driving pro-inflammatory signaling and immune modulation, with their balanced interplay being essential for metabolic homeostasis and disease resilience [1]. Traditional sources of PUFAs, such as animal and plant oils, increasingly fall short of meeting growing demand. This shortfall has driven interest in microbial fermentation as an alternative production method, motivated by economic and societal considerations [3–5].

Docosahexaenoic acid (DHA, $C_{22}H_{32}O_2$), an ω-3 long-chain polyunsaturated fatty acid, features six cis-configured double bonds. A colorless and odorless compound, DHA is insoluble in water but soluble in organic solvents and remains in an oily liquid state at room temperature due to its low melting point of −44 °C and inherent chemical instability. Despite its essential role in human metabolism, DHA cannot be synthesized endogenously and must be obtained from dietary sources. Often referred to as "brain gold", DHA is crucial for brain and visual development, constituting a key component of neuronal cell membranes. Inadequate DHA intake during infancy can impair brain development and lead to cognitive delays, while sufficient levels in adults are vital for maintaining cognitive function [6–8].

Traditionally, DHA has been sourced from deep-sea fish and marine animals [9,10]. However, sustainable non-fish alternatives—collectively designated as "DHA algal oil" and primarily derived from heterotrophic marine microorganisms—are rapidly replacing conventional sources. Although derived from microbial fermentation, those DHA products are uniformly marketed as "algal oils" in accordance with global regulatory frameworks and consumer labeling conventions. Critically, this review specifically addresses high-yield DHA production from non-photosynthetic microbes, which constitute >90% of commercial algal DHA supply due to their superior productivity and fish-independent fermentation systems. These microbial platforms provide essential sustainability advantages, including decoupling from marine food webs and compatibility with circular bioeconomy models using waste feedstocks [11]. Despite its advantages, industrial-scale production of this microbial-sourced DHA (marketed as algal oil) faces significant challenges, primarily due to high production costs [10,12] and inconsistencies in product quality [13–15]. To address these issues, this review explores fermentation technologies and downstream processing methods aimed at cost reduction and quality control, highlighting the recent advancements in low-cost fermentation strategies, as well as refining and antioxidant technologies for marine microbial DHA. This integrated approach aims to enable affordable, high-quality DHA algal oil production at scale, reducing pressure on marine ecosystems while meeting growing nutraceutical demand.

2. Sources of DHA

2.1. Traditional Sources of DHA

Many DHA products on the market are labeled as "deep-sea fish oil", reflecting the reliance on marine fish as the primary traditional source of DHA. Historically, fish from deep-sea regions, such as tuna, cod, herring, and sardines, have served as principal raw materials for DHA extraction [16]. However, in recent years, the use of fish oil as a DHA source has faced several challenges, primarily due to environmental and sustainability concerns, including global climate change and pollution [17,18].

Research indicates that DHA in seals, marine fish, and even commercial deep-sea fish oil consumed by indigenous populations, such as the Inuit, originates from DHA accumulation along the food chain. Planktonic microorganisms and microalgae in the ocean are capable of synthesizing DHA, with fish acquiring it by consuming these primary producers. Therefore, marine microbes and their phototrophic counterparts, rather than fish, are the true original producers of DHA [19].

2.2. New Sources of DHA

The limitations of traditional fish oil derived DHA, coupled with the growing demand for high-quality DHA, have highlighted the insufficiency of fish oil to meet societal needs. As a result, there is an urgent need to identify high-quality, safe, reliable, and sustainable alternatives to fish-derived DHA [20,21]. In this context, oil-producing marine microorganisms—the original producers of DHA—have garnered significant global attention as a promising source of DHA.

An innovative approach to DHA production involves utilizing these marine microbial producers, particularly heterotrophic species, fermented under optimized conditions to accumulate lipids rich in DHA within their cells. This method bypasses the complexities associated with the bioaccumulation of DHA through the marine food chain, enabling the direct production of DHA from its microbial origin. Currently, commercially available microbial DHA oils are primarily derived from heterotrophic marine protists. Compared to fish oil-derived DHA, marine microbial DHA (commercially labeled as algal oil) offers several key advantages:

- High productivity in fermentation systems: Under controlled fermentation conditions, selected heterotrophic strains exhibit rapid growth and can accumulate oil at levels as high as 50% of their dry cell weight.
- Closed-tank bioprocessing: Fermentation occurs in sterile bioreactors with defined and uncontaminated culture media. This controlled environment produces DHA oil with fewer impurities, superior quality, and enhanced safety.
- Efficient extraction processes: Microbial fermentation DHA production integrates seamlessly with downstream processing. Following fermentation, the oil can be efficiently extracted and refined without the geographical and seasonal constraints associated with fish-derived DHA, ensuring consistent production quality.
- Utilization of agricultural by-products: These microorganisms can be cultured using low-cost agricultural by-products and waste materials, such as waste molasses and glycerol, as fermentation substrates. This approach not only reduces production costs but also valorizes waste, contributing to environmental sustainability by mitigating pollution from waste disposal.

Marine microbial DHA thus represents a highly promising alternative to traditional sources, with broad industrial application potential [21–24].

2.3. DHA-Producing Strains

The exploration of microbial sources as a source of DHA began in the late 1970s and early 1980s. During this time, scientists identified specific microalgal strains capable of accumulating substantial amounts of DHA, a fatty acid essential for human health, particularly for brain development. This groundbreaking discovery laid the foundation for subsequent research into these microorganisms as a sustainable and environmentally friendly DHA source. Initially, DHA-producing microalgae were cultivated using autotrophic growth in photobioreactors. However, this method was hampered by slow cell growth rates and low DHA yields, which posed challenges for efficient DHA oil production and complicated downstream processing. The 1980s marked a turning point with the introduction of heterotrophic culture techniques, which revolutionized DHA production. By supplementing the culture medium with carbon and nitrogen sources, heterotrophic cultivation facilitated faster cell growth, higher biomass accumulation, and increased oil content. This approach quickly became the dominant method for microbial DHA fermentation.

The first microalgal strain applied to the industrial production of DHA was *Crypthecodinium cohnii*, using heterotrophic culture methods, and the resulting DHA oil was incorporated into infant formulas and health supplements as a nutritional enhancer, establishing the safety and suitability of microalgal DHA for human consumption [25]. Despite its initial success, *Crypthecodinium cohnii* faced significant limitations, including a slow growth rate, an extended fermentation period of approximately 400 h per batch, and relatively low DHA production efficiency. Consequently, efforts have been directed toward identifying more efficient DHA-producing strains. Thraustochytrids, as marine heterotrophic protists, have emerged as a promising group for DHA production [26]. Genera such as *Thraustochytrium* [27], *Schizochytrium* [28], and *Aurantiochytrium* [29] exhibit biotechnologically relevant DHA-producing capabilities, showing great potential for large-scale commercial production [30–32].

3. Marine Microbial DHA Fermentation

3.1. DHA Synthesis Pathway

Under environmental stress such as nutrient limitations (e.g., nitrogen or phosphorus deficiency), oleaginous microorganisms shift toward lipid synthesis and accumulation. DHA synthesis primarily occurs through two pathways: the fatty acid synthase (FAS) pathway and the polyketide synthase (PKS) pathway [26,33]. Both utilize acetyl-CoA as a precursor and rely on NADPH for energy. The FAS pathway, dominant in most photoautotrophic species, generates fatty acids through cyclic reactions including condensation, reduction, dehydration, and a second reduction (a four-step cycle that sequentially extends and processes the fatty acyl chain). The PKS pathway, employed by heterotrophic marine protists like Thraustochytrids, is a multi-enzyme complex system. Mechanistically distinct from FAS, PKS introduces unsaturated bonds during fatty acyl chain elongation [34].

Therefore, the preference for either pathway is largely strain-specific and influenced by environmental and nutritional conditions. Photoautotrophic microalgae such as *Nannochloropsis* and *Chlorella* typically utilize the FAS pathway, in which unsaturated fatty acids are synthesized through a series of elongation and desaturation reactions of saturated fatty acids. This process is favored under nitrogen-limited conditions and sufficient oxygen availability, which activate desaturase enzymes and promote lipid accumulation [35,36]. In contrast, heterotrophic Thraustochytrids such as *Schizochytrium* and *Aurantiochytrium* predominantly use the PKS pathway, a multi-modular enzyme system that enables the direct formation of DHA without desaturation steps. This pathway is especially active under high carbon-to-nitrogen (C/N) ratios, low oxygen levels, and in the presence of abundant organic carbon sources such as glucose or glycerol [37,38]. Moreover, low cultivation temperatures have been shown to enhance DHA accumulation in PKS-dominant strains, likely by modulating enzyme activity and promoting the entry of relatively large amounts of substrates into the PKS pathway [39]. Overall, the dominance of the FAS or PKS pathway is not only species-dependent but also closely linked to cultivation mode, carbon and nitrogen availability, and oxygen conditions.

3.2. Materials for Microbial DHA Fermentation

In commercial microbial DHA fermentation using heterotrophic strains, the composition of nutrients in the culture medium plays a pivotal role in promoting rapid biomass growth and efficient oil accumulation. The consumption of substrates by the cells is directly linked to cellular activity, growth rates, and the rate of product synthesis.

3.2.1. Traditional Carbon Sources

Carbon sources are essential for the growth and metabolism of microorganisms, serving as the primary energy supply and structural foundation for cellular functions. For DHA, a long-chain fatty acid with 22 carbon atoms, a continuous and efficient supply of carbon is crucial for its synthesis and accumulation. The choice of carbon source directly influences microbial biomass production and DHA oil yield.

Heterotrophic production systems utilize diverse carbon substrates for growth and oil accumulation. Glucose and glycerol are the dominant industrial carbon sources [40–42]. Glucose has been widely recognized as an effective carbon source for DHA production. For example, *Schizochytrium* sp. can efficiently ferment glucose and fructose for DHA synthesis, although it is unable to metabolize sucrose [43]. It is well known that glucose is particularly effective in promoting DHA accumulation during fermentation. Glycerol, another widely used substrate, has also demonstrated excellent potential for DHA production. Additionally, glycerol has been found to enhance the conversion of short-chain saturated fatty acids into unsaturated fatty acids during the oil accumulation phase, further boosting DHA synthesis [40,44].

In addition to traditional carbon sources, C2 compounds such as ethanol and acetic acid have been explored as substrates for DHA fermentation. Research indicates that ethanol and acetic acid are viable carbon sources for DHA production in *Schizochytrium* and other heterotrophic strains, offering alternative options for substrate utilization [41,45].

3.2.2. Traditional Nitrogen Sources

Nitrogen is a vital component for microbial cell synthesis, serving as a building block for proteins, nucleic acids, and enzymes. The choice of nitrogen source significantly influences DHA production during fermentation, with both organic and inorganic nitrogen sources commonly employed.

Organic nitrogen sources such as yeast extract, corn steep liquor, monosodium glutamate (MSG), and peptone are frequently used in DHA production [46]. These sources often provide additional growth factors, including amino acids, vitamins, and fatty acids, which support cellular metabolism and catalysis [47]. For instance, MSG has been identified as an optimal nitrogen source for promoting both biomass accumulation and DHA synthesis [46]. Corn steep liquor, a cost-effective organic nitrogen source, has also demonstrated effectiveness in enhancing DHA production [48]. In comparative studies, MSG enhanced glucose consumption and biomass accumulation in *Schizochytrium* sp., whereas ammonium sulfate, an inorganic nitrogen source, favored oil accumulation [49].

Inorganic nitrogen sources such as ammonium sulfate, ammonia, and sodium nitrate are widely used due to their cost-effectiveness and fast cellular uptake [46,50]. Ammonium sulfate is particularly notable for its dual function: it not only serves as a nitrogen source but also helps regulate pH during fermentation by lowering the medium's pH. Similarly, ammonia is often used for its dual role in providing nitrogen and maintaining a stable pH during fermentation, which enhances process efficiency [51,52].

Carbon and nitrogen sources are critical for optimizing DHA production during microbial fermentation. Traditional carbon and nitrogen sources are usually more expensive. By carefully selecting and balancing carbon and nitrogen sources, fermentation processes can be optimized for higher DHA yields and cost efficiency. Table 1 summarizes the traditional carbon and nitrogen sources and their effects in marine microbial DHA production.

Table 1. Conventional carbon and nitrogen sources for DHA production via fermentation.

Strains	Carbon Sources	Nitrogen Sources	Fermentation Conditions	DHA Content (g/L)	DHA Productivity (mg/L/h)	Ref.
Thraustochytrium sp.	Glucose	Yeast extract, peptone	Batch fermentation (shake flasks, 26 °C, 150 rpm, 120 h)	1.34	11.17	[53]
Thraustochytrium sp. ONC-T18	Glucose	Yeast extract, MSG	Batch fermentation (5 L bioreactor, 25 °C, 120 rpm, 168 h)	4.6	38.33	[54]
Aurantiochytrium sp.	Glucose	Yeast extract, peptone	Batch fermentation (shake flasks, 26 °C, 150 rpm, 120 h)	1.34	11.17	[55]
Aurantiochytrium sp. AF0043	Glucose, glycerol	MSG, Corn steep powder	Fed-batch fermentation (shake flasks, 28 °C, 150 rpm, 120 h)	2.75	22.92	[56]
Aurantiochytrium sp. PKU#SW8	Glucose	MSG	Batch fermentation (shake flasks, 28 °C, 170 rpm, 96 h)	3.64	37.92	[57]
Aurantiochytrium SW1	Fructose	MSG	Batch fermentation (shake flasks, 30 °C, 250 rpm, 120 h)	4.75	39.58	[58]
Aurantiochytrium limacinum SR21	Glucose, glycerol	Yeast extract, MSG	Fed-batch fermentation (5 L bioreactor, 25 °C, 300–400 rpm, 96 h)	32.36	337.1	[40]
Schizochytrium sp. I-F-9	Glucose, glycerol	Peptone, MSG	Fed-batch fermentation (shake flasks, 28 °C, 200 rpm, 120 h)	8.33	69.41	[38]
Schizochytrium sp. ATCC 20888	Glucose	Yeast extract, MSG	Batch fermentation (shake flasks, 25 °C, 200 rpm, 96 h)	6.95	72.4	[59]
Schizochytrium sp. HX-308	Glucose	Yeast extract, MSG	Three stage continuous fermentation (50 L bioreactor, 30 °C, 300 rpm, 147 h)	23.0	156.46	[60]
Schizochytrium sp. ABC101	Glucose	Yeast extract, corn steep liquor	Fed-batch fermentation (5 L bioreactor, 28 °C, 200 rpm, 84 h)	16.7	183.3	[61]

3.2.3. Non-Traditional Low-Cost Materials

Glucose currently serves as the primary carbon source for microbial DHA production, with yeast extract commonly acting as the organic nitrogen source. However, glucose represents one of the most expensive components in industrial fermentation media, often exceeding 60% of total fermentation costs in microbial fermentations when used as the primary carbon source [62]. Despite significant advancements in marine microbial DHA production, challenges remain in scaling up the process and achieving practical implementation. Compared to fish oil-derived DHA, fermentation-sourced DHA oil production costs remain higher, limiting market competitiveness [46,59,63]. Among the various factors contributing to this, the cost of fermentation raw materials is a crucial consideration. The use of low-cost substrates in fermentation systems often compromises DHA yields [48]. To overcome this challenge, further optimization of raw material processing methods and increased efficiency in cell utilization through fermentation regulation are required.

Cheap Carbon Sources

To reduce production costs, there has been growing interest in utilizing waste products and low-cost agricultural by-products as carbon sources for microbial DHA fermentation. This strategy not only helps recycle resources but also minimizes environmental waste while lowering the overall cost of DHA production.

Several agricultural by-products have been explored as viable carbon sources for DHA production. Coconut water, a significant waste product in Southeast Asia, contains sugars like glucose and fructose, which can support *Schizochytrium* fermentation for DHA production [64]. The techno-economic analysis by Anni's research team revealed that glucose replacement with cassava-derived hydrolysate decreased overall DHA manufacturing expenses by 63%, demonstrating the viability of cheap lignocellulosic biomass utilization in microbial lipid production systems [63]. Nguyen et al. demonstrated that utilizing sugarcane bagasse hydrolysate as a carbon source achieved DHA productivity of 17.35% lipid content, with the carbon source cost dramatically reduced to USD 56.1 per ton compared to conventional glucose-based systems (USD 903.9 per ton), representing a 93.8% cost reduction in carbon substrate expenditure [65,66]. Similarly, sweet sorghum straw juice, when used at a 50% addition ratio, achieved DHA productivity comparable to pure glucose fermentation [67]. Cassava pulp has also been identified as an effective substitute for glucose, with a higher economic yield of DHA [63]. Corn syrup, another agricultural by-product, successfully supported *Aurantiochytrium* sp., yielding up to 20.1 g/L of DHA [68]. Additionally, bean dregs from tofu production have been used as a carbon source for *Schizochytrium*, yielding significant DHA production [69].

Crude glycerol, a by-product from the biodiesel industry, has been recognized as a promising carbon source for DHA fermentation. Studies have shown that crude glycerol can support *Schizochytrium* growth, resulting in high biomass and DHA yields [70]. It was reported that cultivation of *Schizochytrium* sp. S31 using crude glycerol achieved DHA productivity of 23.97% lipid content [71]. Since the cost of crude glycerol is only one-ninth that of glucose (USD 100 vs. 903.9 per ton) [66], the substrate cost analysis revealed that crude glycerol expenditure represented merely 11% of the equivalent glucose-based carbon source requirement, demonstrating an 89% cost reduction potential in bulk DHA production systems.

Molasses hydrolysates, derived from sugar production, have also been successfully used as an alternative to glucose, significantly lowering raw material costs [72,73]. Similarly, spruce hydrolysate, an industrial by-product, has demonstrated potential as a sustainable carbon source for DHA fermentation, demonstratings significantly lower production costs than pure glucose-based carbon sources, with substitution achieving an estimated 70% cost reduction [74]. Other waste materials, such as sugarcane top hydrolysates and food waste hydrolysates, have been explored for their low-cost potential, and algae residues after oil extraction have proven effective in stimulating biomass growth during DHA fermentation [75,76].

Using waste products and agricultural by-products as carbon sources for DHA fermentation presents an effective strategy to reduce production costs and address environmental concerns. These low-cost materials not only reduce the financial burden of DHA production but also promote sustainability by recycling waste into valuable products.

Cheap Nitrogen Sources

In addition to exploring low-cost carbon sources, researchers have focused on affordable organic nitrogen sources to further reduce production costs in marine microbial DHA fermentation.

Spent brewery yeast, a by-product of brewing, contains a high protein content (23.6%) and has been explored as a nitrogen source in DHA fermentation. The hydrolysate derived from spent brewery yeast has been successfully used as a substitute for yeast extract in cultivating *Aurantiochytrium* sp. for DHA production [77]. It was also demonstrated that spent brewery yeast is a promising, recyclable material for DHA fermentation [78]. Similarly, algae residues, which remain after oil extraction from microalgae, have shown potential as nitrogen sources. These residues, rich in protein, are commonly repurposed as livestock feed [79], but the non-lipid biomass left after oil extraction can also be utilized in microbial fermentation. It was found that enzymatically hydrolyzed residues from *Chlorella* after oil extraction were effective for cultivating *Chlorella vulgaris* [80]. Likewise, *Schizochytrium* sp. residues after oil extraction could serve as a nitrogen source in DHA fermentation. Strategic substitution of 80% conventional yeast extract (YE) with algal residue extract in DHA bioproduction achieved an 80% cost reduction in nitrogen source expenditure. Similarly, *Schizochytrium* sp. residues after lipid extraction demonstrate viability as an alternative nitrogen substrate for DHA biosynthesis. Implementation of algal-derived residue extract enabled substitution of 80% conventional yeast extract, achieving a proportional 80% cost reduction in nitrogen source [73]. A parallel investigation demonstrated that strategic recycling of cellular residues as nitrogen substrates achieved a 67.31% reduction in nitrogen source expenditure for DHA biosynthesis, concomitant with a 12.75% increase in DHA yield [81]. These strategies achieve dual industrial objectives by simultaneously addressing cost reduction and productivity enhancement in industrial applications.

Other organic nitrogen sources, such as rapeseed meal hydrolysates, which are rich in organic nitrogen, have also been identified as a promising nitrogen source for DHA fermentation, particularly in *Crypthecodinium* cohnii [72]. A techno-economic evaluation by Wang's team revealed that valorization of tofu processing wastewater as both a nitrogen source and a culture medium not only boosted DHA output by 3.6-fold (361.54% increase) in *Thraustochytrid* fermentation but also established a low-cost nutrient supply model through complete elimination of traditional nitrogen substrate inputs [82].

Utilizing non-traditional cheap organic nitrogen sources offers a sustainable and cost-effective approach to reducing DHA production costs. Algae residues, in particular, represent a viable alternative to traditional yeast extract, providing an efficient way to recycle waste materials into valuable resources for fermentation processes. These alternatives help promote both economic and environmental sustainability in DHA production.

Wastewater

Wastewater refers to the liquid produced in various industrial, agricultural, and household activities that contains water-soluble substances, often rich in organic and inorganic components such as nutrients, nitrogen, phosphorus, and sugars. The recycling and utilization of wastewater have become essential strategies for resource conservation, cost reduction, and promoting sustainability, especially in microbial fermentation processes. The reuse of wastewater not only helps reduce waste emissions but also significantly lowers production costs [83,84].

After microbial fermentation, a significant amount of wastewater is often generated, containing inorganic salts that help maintain the osmotic pressure of the fermentation broth, as well as nutrient substances and growth-promoting factors [84]. For example, brewery wastewater, containing essential salts and nutrients, has been successfully used in DHA fermentation. Yamasaki et al. utilized brewery wastewater to achieve a DHA concentration of 3.4 g/L [85]. Similarly, Song et al. repurposed wastewater from *Mortierella alpina* fermentation to replace pure water in DHA production by *Aurantiochytrium*, obtaining a DHA yield of 28.7 g/L, comparable to that using pure water [86]. Additionally, tofu whey wastewater, a by-product of tofu production, has shown promise as a nitrogen source for DHA fermentation. Wang et al. cultured *Schizochytrium* sp. S31 using tofu whey wastewater, achieving significant biomass and DHA productivity [82]. A mixture of saline wastewater and tofu whey wastewater further enhanced biomass and DHA yields [87]. Tofu whey wastewater can even replace commercial nutrients, such as glucose and peptone, leading to a fourfold increase in DHA production [88].

Furthermore, dairy wastewater, such as mozzarella stretching water, has also been successfully applied in DHA fermentation, resulting in a DHA yield of 1.21 g/L [78]. Kitchen wastewater, typically rich in nitrogen, phosphorus, and carbohydrates, has been used for DHA fermentation by *Aurantiochytrium* sp., yielding 7.23 g/L of DHA [89].

The use of agricultural by-products, food processing by-products, fermentation wastewater, and waste residues as raw materials for DHA fermentation offers both environmental and economic benefits. This approach reduces production costs by replacing expensive commercial media, supports the commercialization of marine microbial-sourced

DHA oil, and helps minimize waste and environmental pollution. By recycling these waste streams, resource conservation is promoted, and a more sustainable, circular economy in the biotechnological industry is fostered, enhancing DHA production efficiency and sustainability. Table 2 summarizes the non-traditional low-cost materials and their effects on marine microbial DHA production.

Table 2. DHA fermentation performance using alternative low-cost substrates.

Strains	Fermented Raw Materials	Biomass (g/L)	Lipid (g/L)	DHA (g/L)	Ref.
Schizochytrium limacinum SR21	Crude glycerol	7.89	4.94	1.84	[70]
Schizochytrium limacinum SR21	Sorghum straw sweat	9.38	6.90	2.35	[67]
Schizochytrium limacinum PA-968	Saline wastewater	28.40	9.82	3.1	[90]
Schizochytrium mangrovei Sk-02	Coconut wastewater	28.6	14.13	5.5	[64]
Schizochytrium sp. HX-308	Algal residues and cane molasses	78.26	35.54	15.22	[73]
Schizochytrium limacinum OUC88	Soybean meal hydrolysate	81.84	44.68	19.2	[91]
Aurantiochytrium sp. KRS101	Orange peel extract	5.5	2.85	0.78	[92]
Aurantiochytrium sp. KRS101	Spent yeast	31.8	12.12	10.4	[77]
Aurantiochytrium sp. SW1	Waste fruit extract	41.5	25.6	12.67	[93]
Aurantiochytrium sp. TZ209	Waste cellular residues	70.12	40.55	17.78	[81]
Aurantiochytrium sp. YLH70	Corn syrup	78.5	51	20.1	[68]
Crypthecodinium cohnii ATCC 30772	Crude glycerol	5.34	1.31	1.34	[94]
Thraustochytrium sp. (T18)	Lipid-extracted hydrolysate	14.86	6.43	2.07	[95]

3.3. *pH Control in DHA Fermentation Processes*

In heterotrophic DHA fermentation, pH is a critical parameter that influences both cell growth and metabolism [42,96]. Production strains exhibit varying optimal pH values at different stages of growth and product accumulation. In some cases, the optimal pH for cell growth may differ from that required for target product formation. Several studies have shown that the optimal pH for product formation does not always coincide with the ideal pH for cell growth [97]. Moreover, pH fluctuations during fermentation, caused by substrate consumption (particularly nitrogen sources), can impact intracellular and extracellular ion balances and enzyme activity, further affecting overall productivity, and these fluctuations can impact the intracellular and extracellular ion balance, as well as the activity of enzymes involved in the process [98].

Given its importance, pH control is essential for achieving optimal DHA production. This can be accomplished by carefully designing the initial pH [99] or by maintaining a constant pH throughout the fermentation process [100]. Furthermore, different pH values could significantly affect cell morphology [42]. Therefore, carefully designing the pH has been widely studied to optimize DHA production. For instance, Zhao et al. demonstrated that maintaining an initial pH of 7.0 during the growth phase and then adjusting to pH 5.0 for the production phase significantly enhanced DHA content, achieving a maximum yield of 11.44 g/L [101]. Similarly, Yin et al. applied a two-stage pH strategy using ammonia and citric acid as pH regulators in *Schizochytrium* sp. fermentations. They found that a pH of 7.0 was optimal for cell growth, while a lower pH of 5.0 was favorable for DHA synthesis, resulting in a high DHA yield of 32.75 g/L [42].

Both single-phase and two-phase pH regulation strategies offer effective means of optimizing DHA production during microbial fermentation. Single-phase strategies are simpler to implement and can balance cell growth and product accumulation, whereas two-phase strategies provide greater flexibility to meet the distinct pH needs of different fermentation stages, often resulting in enhanced DHA yields. Selecting the appropriate pH strategy depends on the specific marine microbial strain, process requirements, and production goals.

3.4. Osmotic Control in DHA Fermentation Processes

Osmotic pressure is a key factor influencing cell growth and metabolism, especially for marine heterotrophic strains, as a stable culture environment is essential for optimal microbial performance [102]. Inorganic salts play a crucial role in regulating osmotic pressure and maintaining cellular integrity by serving as nutrients and enzyme activators. Excessive osmotic stress can disrupt cellular metabolism: hypertonic conditions may cause cell death due to water loss, while hypotonic conditions can lead to cell swelling and potential lysis [103]. For example, under hypertonic conditions, *Schizochytrium* sp. synthesizes compatible solutes like cyclohexanol and betaine to maintain metabolic functions [104], and *Chlamydomonas reinhardtii* forms irregular "ghost" cells under high osmotic pressure [105]. Additionally, mineral elements such as Na^+, K^+, Ca^{2+}, and Cl^- often remain in the fermentation broth after DHA production, helping to stabilize osmotic pressure and support marine microbial growth [86].

Osmotic pressure also affects metabolite production, with some microorganisms accumulating specific metabolites under certain osmotic conditions. For instance, *Torulopsis glabrata* accumulates pyruvate under hypertonic conditions [106], while *Thraustochytrium* sp. shows a reduction in oil content without inorganic salts [107]. The stage of marine microbial growth and oil accumulation can also influence the effect of osmotic pressure. High osmotic conditions favor seed proliferation, while low osmotic conditions promote PUFA accumulation [108]. Strategies that manipulate osmotic pressure during fermentation, such as partial reuse of fermentation wastewater, can optimize biomass accumulation and DHA productivity [84]. Additionally, substrate concentration also plays a role in maintaining osmotic balance, with continuous feeding strategies helping to sustain microbial physiological conditions and improve DHA yields [102].

Osmotic pressure, influenced by inorganic salts and mineral elements, is essential for maintaining cell integrity and supporting microbial metabolism, especially during DHA fermentation. Adjusting osmotic pressure at different stages of marine microbial growth and oil accumulation can enhance DHA productivity, making osmotic regulation an important factor in optimizing fermentation processes.

3.5. Two-Stage Regulation Strategy for DHA Fermentation Processes

As discussed, microbial DHA oil synthesis occurs in distinct phases, with cell biomass proliferation and lipid accumulation taking place at different stages of fermentation [109,110]. During the initial stage, when nutrients are abundant, cells prioritize division and growth, leading to an increase in non-lipid biomass. As nutrients, particularly nitrogen sources, become depleted, the cells shift their metabolic focus. The carbon source is redirected toward lipid synthesis, initiating the oil accumulation phase. During this phase, cell proliferation halts, and lipid accumulation becomes the primary metabolic activity, with non-lipid biomass levels remaining relatively constant [111,112]. The asynchronous mechanism of biomass growth and lipid accumulation provides a theoretical basis for the development of a two-stage regulation strategy.

3.5.1. Two-Stage Dissolved Oxygen Regulation

The two-stage dissolved oxygen regulation strategy is designed to optimize the growth and oil accumulation phases of microbial fermentation, addressing the high aerobic oxygen demand during the initial growth phase and the relatively anaerobic conditions needed for DHA accumulation in the later stages. Qu et al. used the oxygen transfer coefficient (KLa) as a benchmark for oxygen supply, maintaining KLa at 150.1 h^{-1} for the first 40 h of fermentation, then reducing it to 88.5 h^{-1} for the remainder of the process. This two-stage KLa strategy resulted in 43.83% and 63.88% higher biomass and DHA content, respectively, compared to constant KLa conditions [109]. Zhang et al. addressed the insufficient aeration in traditional shake flasks by introducing aeration membranes and adjusting the shaking table speed. This modification increased DHA production efficiency by 60% and shortened the fermentation cycle [113]. Zhao et al. designed various agitator combinations to enhance dissolved oxygen and mixing, and their computational fluid dynamics (CFD) simulations revealed that a configuration with straight-blade, arrow-blade, and flat-blade impellers was most effective for both cell growth and DHA accumulation [114]. Guo et al. developed a membrane-pore-material agitator to enhance the KLa value and gas content, incorporating a two-stage ventilation strategy to achieve higher biomass and DHA yields—87.34% and 83.77% higher than conventional reactors [111]. Additionally, optimizing bubble size in bubble column bioreactors could replicate the benefits of a two-stage oxygen supply strategy, improving both cell growth and DHA production [115].

3.5.2. Two-Stage Temperature Regulation

Temperature plays a crucial role in cell growth and metabolism, especially in oil-accumulating microorganisms. While low temperatures promote the synthesis of polyunsaturated fatty acids (PUFAs) by maintaining cellular fluidity, they can also slow down enzymatic activity, which may reduce the rate of oil accumulation. Conversely, higher temperatures favor cell growth and biomass accumulation but may reduce PUFA synthesis [39,116]. Such as a temperature of 30 °C during the cell growth phase, followed by a reduction to 20 °C for the DHA accumulation phase, led to a DHA content of 52% of the total fatty acids [117].

3.5.3. Two-Stage Substrate Regulation

Substrate selection is crucial for both cell growth and metabolic flux partitioning, with nitrogen limitation being a key factor in driving cellular metabolism toward lipid accumulation. A higher carbon-to-nitrogen (C/N) ratio in the culture medium is favorable for triacylglycerol biosynthesis in production phases. In a two-stage fermentation strategy, the first stage supports biomass growth without nutrient restrictions, while the second stage introduces nitrogen limitation to promote lipid accumulation. For example, Silvina et al. used a two-stage medium strategy for *Aurantiochytrium limacinum*, where the cells were first cultured in a medium optimized for biomass production, followed by transfer to a high C/N ratio medium (55:1) for lipid accumulation. This strategy resulted in a DHA productivity of 3.7 g/(L·day) [118]. Li et al. also explored the use of glucose and glycerol as carbon sources for DHA fermentation by *Schizochytrium* sp. and found that glucose supported biomass growth, while glycerol promoted DHA accumulation [40]. Based on this, they implemented a two-stage substrate fermentation strategy where glucose was used for biomass growth and glycerol for DHA synthesis, improving both production efficiency and reducing costs. In addition, Yin et al. applied this two-stage strategy using glucose for biomass accumulation in *Schizochytrium* sp., followed by hydrolyzed molasses for DHA production. This strategy increased the utilization of inexpensive molasses, enhancing DHA production efficiency while reducing raw material costs [73]. Furthermore, the two-stage feeding control strategy has also been successfully applied using alternative nitrogen sources, such as waste *Pichia pastoris*, in industrial DHA fermentation [119].

In summary, during marine microbial DHA fermentation, selecting suitable low-cost raw materials (such as agricultural waste or industrial by-products) to replace traditional expensive substrates can significantly reduce the economic cost of fermentation while promoting resource recycling and sustainable development. Meanwhile, optimizing key parameters in the fermentation process (such as temperature, pH, dissolved oxygen levels, and nutrient composition) and precisely controlling the fermentation conditions can enhance cell growth and DHA accumulation efficiency. This strategy provides essential technical support for achieving large-scale and cost-effective production of marine microbial-sourced DHA oil (Figure 1).

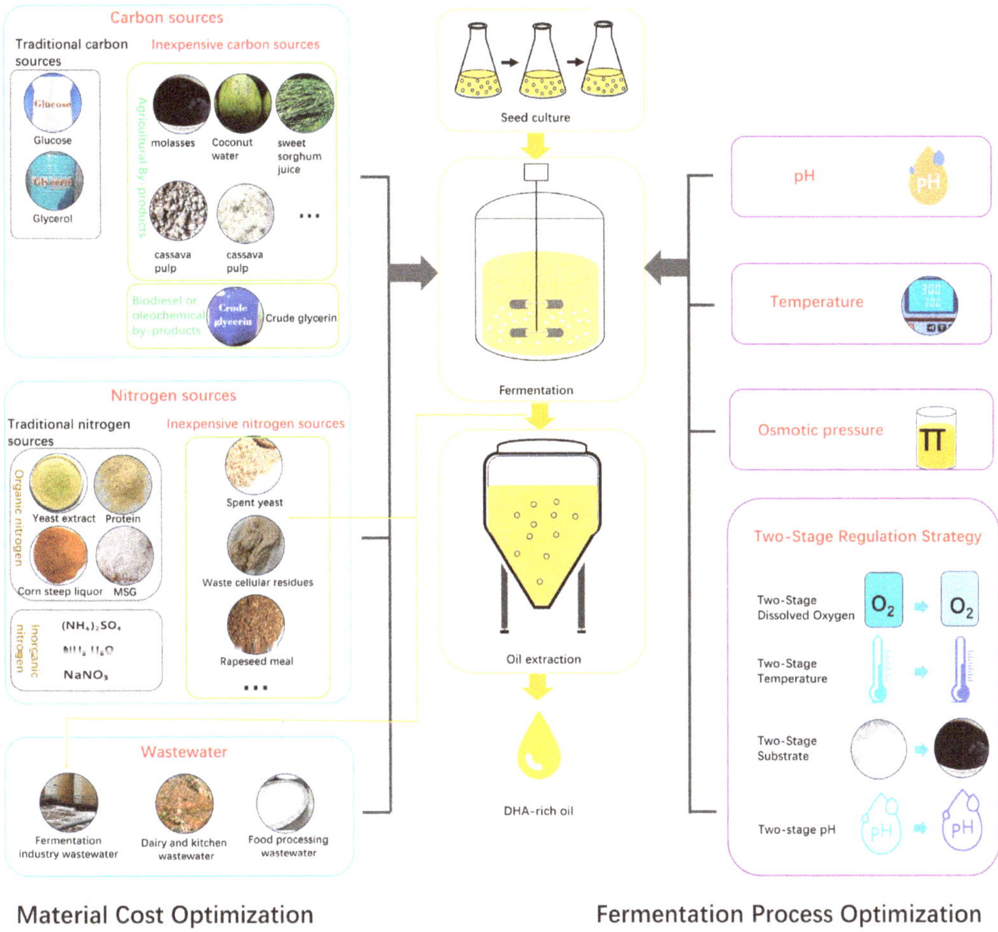

Figure 1. Schematic of marine microbial DHA fermentation: Cost-control and efficiency-enhancement strategies.

4. Post-Treatment of Marine Microbial DHA Oils

Marine microbial-sourced DHA oil exists both within the cell membrane as a functional lipid and in the cytoplasm as a storage lipid. To obtain DHA oil, several post-processing steps are necessary, including cell disruption and oil extraction. The resulting crude microbial oil contains various impurities such as pigments, proteins, and colloidal substances. Furthermore, during the extraction process, oxidation of certain lipids can lead to the formation of undesirable compounds, which can negatively impact the sensory qualities and fluidity of the DHA oil. Consequently, after the initial extraction, further refining treatments are required to remove these impurities and enhance the overall quality of the DHA oil.

4.1. Extraction of Oils

4.1.1. Cell Lysis

DHA oil resides within microbial cells, necessitating cell disruption before oil extraction. Common methods for cell lysis include chemical, mechanical, and enzymatic approaches.

- Chemical lysis: This method uses strong acids or bases to break down the cell wall by dissolving glycoproteins, cellulose, and other structural components, thereby releasing the oil. While chemical lysis is relatively simple and does not require extensive sample pretreatment, the hot acid method is ineffective for extracting oils from cell membranes. Moreover, the use of acids and alkalis is highly corrosive and poses significant safety risks, making it unsuitable for large-scale production.
- Mechanical lysis: This technique utilizes high shear force from a high-pressure homogenizer to rupture cell walls and membrane components. Unlike chemical methods, mechanical lysis does not introduce chemicals, minimizing the risk of damaging the oil components and improving extraction efficiency. However, this method is energy-intensive and involves complex operational procedures, leading to higher labor and energy costs.
- Enzymatic lysis: This method uses specific enzymes tailored to the composition of the cell wall. The enzymes interact with the structural components of the cell wall and membrane, breaking them down and releasing the intracellular contents. Due to the complexity of the cell wall structures, a combination of enzymes is often required to achieve optimal lysis. With advancements in biotechnological enzymes and the increasing demand for efficient production methods, enzymatic lysis has become the predominant technique for disrupting DHA-producing microbial cells [120–122].

4.1.2. Oils Extraction

Oil extraction from microbial biomass is commonly carried out using organic solvents due to the fat-soluble nature of oils. Organic solvents like n-hexane, cyclohexane, acetone, and chloroform are widely used for their ability to dissolve oils efficiently, as they are insoluble in water and have low boiling points that facilitate easy solvent removal through distillation [123,124]. However, organic solvent extraction has significant drawbacks. One of the primary concerns is the potential for solvent residues to remain in the extracted oil, which can affect oil quality and complicate the reuse of microbial residue. Furthermore, the flammability and volatility of organic solvents necessitate stringent safety measures during the extraction process, posing environmental and health risks to operators. These factors make the process less sustainable and raise concerns about its long-term viability.

To overcome these issues, supercritical fluid extraction, especially using supercritical carbon dioxide (CO_2), has emerged as a promising alternative [125,126]. This method offers high purity and rapid separation, without the risk of solvent residues, making it an attractive choice for food-grade oil extraction [127,128]. Supercritical CO_2 is particularly beneficial due to its ease of removal after extraction. However, the method's sensitivity to the moisture content of the sample presents a challenge, as high-water content can hinder the extraction efficiency, requiring pre-drying of the biomass. Despite its advantages, supercritical fluid extraction is not yet widely adopted for marine microbial DHA extraction on an industrial scale. The high cost of equipment and operational complexities limit its application from laboratory production to industrial scale [124]. Consequently, researchers and manufacturers are increasingly exploring physical extraction methods, such as centrifugation, which offer safer and more efficient alternatives [83]. Marine microbial DHA-rich oil, composed of various fatty acids including palmitic acid (C16:0), docosapentaenoic acid (DPA, C22:5), and DHA, has a lower density than water due to the large molecular spacing between these fatty acids, making it well-suited for separation via centrifugation.

The three-phase centrifuge is a particularly promising method for extracting microbial oil. It efficiently separates the oil, water, and solid phases into distinct layers, allowing for continuous and effective oil-water separation with minimal environmental impact, providing a safer, more sustainable extraction option [83,129]. The three-phase centrifuge enables continuous, high-efficiency DHA oil separation from microbial biomass, operating without interruptions to simultaneously isolate oil, nutrient-rich wastewater, and solid residues. The wastewater can be reused in fermentation media [83,84], while residues are repurposed as organic nitrogen sources [81]. This closed-loop process supports sustainable biomanufacturing by minimizing waste and maximizing resource recovery.

4.2. Oils Refining

Crude microbial DHA oil, after extraction, contains various impurities such as proteins, phospholipids, free fatty acids, and pigments, which can affect its safety, stability, and suitability for further processing. To improve the quality, appearance, and shelf life of the oil, refining processes are used to remove these contaminants. Oil refining typically involves several sequential steps, either carried out continuously or batchwise:

- Degumming: This step removes phospholipids and other gum-forming agents that can affect the oil's clarity and quality.
- Neutralization: Free fatty acids are removed, which helps reduce acidity and improve the oil's stability.
- Decolorization: Pigments and certain other contaminants are removed, improving the oil's color and purity.
- Deodorization: Volatile compounds responsible for off-flavors and undesirable odors are eliminated to improve the oil's sensory characteristics.

To enhance the oil's fluidity and expand its potential applications in the healthcare and pharmaceutical industries, winterization is typically performed before deodorization. Winterization helps remove certain saturated fatty acids, ensuring the oil remains fluid even at lower temperatures.

4.2.1. Degumming

Colloidal impurities in crude microbial DHA oils (phospholipids, proteins, glyceryl esters) compromise oil stability and complicate subsequent refining steps—such as causing emulsification during alkali refining or increasing decolorizing agent consumption. Thus, degumming acts as the first critical step in oil refining to address these issues [130].

The most common industrial degumming process involves adding salt water or phosphoric acid under heat: this hydrates and coagulates colloidal substances, which are then removed via sedimentation and separation. In recent years, more efficient technologies have emerged to overcome limitations of traditional methods. One is enzymatic degumming, which uses phospholipases to hydrolyze non-hydrated phospholipids in crude oil; released fatty acids enhance phospholipids' hydrophilicity, enabling their transfer to the aqueous phase for separation, and this environmentally friendly approach also offers operational simplicity [131,132]. And several commercial phospholipases, such as Lecitase Ultra, GumZyme, and Lysomax, have been successfully developed and applied [130]. Another is ultrafiltration membrane separation: while phospholipids and glyceryl esters (with similar molecular weights of 700–900 Da) are difficult to separate with conventional membranes, phospholipids' amphiphilic nature allows them to form reverse micelles ($\geq$2000 Da) in oil-water mixtures—these micelles enable specialized membranes to achieve efficient phospholipid removal, with efficiency reaching up to 89% [133,134].

4.2.2. Neutralization

Free fatty acids (FFAs) in crude microbial DHA oils increase acidity and trigger oxidative rancidity, generating harmful compounds and degrading oil quality. Thus, alkali refining acts as a critical deacidification step to address these issues.

The process involves adding alkaline agents (e.g., NaOH, KOH, Na_2CO_3, LiOH, $Ca(OH)_2$, $CaCO_3$, and NH_4OH) to neutralize FFAs [134], with NaOH the most widely used. When NaOH reacts with FFAs, water-soluble sodium soaps form; these soaps are then removed by water washing, reducing acidity and the risk of rancidity. Precise alkali dosage is vital: insufficient alkali leaves residual FFAs (keeping the oil acidic), while excess alkali triggers saponification of neutral oils (increasing oil loss). Thus, accurate acid value determination is essential to calculate the optimal alkali amount. When properly controlled, alkali refining enhances oil stability, shelf life, and quality by neutralizing FFAs and inhibiting rancidity.

Beyond traditional alkaline methods, enzymatic deacidification offers an alternative [130], such as lipases (e.g., Novozym 435) selectively removing FFAs under mild conditions while minimizing DHA loss [135].

4.2.3. Decolorization

Crude microbial DHA oil appears dark yellow after extraction, primarily due to pigments, notably carotenoids, the dominant pigments in marine microbial-sourced DHA [136]. These pigments degrade the oil's appearance and functionality, so decolorization is essential to improve visual and functional quality.

Decolorization employs adsorbents (e.g., activated carbon, activated clay) to remove pigments, reduce color intensity, and eliminate impurities like metal ions and saponins. Choosing the right adsorbent is key to achieving effective decolorization [129,137]. Activated carbon (with a large surface area and micropores) effectively removes pigments but causes high oil loss due to strong oil adsorption. Activated clay (often derived from bentonite) offers high adsorptivity and chemical activity, enabling efficient decolorization with lower oil loss and filter residue. In some cases, combining multiple decolorants enhances pigment removal [138]. However, decolorization may deplete beneficial bioactives (e.g., vitamin E) and increase peroxide value [139]; thus, strict process control is vital to balance visual improvement with minimizing nutrient loss and preserving oil quality.

4.2.4. Deodorization

Deodorization acts as the final critical step in oil refining, aiming to remove volatile odor-causing compounds (e.g., aldehydes, ketones, and free fatty acids) via steam distillation. This process not only eliminates unwanted flavors and odors but also enhances the oil's stability, quality, and sensory properties [139]. Key parameters like temperature, pressure, duration, and stripping steam volume shape the process outcomes with inherent trade-offs: higher temperatures boost the removal of FFAs but increase the risk of trans fatty acid formation, which compromises the oil's nutritional quality [140]; excessive stripping steam improves impurity removal yet leads to greater oil losses. For batch deodorization, typical conditions are 230–260 °C and 3–5 mbar, with the steam amount ranging from 5 to 15% of the oil's weight. In continuous or semi-continuous systems, the steam volume is reduced to 0.5–2% of the oil's weight to minimize losses [141].

The effectiveness of deodorization is often measured through the p-anisidine value, which quantifies secondary oxidation products like aldehydes and ketones [142]. Elevated levels of these compounds can adversely affect health by potentially raising blood pressure, interfering with the absorption of fat-soluble vitamins, and, in some cases, having carcinogenic properties. China's DHA Algal Oil LS/T 3243-2015 requires the p-anisidine value to not exceed 15, and the industry often enforces even stricter standards. Case studies, like Yin et al.'s use of deoxygenated steam as a stripping gas, achieved a peroxide value of 0 meq/kg, a p-anisidine value of 3.50, and an acid value of 0.37 mg/g, producing DHA-rich oil that met high-quality market standards [15].

Beyond traditional steam distillation, innovations include using nitrogen as a stripping gas, which preserves heat-sensitive omega–3 fatty acids and reduces oxidative degradation [143], and nanofiltration membrane technology that can selectively remove volatile odorants under mild conditions, reducing thermal degradation compared to conventional high-temperature deodorization [135].

In summary, the post-treatment of marine microbial-sourced DHA oils is a critical chain for converting biomass into high-quality DHA-enriched products. Table 3 summarizes the refining processes and the specific impurities they target, visually illustrating the pathways for quality improvement. However, it should be noticed that most studies in this field currently prioritize technical performance indicators (e.g., extraction yield, purification efficiency, product quality) over economic evaluations across the entire post-treatment chain. Systematic cost data—which should cover raw material costs, equipment amortization, energy consumption, and other dimensions of multiple unit operations—are rarely reported in the literature. Without a unified quantitative model or validated data from industrial-scale practices, accurate cost calculations would remain speculative. Given this context, this section only focuses on the technical principles, key parameter optimization, and industrialization bottlenecks of each post-treatment unit (extraction and refining processes), aiming to provide a foundation for subsequent economic analyses and industrial-scale cost modeling in this domain.

Table 3. Oil refining processes and the removal of relevant impurities.

Oil Refining Processes	Major Impurity Components
Degumming	Phospholipids, proteins
Neutralization	Free fatty acids, phospholipids, metal ions, soap stock
Decolorization	Pigment, metal ions, and soap stock
Deodorization	Secondary oxidation products, free fatty acids, pigments, sterols, and squalene
Winterization	Saturated fatty acids

5. Antioxidant Technology of Marine Microbial DHA Oils

5.1. Oxidation Processes in PUFA-Rich Oils

Oils enriched with polyunsaturated fatty acids (PUFAs) like DHA, are inherently vulnerable to oxidation during processing and storage. Exposure to light, heat, and oxygen triggers a series of oxidative reactions that not only reduce shelf life via rancidity development but also generate toxic by-products with potential health implications. The susceptibility of PUFAs to oxidation stems from the instability of their multiple double bonds, which render them highly reactive toward environmental stressors (e.g., light, heat, oxygen).

5.1.1. Autoxidation of Oils

Autoxidation, a free-radical-mediated chain reaction, unfolds in three sequential stages: initiation, propagation, and termination.

Initiation commences with the abstraction of hydrogen atoms from unsaturated fatty acids or glycerides, yielding lipid radicals. Light, heat, and transition metals accelerate this step, with hydrogen atoms adjacent to double bonds (especially conjugated double bonds) being preferentially abstracted due to their lower bond energy.

Propagation stage: In this stage, lipid radicals react with molecular oxygen to form peroxyl radicals. These peroxyl radicals act as active chain carriers in the free radical chain reaction. They can attack new lipid molecules, leading to the formation of hydroperoxides and the generation of additional free radicals. The oxidation process continues as long as there is a source of hydrogen or until the chain reaction is interrupted by antioxidants or other inhibitors [144].

Termination neutralizes free radicals through radical–radical coupling or other interactions, forming stable non-radical products (e.g., free-radical polymers) and halting the chain reaction.

5.1.2. Photo-Oxidation of Oils

Photo-oxidation diverges from autoxidation as it does not rely on free-radical initiation. Instead, triplet oxygen (in an excited state) directly abstracts hydrogen from unsaturated fatty acids, driving lipid oxidation. This process occurs via two mechanism-dependent pathways.

Type I (low oxygen concentration): Excited photosensitizers independently extract hydrogen from fatty acids, generating lipid free radicals analogous to those in autoxidation [145]. The excited photosensitizers then return to the ground state by reacting with these radicals, sustaining the oxidative process.

Type II (sufficient oxygen): Photosensitizers transfer energy to molecular oxygen, elevating it to the singlet state ($^1O_2$). This highly reactive $^1O_2$ attacks electron-rich double bonds in unsaturated fatty acids at a rate ~1500-fold faster than ground-state oxygen, forming hydroperoxides [146].

5.2. Antioxidants for Marine Microbial DHA Oils

Oxidation not only shortens the shelf life of marine microbial-derived DHA-rich oils but also compromises their safety and nutritional quality by generating harmful by-products. Antioxidants are thus indispensable for inhibiting oxidation, extending shelf life, and preserving nutrient integrity. By definition, antioxidants interrupt oxidative chain reactions either by suppressing lipid free-radical formation or by reducing their concentration. Given DHA's six unsaturated double bonds, hydrogen atoms adjacent to these bonds are particularly prone to abstraction, making DHA-rich oils exceptionally susceptible to oxidation [147] and necessitating robust antioxidant protection during storage.

Phenolic antioxidants, a prominent class of inhibitors, exert their effect by donating hydrogen atoms to lipid radicals, terminating the oxidative chain reaction. The resultant antioxidant radicals exhibit low reactivity and do not propagate new radicals; moreover, they can form stable non-radical products via interaction with lipid radicals, further quenching oxidation.

5.3. Types of Antioxidants

5.3.1. Synthetic Antioxidants

Synthetic antioxidants are chemically engineered to inhibit oxidation at low concentrations (<200 ppm). Common examples include tert-butylhydroquinone (TBHQ), butylated hydroxytoluene (BHT), butylated hydroxyanisole (BHA), and ethylenediaminetetraacetic acid (EDTA).

TBHQ, a polar phenolic compound, demonstrates potent antioxidant activity—especially during the early stages of oxidation—and effectively extends the shelf life of DHA-rich fish oils, outperforming other synthetic analogs. EDTA, originally a chelating agent, mitigates oxidation by sequestering metal ions that catalyze hydroperoxide decomposition, thus preventing secondary oxidation. Despite their efficacy, synthetic antioxidants face scrutiny over potential health risks, driving consumer demand for natural alternatives [148].

5.3.2. Natural Antioxidants

Natural antioxidants, abundant in plants, marine organisms, and microorganisms, protect host organisms from oxidative damage and offer applications in food preservation and nutraceuticals. Table 4 provides examples of natural antioxidants and their sources, highlighting the diversity of antioxidants found in nature and their broad potential applications.

Table 4. Natural antioxidants and their sources.

Antioxidants	Examples	Sources
Tocopherol	α-, β-, γ-, δ-tocopherol	Seeds, grains, nuts, vegetable oils, etc.
Trienyltocopherol	α-, β-, γ-, δ- triene tocopherols	Palm oil, rice bran oil
Ascorbic acid	Vitamin C, ascorbate derivatives	Fruits, vegetables, etc.
Carotenoids	β-carotene, lycopene, lutein, astaxanthin	Carrots, tomatoes, microalgae, etc.
Phenols	Flavonoids, phenolic acids, tannins, lignans	Fruits, vegetables, grains, etc.
Peptides	Glutathione, metallothioneins, antioxidant peptides	Animal liver, eggs, milk, etc.
Enzymes	Superoxide dismutase, catalase, glutathione peroxidase	Plant and animal tissues

- Tocopherol: A fat-soluble compound, tocopherol enhances DHA oil stability [149], though its efficacy is concentration-dependent. Structural isomers (α-, γ-, δ-tocopherols) exhibit varying activities; while tocopherol scavenges singlet oxygen and donates hydrogen to lipid radicals, excessive concentrations (>740 mg/kg) can paradoxically promote oxidation (e.g., by generating free radicals during decomposition) [150]).
- Rosemary extract: Derived from Rosmarinus officinalis, rosemary extract contains bioactive compounds (e.g., rosmarinic acid, carnosic acid, carnosol) and exists as water- or fat-soluble fractions. It is widely used in food preservation for its strong antioxidant properties [151,152], where a 50 mg/kg dose effectively controls oxidation rates [153], and it preserves the sensory quality of sardine oil during cold storage [154].
- Ascorbic acid: A water-soluble antioxidant, ascorbic acid neutralizes reactive oxygen species (ROS) via redox reactions [155] but exhibits limited solubility in oils. Esterification with fatty acids (e.g., ascorbyl palmitate) improves lipid compatibility while retaining antioxidant activity, making it suitable for DHA oil systems.
- Tea polyphenols: Extracted from tea leaves, tea polyphenols (e.g., catechins) are water-soluble and exhibit strong antioxidant activity. O'Sullivan et al. showed that 0.04% (*w*/*w*) tea polyphenols inhibit thermal degradation in frying oil [156]. Recent studies have further validated the antioxidant properties of tea polyphenols and demonstrated their successful application in protecting DHA algae oil [12,157,158].

5.4. Synergistic Effects of Antioxidants

While individual antioxidants target specific oxidation components or stages, their efficacy is often limited in isolation due to rapid activity loss. Combining antioxidants thus emerges as a strategic approach: synergistic interactions enhance overall antioxidant capacity, reduce dosage requirements, and lower costs [159]. These interactions amplify protective effects across multiple oxidative pathways and stages, providing comprehensive defense against oxidation [13].

5.4.1. Vitamins and Polyphenols

Polyphenols exhibit amphiphilic properties, allowing them to dissolve in both water and oil, which enhances their ability to interact effectively with vitamin antioxidants in complex systems. Phenolic compounds with redox activity can also facilitate the regeneration of vitamins. For example, research by Dai et al. demonstrated that polyphenols could regenerate vitamin C [160]. In a model system using linoleic acid methyl ester peroxides, the flavonoid quercetin in catechins exhibited a synergistic effect with α-tocopherol by preventing chain oxidation of oils and regenerating α-tocopherol [161]. Specific bioactive compounds in rosemary extract (e.g., phenolic acids, diterpenoids), acting as hydrogen donors, can donate a hydrogen atom to the α-tocopherol radical, converting it back to its active form. Similarly, ascorbic acid can recycle oxidized α- and γ-tocopherol radicals back to their native state through direct reduction, thereby delaying tocopherol depletion [162]. The combination of green tea polyphenols (GTP), α-tocopherol, and ascorbate demonstrated cooperative inhibition of lipid peroxidation via a sequential redox cycling mechanism: GTPs reduced α-tocopheroxyl radicals to regenerate α-tocopherol, while ascorbate subsequently restored GTPs' antioxidant capacity by neutralizing their oxidized derivatives [163].

A study on various phenolic compounds (e.g., chlorogenic acid, gallic acid, protocatechuic acid, and vanillic acid) in mango pulp revealed that most combinations, including all four compounds, exhibited significant synergistic effects [164]. Additionally, quercetin and resveratrol have also been proven to have a synergistic effect [165].

5.4.2. Vitamins and Carotenoids

Carotenoids are effective quenchers of singlet oxygen, a highly reactive form of oxygen. They achieve this by absorbing excess energy and returning singlet oxygen to its ground state [147]. Marine microbial-sourced DHA oils naturally contains carotenoids, which are primary pigments contributing to the oil's color [31,166]. Early studies have shown that β-carotene and α-tocopherol can offer mutual protection against the formation of linoleic acid peroxides, effectively preventing oxidation [167]. Carotenoids naturally present in palm oil could protect tocopherols from decomposition at high temperatures by undergoing their own oxidation. In turn, tocopherols can reduce some carotenoid free radicals back to their active forms [168]. However, carotenoids present a challenge in practical applications due to their dark color, which contradicts the goal of oil decolorization. As a result, carotenoids are typically removed during the decolorization process of DHA oil, limiting their use as antioxidants in this context.

5.4.3. Application of Synergistic Antioxidants in DHA Oils

DHA oil, with its unique structure featuring six unsaturated double bonds, is highly susceptible to oxidation, influenced by both intrinsic and extrinsic factors. Therefore, selecting an appropriate antioxidant system is crucial to preserving the oil's stability. Due to the synergistic effects and concentration dependencies among antioxidants, selecting the most effective antioxidant combination is not straightforward. Optimizing the combination and dosage often requires methodologies such as response surface methodology [157, 169]. For instance, Shen et al. found that a combination of 80 mg/kg ascorbyl palmitate, 80 mg/kg vitamin E, 40 mg/kg phytic acid, and 80 mg/kg tea polyphenols provided better antioxidant protection than each antioxidant used individually, extending the shelf life of DHA-rich algal oils from 7.5 to 28.2 days [13]. Similarly, a combination of 53.20 mg/kg octyl gallate and 360 mg/kg tea polyphenol palmitate resulted in the highest antioxidant capacity, extending the shelf life of DHA algae oil by a factor of 4.24 [170]. In another study, Yin et al. optimized an antioxidant combination consisting of 0.0259% rosemary extract, 0.0224% vitamin E, and 0.0166% ascorbyl palmitate, which extended the oxidation induction time of marine microbial DHA-rich oils to 20.21–10.47 days longer than the control group—thus enhancing the antioxidant capacity and prolonging the shelf life of DHA oil [157]. While a study has suggested that the most effective protection for DHA algae oil involves four antioxidants, consisting of 80 mg/kg ascorbyl palmitate, 80 mg/kg vitamin E, 40 mg/kg phytic acid, and 80 mg/kg tea polyphenols [13].

In summary, crude marine microbial DHA oil obtained after oil extraction can undergo a series of refining processes, such as degumming, neutralization, decolorization, and deodorization, to remove impurities and achieve high-purity, high-quality DHA oil. Furthermore, the refined oil can be enhanced by adding appropriate antioxidants, such as vitamin E, polyphenols, or natural plant extracts, to effectively increase the oil's antioxidant activity, reduce oxidation, and thus extend its shelf life (Figure 2). These measures not only enhance the quality of marine microbial-sourced DHA oil, but also broaden its applications across diverse sectors, including food and nutritional fortification, pharmaceuticals and healthcare, and animal feed and aquaculture industries. This progress aligns with the market demand for high-quality, stable DHA oils with extended shelf life.

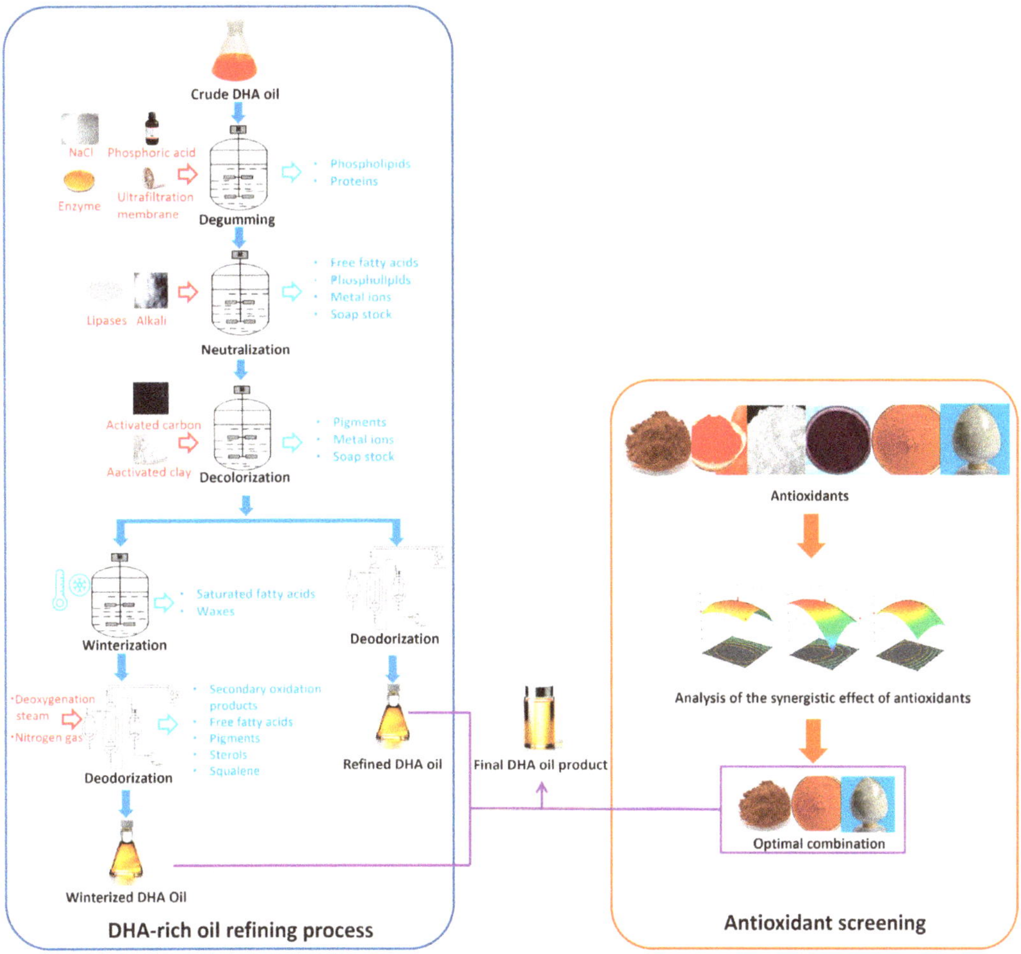

Figure 2. Quality enhancement framework for marine microbial DHA oil: Refining workflows and antioxidant selection.

6. Conclusions

The industrial production of marine microbial DHA oil has achieved significant advancements, particularly in enhancing production efficiency and improving oil quality. The shift toward utilizing heterotrophic microorganisms as a primary source of DHA presents a promising avenue for reducing production costs, primarily through the adoption of low-cost substrates and the optimization of fermentation processes. These developments not only lower the economic barriers to DHA production but also align with the broader goals of sustainable and environmentally friendly manufacturing practices. Ensuring the high quality of marine microbial-sourced DHA oil (commercially designated as algal oils) remains a critical priority, particularly in terms of its purity, stability, and nutritional value. Meeting these standards is essential for addressing market demands and maintaining consumer trust in the product. To this end, the implementation of robust quality preservation strategies, such as antioxidant treatments to mitigate oxidation, has proven effective. These measures are vital for extending the shelf life and maintaining the functional integrity of DHA oil, thereby enhancing its applicability in various industries, including food, pharmaceuticals, and nutraceuticals. Looking ahead, future research should focus on advancing the scalability of marine microbial DHA production processes to meet the escalating global demand for high-quality, cost-effective DHA oil. This includes the exploration of novel biotechnological approaches, such as genetic engineering and metabolic pathway optimization, to further enhance marine microbial productivity and DHA yield. Additionally, the development of innovative downstream processing techniques will be crucial for improving extraction efficiency and reducing energy consumption. Moreover, the integration of circular economy principles, such as the utilization of waste streams as substrates and the recycling of by-products, could further enhance the sustainability of marine microbial DHA production. Collaborative efforts among researchers, industry stakeholders, and policymakers will be essential to drive these innovations and ensure the widespread adoption of marine microbial DHA as a viable and sustainable alternative to traditional fish oil sources.

Author Contributions: F.Y.: conceptualization, methodology, investigation, writing—original; X.S.: formal analysis, data curation, funding acquisition; X.L.: validation, methodology; W.Z.: visualization, investigation, methodology; L.Y.: visualization, investigation; Y.Z.: formal analysis, investigation; Y.F.: supervision, project administration, writing—review and editing. All authors have read and agreed to the published version of the manuscript.

Funding: This work was supported by the Taizhou Science and Technology Plan Project (grant number: 23gyb07).

Institutional Review Board Statement: Not applicable.

Data Availability Statement: Not applicable.

Conflicts of Interest: The authors declare no conflicts of interest.

References

1. Liput, K.P.; Lepczyński, A.; Ogłuszka, M.; Nawrocka, A.; Poławska, E.; Grzesiak, A.; Ślaska, B.; Pareek, C.S.; Czarnik, U.; Pierzchała, M. Effects of Dietary n-3 and n-6 Polyunsaturated Fatty Acids in Inflammation and Cancerogenesis. *Int. J. Mol. Sci.* **2021**, *22*, 6965. [CrossRef] [PubMed]
2. Goff, Z.D.; Nissen, S.E. N-3 polyunsaturated fatty acids for cardiovascular risk. *Curr. Opin. Cardiol.* **2022**, *37*, 356–363. [CrossRef] [PubMed]
3. Bellou, S.; Triantaphyllidou, I.-E.; Aggeli, D.; Elazzazy, A.M.; Baeshen, M.N.; Aggelis, G. Microbial oils as food additives: Recent approaches for improving microbial oil production and its polyunsaturated fatty acid content. *Curr. Opin. Biotechnol.* **2016**, *37*, 24–35. [CrossRef] [PubMed]

4. Shah, A.M.; Yang, W.; Mohamed, H.; Zhang, Y.; Song, Y. Microbes: A Hidden Treasure of Polyunsaturated Fatty Acids. *Front. Nutr.* **2022**, *9*, 827837. [CrossRef]
5. Yan, C.X.; Zhang, Y.; Yang, W.Q.; Ma, W.; Sun, X.M.; Huang, H. Universal and unique strategies for the production of polyunsaturated fatty acids in industrial oleaginous microorganisms. *Biotechnol. Adv.* **2024**, *70*, 108298. [CrossRef]
6. Cardoso, C.; Afonso, C.; Bandarra, N.M. Dietary DHA and health: Cognitive function ageing. *Nutr. Res. Rev.* **2016**, *29*, 281–294. [CrossRef]
7. Díaz, M.; Mesa-Herrera, F.; Marín, R. DHA and Its Elaborated Modulation of Antioxidant Defenses of the Brain: Implications in Aging and AD Neurodegeneration. *Antioxidants* **2021**, *10*, 907. [CrossRef]
8. Li, J.; Pora, B.L.R.; Dong, K.; Hasjim, J. Health benefits of docosahexaenoic acid and its bioavailability: A review. *Food Sci. Nutr.* **2021**, *9*, 5229–5243. [CrossRef]
9. Santos, H.O.; Price, J.C.; Bueno, A.A. Beyond Fish Oil Supplementation: The Effects of Alternative Plant Sources of Omega-3 Polyunsaturated Fatty Acids upon Lipid Indexes and Cardiometabolic Biomarkers—An Overview. *Nutrients* **2020**, *12*, 3159. [CrossRef]
10. Xu, X.; Huang, C.; Xu, Z.; Xu, H.; Wang, Z.; Yu, X. The strategies to reduce cost and improve productivity in DHA production by *Aurantiochytrium* sp.: From biochemical to genetic respects. *Appl. Microbiol. Biotechnol.* **2020**, *104*, 9433–9447. [CrossRef]
11. Jesionowska, M.; Ovadia, J.; Hockemeyer, K.; Clews, A.C.; Xu, Y. EPA and DHA in microalgae: Health benefits, biosynthesis, and metabolic engineering advances. *J. Am. Oil Chem. Soc.* **2023**, *100*, 831–842. [CrossRef]
12. Kumari, A.; Pabbi, S.; Tyagi, A. Recent advances in enhancing the production of long chain omega-3 fatty acids in microalgae. *Crit. Rev. Food Sci. Nutr.* **2024**, *64*, 10564–10582. [CrossRef] [PubMed]
13. Shen, Y.; Lu, T.; Liu, X.Y.; Zhao, M.T.; Yin, F.W.; Rakariyatham, K.; Zhou, D.Y. Improving the oxidative stability and lengthening the shelf life of DHA algae oil with composite antioxidants. *Food Chem.* **2020**, *313*, 126139. [CrossRef] [PubMed]
14. Wang, J.; Han, L.; Wang, D.; Sun, Y.; Huang, J.; Shahidi, F. Stability and stabilization of omega-3 oils: A review. *Trends Food Sci. Technol.* **2021**, *118*, 17–35. [CrossRef]
15. Yin, F.; Sun, X.; Zheng, W.; Luo, X.; Peng, C.; Jia, Q.; Fu, Y. Improving the quality of microalgae DHA-rich oil in the deodorization process using deoxygenated steam. *J. Food Process. Preserv.* **2020**, *44*, e14602. [CrossRef]
16. Singer, P.; Richter, V.; Singer, K.; Loehlein, I. Analyses and Declarations of Omega-3 Fatty Acids in Canned Seafood May Help to Quantify Their Dietary Intake. *Nutrients* **2021**, *13*, 2970. [CrossRef]
17. Liu, Y.; Ren, X.; Fan, C.; Wu, W.; Zhang, W.; Wang, Y. Health Benefits, Food Applications, and Sustainability of Microalgae-Derived N-3 PUFA. *Foods* **2022**, *11*, 1883. [CrossRef]
18. Jiang, M.; Hu, Z.; Huang, Y.; Chen, X.D.; Wu, P. Impact of wall materials and DHA sources on the release, digestion and absorption of DHA microcapsules: Advancements, challenges and future directions. *Food Res. Int.* **2024**, *191*, 114646. [CrossRef]
19. Gladyshev, M.I.; Sushchik, N.N.; Makhutova, O.N. Production of EPA and DHA in aquatic ecosystems and their transfer to the land. *Prostaglandins Other Lipid Mediat.* **2013**, *107*, 117–126. [CrossRef]
20. Lenihan-Geels, G.; Bishop, K.S.; Ferguson, L.R. Alternative sources of omega-3 fats: Can we find a sustainable substitute for fish? *Nutrients* **2013**, *5*, 1301–1315. [CrossRef]
21. Magoni, C.; Bertacchi, S.; Giustra, C.M.; Guzzetti, L.; Cozza, R.; Ferrari, M.; Torelli, A.; Marieschi, M.; Porro, D.; Branduardi, P.; et al. Could microalgae be a strategic choice for responding to the demand for omega-3 fatty acids? A European perspective. *Trends Food Sci. Technol.* **2022**, *121*, 142–155. [CrossRef]
22. Veerasamy, V.; Neethirajan, V.; Singarayar, M.S.; Balasundaram, D.; Dharmar, P.; Thilagar, S. Microalgal biomass and lipid synergy for omega fatty acid enrichment: A sustainable source for food supplements & nutraceuticals. *Algal Res. Biomass Biofuels Bioprod.* **2024**, *80*, 103514. [CrossRef]
23. Zhang, H.; Xiangying, Z.; Jiaxiang, Z.; Liping, L.; Liu, J. Overview of Docosahexaenoic Acid (DHA) Biosynthesis by Aurantiochytrium: Biosynthetic Pathway, High-Yield Mechanism, and Metabolic Engineering Strategies. *Food Rev. Int.* **2025**, *41*, 883–921. [CrossRef]
24. Sharma, T.; Nisha, D.; Preeti, M.K.; Kumar, M.R.; Sudheer, P.; Kant, S.R.; Manish, K.; Abhishek, G.; Nayak, M. Microalgae as an emerging alternative raw material of docosahexaenoic acid and eicosapentaenoic acid—A review. *Crit. Rev. Food Sci. Nutr.* **2025**, *31*, 1–20. [CrossRef]
25. Barclay, W.; Meager, K.; Abril, J. Heterotrophic production of long chain omega-3 fatty acids utilizing algae and algae-like microorganisms. *J. Appl. Phycol.* **1994**, *6*, 123–129. [CrossRef]
26. Sun, X.M.; Xu, Y.S.; Huang, H. Thraustochytrid Cell Factories for Producing Lipid Compounds. *Trends Biotechnol.* **2021**, *39*, 648–650. [CrossRef]
27. Bajpai, P.; Bajpai, P.K.; Ward, O.P. Production of docosahexaenoic acid by *Thraustochytrium aureum*. *Appl. Microbiol. Biotechnol.* **1991**, *35*, 706–710. [CrossRef]
28. Barclay, W.R. Process for the Heterotrophic Production of Microbial Products with High Concentrations of Omega-3 Highly Unsaturated Fatty Acids. U.S. Patent 5,130,242, 14 July 1992.

29. Yokoyama, R.; Honda, D. Taxonomic rearrangement of the genus *Schizochytrium* sensu lato based on morphology, chemotaxonomic characteristics, and 18S rRNA gene phylogeny (Thraustochytriaceae, Labyrinthulomycetes): Emendation for *Schizochytrium* and erection of *Aurantiochytrium* and *Oblongichytrium* gen. nov. *Mycoscience* **2007**, *48*, 199–211. [CrossRef]
30. Lee Chang, K.J.; Nichols, C.M.; Blackburn, S.I.; Dunstan, G.A.; Koutoulis, A.; Nichols, P.D. Comparison of Thraustochytrids *Aurantiochytrium* sp., *Schizochytrium* sp., *Thraustochytrium* sp., and *Ulkenia* sp. for production of biodiesel, long-chain omega-3 oils, and exopolysaccharide. *Mar. Biotechnol.* **2014**, *16*, 396–411. [CrossRef]
31. Yin, F.W.; Zhan, C.T.; Huang, J.; Sun, X.L.; Yin, L.F.; Zheng, W.L.; Luo, X.; Zhang, Y.Y.; Fu, Y.Q. Efficient Co-production of Docosahexaenoic Acid Oil and Carotenoids in *Aurantiochytrium* sp. Using a Light Intensity Gradient Strategy. *Appl. Biochem. Biotechnol.* **2023**, *195*, 623–638. [CrossRef]
32. Han, X.; Li, Z.; Wen, Y.; Chen, Z. Overproduction of docosahexaenoic acid in *Schizochytrium* sp. through genetic engineering of oxidative stress defense pathways. *Biotechnol. Biofuels* **2021**, *14*, 70. [CrossRef] [PubMed]
33. Blasio, M.; Balzano, S. Fatty Acids Derivatives From Eukaryotic Microalgae, Pathways and Potential Applications. *Front. Microbiol.* **2021**, *12*, 718933. [CrossRef] [PubMed]
34. Metz, J.G.; Roessler, P.; Facciotti, D.; Levering, C.; Dittrich, F.; Lassner, M.; Valentine, R.; Lardizabal, K.; Domergue, F.; Yamada, A.; et al. Production of polyunsaturated fatty acids by polyketide synthases in both prokaryotes and eukaryotes. *Science* **2001**, *293*, 290–293. [CrossRef] [PubMed]
35. Li, J.; Han, D.; Wang, D.; Ning, K.; Jia, J.; Wei, L.; Jing, X.; Huang, S.; Chen, J.; Li, Y. Choreography of transcriptomes and lipidomes of Nannochloropsis reveals the mechanisms of oil synthesis in microalgae. *Plant Cell* **2014**, *26*, 1645–1665. [CrossRef]
36. Bellou, S.; Baeshen, M.N.; Elazzazy, A.M.; Aggeli, D.; Sayegh, F.; Aggelis, G. Microalgal lipids biochemistry and biotechnological perspectives. *Biotechnol. Adv.* **2014**, *32*, 1476–1493. [CrossRef]
37. Diao, J.; Li, X.; Pei, G.; Liu, L.; Chen, L. Comparative metabolomic analysis of *Crypthecodinium cohnii* in response to different dissolved oxygen levels during docosahexaenoic acid fermentation. *Biochem. Biophys. Res. Commun.* **2018**, *499*, 941–947. [CrossRef]
38. Ding, J.; Fu, Z.; Zhu, Y.; He, J.; Ma, L.; Bu, D. Enhancing docosahexaenoic acid production of *Schizochytrium* sp. by optimizing fermentation using central composite design. *BMC Biotechnol.* **2022**, *22*, 39. [CrossRef]
39. Hu, F.; Clevenger, A.L.; Zheng, P.; Huang, Q.; Wang, Z. Low-temperature effects on docosahexaenoic acid biosynthesis in *Schizochytrium* sp. TIO01 and its proposed underlying mechanism. *Biotechnol. Biofuels* **2020**, *13*, 172. [CrossRef]
40. Li, J.; Liu, R.; Chang, G.; Li, X.; Chang, M.; Liu, Y.; Jin, Q.; Wang, X. A strategy for the highly efficient production of docosahexaenoic acid by *Aurantiochytrium limacinum* SR21 using glucose and glycerol as the mixed carbon sources. *Bioresour. Technol.* **2015**, *177*, 51–57. [CrossRef]
41. Sahin, D.; Tas, E.; Altindag, U.H. Enhancement of docosahexaenoic acid (DHA) production from *Schizochytrium* sp. S31 using different growth medium conditions. *AMB Express* **2018**, *8*, 7. [CrossRef]
42. Yin, F.-W.; Zhang, Y.T.; Jiang, J.-Y.; Guo, D.-S.; Gao, S.; Gao, Z. Efficient docosahexaenoic acid production by *Schizochytrium* sp. via a two-phase pH control strategy using ammonia and citric acid as pH regulators. *Process Biochem.* **2019**, *77*, 1–7. [CrossRef]
43. Ma, W.; Zhang, Z.; Yang, W.; Huang, P.; Gu, Y.; Sun, X.; Huang, H. Enhanced docosahexaenoic acid production from cane molasses by engineered and adaptively evolved *Schizochytrium* sp. *Bioresour. Technol.* **2023**, *376*, 128833. [CrossRef] [PubMed]
44. Chang, G.; Luo, Z.; Gu, S.; Wu, Q.; Chang, M.; Wang, X. Fatty acid shifts and metabolic activity changes of *Schizochytrium* sp. S31 cultured on glycerol. *Bioresour. Technol.* **2013**, *142*, 255–260. [CrossRef] [PubMed]
45. de Swaaf, M.E.; Pronk, J.T.; Sijtsma, L. Fed-batch cultivation of the docosahexaenoic-acid-producing marine alga *Crypthecodinium cohnii* on ethanol. *Appl. Microbiol. Biotechnol.* **2003**, *61*, 40–43. [CrossRef]
46. Sun, L.; Ren, L.; Zhuang, X.; Ji, X.; Yan, J.; Huang, H. Differential effects of nutrient limitations on biochemical constituents and docosahexaenoic acid production of *Schizochytrium* sp. *Bioresour. Technol.* **2014**, *159*, 199–206. [CrossRef]
47. Li, M.; Liao, X.; Zhang, D.; Du, G.; Chen, J. Yeast extract promotes cell growth and induces production of polyvinyl alcohol-degrading enzymes. *Enzyme Res.* **2011**, *2011*, 179819. [CrossRef]
48. Ling, X.; Guo, J.; Liu, X.; Zhang, X.; Wang, N.; Lu, Y.; Ng, I.S. Impact of carbon and nitrogen feeding strategy on high production of biomass and docosahexaenoic acid (DHA) by *Schizochytrium* sp. LU310. *Bioresour. Technol.* **2015**, *184*, 139–147. [CrossRef]
49. Ren, L.J.; Sun, L.N.; Zhuang, X.Y.; Qu, L.; Ji, X.J.; Huang, H. Regulation of docosahexaenoic acid production by *Schizochytrium* sp.: Effect of nitrogen addition. *Bioprocess Biosyst. Eng.* **2014**, *37*, 865–872. [CrossRef]
50. Li, S.; Hu, Z.; Yang, X.; Li, Y. Effect of Nitrogen Sources on Omega-3 Polyunsaturated Fatty Acid Biosynthesis and Gene Expression in *Thraustochytriidae* sp. *Mar. Drugs* **2020**, *18*, 612. [CrossRef]
51. Ganuza, E.; Anderson, A.J.; Ratledge, C. High-cell-density cultivation of *Schizochytrium* sp. in an ammonium/pH-auxostat fed-batch system. *Biotechnol. Lett.* **2008**, *30*, 1559–1564. [CrossRef]
52. Xie, D.; Jackson, E.N.; Zhu, Q. Sustainable source of omega-3 eicosapentaenoic acid from metabolically engineered *Yarrowia lipolytica*: From fundamental research to commercial production. *Appl. Microbiol. Biotechnol.* **2015**, *99*, 1599–1610. [CrossRef] [PubMed]

53. Chauhan, A.S.; Chen, C.-W.; Tambat, V.S.; Singhania, R.R.; Chang, J.-S.; Dong, C.-D.; Patel, A.K. Bioprocess engineering to produce essential polyunsaturated fatty acids from *Thraustochytrium* sp. *Bioresour. Technol.* **2023**, *383*, 129209. [CrossRef] [PubMed]
54. Burja, A.M.; Radianingtyas, H.; Windust, A.; Barrow, C.J. Isolation and characterization of polyunsaturated fatty acid producing *Thraustochytrium* species: Screening of strains and optimization of omega-3 production. *Appl. Microbiol. Biotechnol.* **2006**, *72*, 1161–1169. [CrossRef] [PubMed]
55. Chauhan, A.S.; Patel, A.K.; Chen, C.-W.; Chang, J.-S.; Michaud, P.; Dong, C.-D.; Singhania, R.R. Enhanced production of high-value polyunsaturated fatty acids (PUFAs) from potential thraustochytrid *Aurantiochytrium* sp. *Bioresour. Technol.* **2023**, *370*, 128536. [CrossRef]
56. Trovao, M.; Pereira, H.; Costa, M.; Machado, A.; Barros, A.; Soares, M.; Carvalho, B.; Silva, J.T.; Varela, J.; Silva, J. Lab-Scale Optimization of *Aurantiochytrium* sp. Culture Medium for Improved Growth and DHA Production. *Appl. Sci.* **2020**, *10*, 2500. [CrossRef]
57. Chen, X.; He, Y.; Liu, L.; Zhu, X.; Sen, B.; Wang, G. Nitrogen Starvation Enhances the Production of Saturated and Unsaturated Fatty Acids in *Aurantiochytrium* sp. PKU#SW8 by Regulating Key Biosynthetic Genes. *Mar. Drugs* **2022**, *20*, 621. [CrossRef]
58. Nazir, Y.; Shuib, S.; Kalil, M.S.; Song, Y.; Hamid, A.A. Optimization of Culture Conditions for Enhanced Growth, Lipid and Docosahexaenoic Acid (DHA) Production of *Aurantiochytrium* SW1 by Response Surface Methodology. *Sci. Rep.* **2018**, *8*, 8909. [CrossRef]
59. Jiang, X.; Zhang, J.; Zhao, J.; Gao, Z.; Zhang, C.; Chen, M. Regulation of lipid accumulation in *Schizochytrium* sp ATCC 20888 in response to different nitrogen sources. *Eur. J. Lipid Sci. Technol.* **2017**, *119*, 1700025. [CrossRef]
60. Guo, D.-S.; Ji, X.-J.; Ren, L.-J.; Yin, F.-W.; Sun, X.-M.; Huang, H.; Zhen, G. Development of a multi-stage continuous fermentation strategy for docosahexaenoic acid production by *Schizochytrium* sp. *Bioresour. Technol.* **2018**, *269*, 32–39. [CrossRef]
61. Ju, J.-H.; Ko, D.-J.; Heo, S.-Y.; Lee, J.-J.; Kim, Y.-M.; Lee, B.-S.; Kim, M.-S.; Kim, C.-H.; Seo, J.-W.; Oh, B.-R. Regulation of lipid accumulation using nitrogen for microalgae lipid production in *Schizochytrium* sp. ABC101. *Renew. Energy* **2020**, *153*, 580–587. [CrossRef]
62. Fei, Q.; Chang, H.N.; Shang, L.; Choi, J.D.; Kim, N.; Kang, J. The effect of volatile fatty acids as a sole carbon source on lipid accumulation by *Cryptococcus albidus* for biodiesel production. *Bioresour. Technol.* **2011**, *102*, 2695–2701. [CrossRef] [PubMed]
63. Aini, U.N.; Lunprom, S.; Reungsang, A.; Salakkam, A. Docosahexaenoic acid (DHA) production by *Aurantiochytrium limacinum* using cassava pulp hydrolysate as an alternative low-cost carbon source. *Front. Mar. Sci.* **2022**, *9*, 985119. [CrossRef]
64. Unagul, P.; Assantachai, C.; Phadungruengluij, S.; Suphantharika, M.; Tanticharoen, M.; Verduyn, C. Coconut water as a medium additive for the production of docosahexaenoic acid (C22:6 n3) by *Schizochytrium mangrovei* Sk-02. *Bioresour. Technol.* **2007**, *98*, 281–287. [CrossRef] [PubMed]
65. Nguyen, H.C.; Su, C.-H.; Yu, Y.-K.; Huong, D.T.M. Sugarcane bagasse as a novel carbon source for heterotrophic cultivation of oleaginous microalga *Schizochytrium* sp. *Ind. Crops Prod.* **2018**, *121*, 99–105. [CrossRef]
66. Zhang, Z.-X.; Xu, Y.-S.; Li, Z.-J.; Xu, L.-W.; Ma, W.; Li, Y.-F.; Guo, D.-S.; Sun, X.-M.; Huang, H. Turning waste into treasure: A new direction for low-cost production of lipid chemicals from *Thraustochytrids*. *Biotechnol. Adv.* **2024**, *73*, 108354. [CrossRef]
67. Liang, Y.; Sarkany, N.; Cui, Y.; Yesuf, J.; Trushenski, J.; Blackburn, J.W. Use of sweet sorghum juice for lipid production by *Schizochytrium limacinum* SR21. *Bioresour. Technol.* **2010**, *101*, 3623–3627. [CrossRef]
68. Yu, X.J.; Yu, Z.Q.; Liu, Y.L.; Sun, J.; Zheng, J.Y.; Wang, Z. Utilization of High-Fructose Corn Syrup for Biomass Production Containing High Levels of Docosahexaenoic Acid by a Newly Isolated *Aurantiochytrium* sp. YLH70. *Appl. Biochem. Biotechnol.* **2015**, *177*, 1229–1240. [CrossRef]
69. Fan, K.W.; Chen, F.; Jones, E.B.; Vrijmoed, L.L. Eicosapentaenoic and docosahexaenoic acids production by and okara-utilizing potential of thraustochytrids. *J. Ind. Microbiol. Biotechnol.* **2001**, *27*, 199–202. [CrossRef]
70. Lung, Y.-T.; Tan, C.H.; Show, P.L.; Ling, T.C.; Lan, J.C.-W.; Lam, H.L.; Chang, J.-S. Docosahexaenoic acid production from crude glycerol by *Schizochytrium limacinum* SR21. *Clean Technol. Environ. Policy* **2016**, *18*, 2209–2216. [CrossRef]
71. Kujawska, N.; Talbierz, S.; Dębowski, M.; Kazimierowicz, J.; Zieliński, M. Optimizing Docosahexaenoic Acid (DHA) Production by *Schizochytrium* sp. Grown on Waste Glycerol. *Energies* **2021**, *14*, 1685. [CrossRef]
72. Gong, Y.; Liu, J.; Jiang, M.; Liang, Z.; Jin, H.; Hu, X.; Wan, X.; Hu, C. Improvement of Omega-3 Docosahexaenoic Acid Production by Marine Dinoflagellate *Crypthecodinium cohnii* Using Rapeseed Meal Hydrolysate and Waste Molasses as Feedstock. *PLoS ONE* **2015**, *10*, e0125368. [CrossRef] [PubMed]
73. Yin, F.W.; Zhu, S.Y.; Guo, D.S.; Ren, L.J.; Ji, X.J.; Huang, H.; Gao, Z. Development of a strategy for the production of docosahexaenoic acid by *Schizochytrium* sp. from cane molasses and algae-residue. *Bioresour. Technol.* **2019**, *271*, 118–124. [CrossRef] [PubMed]
74. Olsen, P.M.; Kósa, G.; Klüver, M.; Kohler, A.; Shapaval, V.; Horn, S.J. Production of docosahexaenoic acid from spruce sugars using *Aurantiochytrium limacinum*. *Bioresour. Technol.* **2023**, *376*, 128827. [CrossRef] [PubMed]

75. Didrihsone, E.; Dubencovs, K.; Grube, M.; Shvirksts, K.; Suleiko, A.; Suleiko, A.; Vanags, J. Crypthecodinium cohnii Growth and Omega Fatty Acid Production in Mediums Supplemented with Extract from Recycled Biomass. *Mar. Drugs* **2022**, *20*, 68. [CrossRef]
76. Wang, X.; Liu, S.-F.; Wang, Z.-Y.; Wang, Y.; Hao, T.-B.; Balamurugan, S.; Li, D.-W.; He, Y.; Li, H.-Y.; Lin, C.S.K. A waste upcycling loop: Two-factor adaptive evolution of microalgae to increase polyunsaturated fatty acid production using food waste. *J. Clean. Prod.* **2022**, *331*, 130018. [CrossRef]
77. Ryu, B.G.; Kim, K.; Kim, J.; Han, J.I.; Yang, J.W. Use of organic waste from the brewery industry for high-density cultivation of the docosahexaenoic acid-rich microalga, *Aurantiochytrium* sp. KRS101. *Bioresour. Technol.* **2013**, *129*, 351–359. [CrossRef]
78. Russo, G.L.; Langellotti, A.L.; Verardo, V.; Martín-García, B.; Di Pierro, P.; Sorrentino, A.; Baselice, M.; Oliviero, M.; Sacchi, R.; Masi, P. Formulation of New Media from Dairy and Brewery Wastes for a Sustainable Production of DHA-Rich Oil by *Aurantiochytrium mangrovei*. *Mar. Drugs* **2021**, *20*, 30. [CrossRef]
79. Bryant, H.L.; Gogichaishvili, I.; Anderson, D.; Richardson, J.W.; Sawyer, J.; Wickersham, T.; Drewery, M.L. The value of post-extracted algae residue. *Algal Res. Biomass Biofuels Bioprod.* **2012**, *1*, 185–193. [CrossRef]
80. Zheng, H.; Gao, Z.; Yin, F.; Ji, X.; Huang, H. Lipid production of *Chlorella vulgaris* from lipid-extracted microalgal biomass residues through two-step enzymatic hydrolysis. *Bioresour. Technol.* **2012**, *117*, 1–6. [CrossRef]
81. Yin, F.-W.; Huang, J.; Zhan, C.-T.; Sun, X.-L.; Zheng, W.-L.; Luo, X.; Zhang, Y.-Y.; Yin, L.-F.; Fu, Y.-Q. Recycling Fermentation Strategy for Waste Cellular Residues in the Production of Polyunsaturated Fatty Acids. *Fermentation* **2024**, *10*, 81. [CrossRef]
82. Wang, S.K.; Wang, X.; Tian, Y.T.; Cui, Y.H. Nutrient recovery from tofu whey wastewater for the economical production of docosahexaenoic acid by *Schizochytrium* sp. S31. *Sci. Total. Environ.* **2020**, *710*, 136448. [CrossRef] [PubMed]
83. Chen, Z.L.; Yang, L.H.; He, S.J.; Du, Y.H.; Guo, D.S. Development of a green fermentation strategy with resource cycle for the docosahexaenoic acid production by *Schizochytrium* sp. *Bioresour. Technol.* **2023**, *385*, 129434. [CrossRef] [PubMed]
84. Yin, F.W.; Guo, D.S.; Ren, L.J.; Ji, X.J.; Huang, H. Development of a method for the valorization of fermentation wastewater and algal-residue extract in docosahexaenoic acid production by *Schizochytrium* sp. *Bioresour. Technol.* **2018**, *266*, 482–487. [CrossRef] [PubMed]
85. Yamasaki, T.; Aki, T.; Shinozaki, M.; Taguchi, M.; Kawamoto, S.; Ono, K. Utilization of *Shochu* distillery wastewater for production of polyunsaturated fatty acids and xanthophylls using thraustochytrid. *J. Biosci. Bioeng.* **2006**, *102*, 323–327. [CrossRef]
86. Song, X.; Ma, Z.; Tan, Y.; Zhang, H.; Cui, Q. Wastewater recycling technology for fermentation in polyunsaturated fatty acid production. *Bioresour. Technol.* **2017**, *235*, 79–86. [CrossRef]
87. Wang, S.K.; Tian, Y.T.; Dai, Y.R.; Wang, D.; Liu, K.C.; Cui, Y.H. Development of an alternative medium via completely replaces the medium components by mixed wastewater and crude glycerol for efficient production of docosahexaenoic acid by *Schizochytrium* sp. *Chemosphere* **2022**, *291*, 132868. [CrossRef]
88. Lee, G.-I.; Shin, W.-S.; Jung, S.M.; Kim, W.; Lee, C.; Kwon, J.-H. Effects of soybean curd wastewater on growth and DHA production in *Aurantiochytrium* sp. *LWT-Food Sci. Technol.* **2020**, *134*, 110245. [CrossRef]
89. Wang, Q.; Jin, W.; Qin, Y.; Zhou, X.; Han, W.; Gao, S.; Li, X.; Naushad, M.; Jiang, G.; Liu, H. Development of an alternative low-cost culture medium for a new isolated high-production DHA strain using kitchen wastewater. *Process Saf. Environ. Prot.* **2024**, *183*, 698–707. [CrossRef]
90. Humhal, T.; Kastanek, P.; Jezkova, Z.; Cadkova, A.; Kohoutkova, J.; Branyik, T. Use of saline waste water from demineralization of cheese whey for cultivation of *Schizochytrium limacinum* PA-968 and *Japonochytrium marinum* AN-4. *Bioprocess Biosyst. Eng.* **2017**, *40*, 395–402. [CrossRef]
91. Song, X.; Zang, X.; Zhang, X. Production of High Docosahexaenoic Acid by *Schizochytrium* sp Using Low-cost Raw Materials from Food Industry. *J. Oleo Sci.* **2015**, *64*, 197–204. [CrossRef]
92. Park, W.-K.; Moon, M.; Shin, S.-E.; Cho, J.M.; Suh, W.I.; Chang, Y.K.; Lee, B. Economical DHA (Docosahexaenoic acid) production from *Aurantiochytrium* sp KRS101 using orange peel extract and low cost nitrogen sources. *Algal Res. Biomass Biofuels Bioprod.* **2018**, *29*, 71–79. [CrossRef]
93. Nazir, Y.; Halim, H.; Al-Shorgani, N.K.N.; Manikan, V.; Hamid, A.A.; Song, Y. Efficient conversion of extracts from low-cost, rejected fruits for high-valued Docosahexaenoic acid production by *Aurantiochytrium* sp. SW1. *Algal Res.* **2020**, *50*, 101977. [CrossRef]
94. Moniz, P.; Silva, C.; Oliveira, A.C.; Reis, A.; Lopes da Silva, T. Raw Glycerol Based Medium for DHA and Lipids Production, Using the Marine Heterotrophic Microalga *Crypthecodinium cohnii*. *Processes* **2021**, *9*, 2005. [CrossRef]
95. Lowrey, J.; Armenta, R.E.; Brooks, M.S. Recycling of lipid-extracted hydrolysate as nitrogen supplementation for production of thraustochytrid biomass. *J. Ind. Microbiol. Biotechnol.* **2016**, *43*, 1105–1115. [CrossRef]
96. Jiang, M.; Chen, J.-N.; He, A.-Y.; Wu, H.; Kong, X.-P.; Liu, J.-L.; Yin, C.-Y.; Chen, W.-F.; Chen, P. Enhanced acetone/butanol/ethanol production by *Clostridium beijerinckii* IB4 using pH control strategy. *Process Biochem.* **2014**, *49*, 1238–1244. [CrossRef]
97. Alonso, S.; Rendueles, M.; Díaz, M. Selection method of pH conditions to establish *Pseudomonas taetrolens* physiological states and lactobionic acid production. *Appl. Microbiol. Biotechnol.* **2013**, *97*, 3843–3854. [CrossRef]

98. Zhang, L.; Zhang, J.; Zeng, G.; Dong, H.; Chen, Y.; Huang, C.; Zhu, Y.; Xu, R.; Cheng, Y.; Hou, K.; et al. Multivariate relationships between microbial communities and environmental variables during co-composting of sewage sludge and agricultural waste in the presence of PVP-AgNPs. *Bioresour. Technol.* **2018**, *261*, 10–18. [CrossRef]
99. Kautharapu, K.B.; Rathmacher, J.; Jarboe, L.R. Growth condition optimization for docosahexaenoic acid (DHA) production by *Moritella marina* MP-1. *Appl. Microbiol. Biotechnol.* **2013**, *97*, 2859–2866. [CrossRef]
100. Song, P.; Yuan, K.; Qin, T.; Zhang, K.; Ji, X.J.; Ren, L.; Guan, R.; Wen, J.; Huang, H. Metabolomics profiling reveals the mechanism of increased pneumocandin B(0) production by comparing mutant and parent strains. *J. Ind. Microbiol. Biotechnol.* **2018**, *45*, 767–780. [CrossRef]
101. Zhao, B.; Li, Y.; Mbifile, M.D.; Li, C.; Yang, H.; Wang, W. Improvement of docosahexaenoic acid fermentation from *Schizochytrium* sp. AB-610 by staged pH control based on cell morphological changes. *Eng. Life Sci.* **2017**, *17*, 981–988. [CrossRef]
102. Guo, D.S.; Tong, L.L.; Ji, X.J.; Ren, L.J.; Ding, Q.Q. Development of a Strategy to Improve the Stability of Culture Environment for Docosahexaenoic Acid Fermentation by *Schizochytrium* sp. *Appl. Biochem. Biotechnol.* **2020**, *192*, 881–894. [CrossRef] [PubMed]
103. He, B.; Hou, L.; Dong, M.; Shi, J.; Huang, X.; Ding, Y.; Cong, X.; Zhang, F.; Zhang, X.; Zang, X. Transcriptome Analysis in *Haematococcus pluvialis*: Astaxanthin Induction by High Light with Acetate and Fe^{2+}. *Int. J. Mol. Sci.* **2018**, *19*, 175. [CrossRef] [PubMed]
104. Jakobsen, A.N.; Aasen, I.M.; Strøm, A.R. Endogenously synthesized (–)-proto-quercitol and glycine betaine are principal compatible solutes of *Schizochytrium* sp. strain S8 (ATCC 20889) and three new isolates of phylogenetically related thraustochytrids. *Appl. Environ. Microbiol.* **2007**, *73*, 5848–5856. [CrossRef] [PubMed]
105. Hounslow, E.; Kapoore, R.V.; Vaidyanathan, S.; Gilmour, D.J.; Wright, P.C. The Search for a Lipid Trigger: The Effect of Salt Stress on the Lipid Profile of the Model Microalgal Species *Chlamydomonas reinhardtii* for Biofuels Production. *Curr. Biotechnol.* **2016**, *5*, 305–313. [CrossRef]
106. Liu, L.; Xu, Q.; Li, Y.; Shi, Z.; Zhu, Y.; Du, G.; Chen, J. Enhancement of pyruvate production by osmotic-tolerant mutant of *Torulopsis glabrata*. *Biotechnol. Bioeng.* **2007**, *97*, 825–832. [CrossRef]
107. Weete, J.D.; Kim, H.; Gandhi, S.R.; Wang, Y.; Dute, R. Lipids and ultrastructure of *Thraustochytrium* sp. ATCC 26185. *Lipids* **1997**, *32*, 839–845. [CrossRef]
108. Hu, X.C.; Ren, L.J.; Chen, S.L.; Zhang, L.; Ji, X.J.; Huang, H. The roles of different salts and a novel osmotic pressure control strategy for improvement of DHA production by *Schizochytrium* sp. *Bioprocess Biosyst. Eng.* **2015**, *38*, 2129–2136. [CrossRef]
109. Qu, L.; Ji, X.J.; Ren, L.J.; Nie, Z.K.; Feng, Y.; Wu, W.J.; Ouyang, P.K.; Huang, H. Enhancement of docosahexaenoic acid production by *Schizochytrium* sp. using a two-stage oxygen supply control strategy based on oxygen transfer coefficient. *Lett. Appl. Microbiol.* **2011**, *52*, 22–27. [CrossRef]
110. Jia, L.; Li, T.; Yang, Z.; He, T.; Ding, J.; Li, T.; Huang, A. Eliminating the accumulation of reactive oxygen species through periodic hypoxic stress control for effective DHA production by *Schizochytrium* sp. *Chem. Eng. Sci.* **2023**, *280*, 119040. [CrossRef]
111. Guo, D.-S.; Ji, X.-J.; Ren, L.-J.; Li, G.-L.; Huang, H. Improving docosahexaenoic acid production by *Schizochytrium* sp using a newly designed high-oxygen-supply bioreactor. *AICHE J.* **2017**, *63*, 4278–4286. [CrossRef]
112. Liyanaarachchi, V.C.; Premaratne, M.; Ariyadasa, T.U.; Nimarshana, P.H.V.; Malik, A. Two-stage cultivation of microalgae for production of high-value compounds and biofuels: A review. *Algal Res. Biomass Biofuels Bioprod.* **2021**, *57*, 102353. [CrossRef]
113. Zhang, L.; Zhao, H.; Lai, Y.; Wu, J.; Chen, H. Improving docosahexaenoic acid productivity of *Schizochytrium* sp. by a two-stage AEMR/shake mixed culture mode. *Bioresour. Technol.* **2013**, *142*, 719–722. [CrossRef] [PubMed]
114. Zhao, X.; Ren, L.; Guo, D.; Wu, W.; Ji, X.; Huang, H. CFD investigation of *Schizochytrium* sp. impeller configurations on cell growth and docosahexaenoic acid synthesis. *Bioprocess Biosyst. Eng.* **2016**, *39*, 1297–1304. [CrossRef] [PubMed]
115. Hoseinkhani, N.; Jalili, H.; Ansari, S.; Amrane, A. Impact of bubble size on docosahexaenoic acid production by *Crypthecodinium cohnii* in bubble column bioreactor. *Biomass Convers. Biorefin.* **2021**, *11*, 1137–1144. [CrossRef]
116. Hu, X.; Luo, Y.; Man, Y.; Tang, X.; Bi, Z.; Ren, L. Lipidomic and transcriptomic analysis reveals the self-regulation mechanism of *Schizochytrium* sp. in response to temperature stresses. *Algal Res. Biomass Biofuels Bioprod.* **2022**, *64*, 102664. [CrossRef]
117. Zeng, Y.; Ji, X.J.; Lian, M.; Ren, L.J.; Jin, L.J.; Ouyang, P.K.; Huang, H. Development of a temperature shift strategy for efficient docosahexaenoic acid production by a marine fungoid protist, *Schizochytrium* sp. HX-308. *Appl. Biochem. Biotechnol.* **2011**, *164*, 249–255. [CrossRef]
118. Rosa, S.M.; Soria, M.A.; Vélez, C.G.; Galvagno, M.A. Improvement of a two-stage fermentation process for docosahexaenoic acid production by *Aurantiochytrium limacinum* SR21 applying statistical experimental designs and data analysis. *Bioresour. Technol.* **2010**, *101*, 2367–2374. [CrossRef]
119. Jia, L.; Li, T.; Wang, R.; Ma, M.; Yang, Z. Enhancing docosahexaenoic acid production from *Schizochytrium* sp. by using waste *Pichia pastoris* as nitrogen source based on two-stage feeding control. *Bioresour. Technol.* **2024**, *403*, 130891. [CrossRef]
120. Patel, A.; Mikes, F.; Matsakas, L. An Overview of Current Pretreatment Methods Used to Improve Lipid Extraction from Oleaginous Microorganisms. *Molecules* **2018**, *23*, 1562. [CrossRef]

121. Lin, Y.; Xie, X.; Yuan, B.; Fu, J.; Liu, L.; Tian, H.; Chen, T.; He, D. Optimization of Enzymatic Cell Disruption for Improving Lipid Extraction from *Schizochytrium* sp through Response Surface Methodology. *J. Oleo Sci.* **2018**, *67*, 215–224. [CrossRef]
122. Isa, M.H.; Metin, C.; Ercan, E.; Alparslan, Y. Effect of different cell disruption methods on lipid yield of *Schizochytrium* sp. *J. Am. Oil Chem. Soc.* **2022**, *99*, 129–139. [CrossRef]
123. Saini, R.K.; Prasad, P.; Shang, X.; Keum, Y.S. Advances in Lipid Extraction Methods—A Review. *Int. J. Mol. Sci.* **2021**, *22*, 13643. [CrossRef] [PubMed]
124. Li, X.; Liu, J.; Chen, G.; Zhang, J.; Wang, C.; Liu, B. Extraction and purification of eicosapentaenoic acid and docosahexaenoic acid from microalgae: A critical review. *Algal Res.* **2019**, *43*, 101619. [CrossRef]
125. Dhara, O.; Rani, K.N.P.; Chakrabarti, P.P. Supercritical Carbon Dioxide Extraction of Vegetable Oils: Retrospective and Prospects. *Eur. J. Lipid Sci. Technol.* **2022**, *124*, 2200006. [CrossRef]
126. Gilcher, E.B.; Lane, M.K.M.; Pontious, R.S.; Apatoff, M.B.L.; Ahrens-Viquez, M.M.; Zimmerman, J.B. Sequential Extraction and Purification of Triglycerides and Carotenoids with Supercritical Carbon Dioxide for Valorization of the Integrated Algal Biorefinery. *ACS Sustain. Chem. Eng.* **2025**, *13*, 1667–1676. [CrossRef]
127. Crampon, C.; Nikitine, C.; Zaier, M.; Lepine, O.; Tanzi, C.D.; Vian, M.A.; Chemat, F.; Badens, E. Oil extraction from enriched *Spirulina platensis* microalgae using supercritical carbon dioxide. *J. Supercrit. Fluids* **2017**, *119*, 289–296. [CrossRef]
128. Tzima, S.; Georgiopoulou, I.; Louli, V.; Magoulas, K. Recent Advances in Supercritical CO_2 Extraction of Pigments, Lipids and Bioactive Compounds from Microalgae. *Molecules* **2023**, *28*, 1410. [CrossRef]
129. Chen, W.; Li, T.; Du, S.; Chen, H.; Wang, Q. Microalgal polyunsaturated fatty acids: Hotspots and production techniques. *Front. Bioeng. Biotechnol.* **2023**, *11*, 1146881. [CrossRef]
130. Dunford, N.T. Enzyme-aided oil and oilseed processing: Opportunities and challenges. *Curr. Opin. Food Sci.* **2022**, *48*, 100943. [CrossRef]
131. Sampaio, K.A.; Zyaykina, N.; Wozniak, B.; Tsukamoto, J.; De Greyt, W.; Stevens, C.V. Enzymatic degumming: Degumming efficiency versus yield increase. *Eur. J. Lipid Sci. Technol.* **2015**, *117*, 81–86. [CrossRef]
132. Lamas, D.L. Effect of enzymatic degumming process on the physicochemical and nutritional properties of fish byproducts oil. *Appl. Food Res.* **2022**, *2*, 100170. [CrossRef]
133. Kim, I.-C.; Kim, J.-H.; Lee, K.-H.; Tak, T.-M. Phospholipids separation (degumming) from crude vegetable oil by polyimide ultrafiltration membrane. *J. Membr. Sci.* **2002**, *205*, 113–123. [CrossRef]
134. de Moura, J.M.; Gonçalves, L.A.; Petrus, J.C.C.; Viotto, L.A. Degumming of vegetable oil by microporous membrane. *J. Food Eng.* **2005**, *70*, 473–478. [CrossRef]
135. Yi, M.; You, Y.; Zhang, Y.; Wu, G.; Karrar, E.; Zhang, L.; Zhang, H.; Jin, Q.; Wang, X. Highly valuable fish oil: Formation process, enrichment, subsequent utilization, and storage of eicosapentaenoic acid ethyl esters. *Molecules* **2023**, *28*, 672. [CrossRef]
136. Byreddy, A.R. Thraustochytrids as an alternative source of omega-3 fatty acids, carotenoids and enzymes. *Lipid Technol.* **2016**, *28*, 68–70. [CrossRef]
137. Chen, Z.; Zhou, Y.; Tang, J.; Wang, Y.; Xiao, Z.; Chenfeng, J.; Liang, D. Method of Producing Oil Containing Polyunsaturated Fatty Acids by Using Schizochytrium sp. Chinese Patent Application No. 201910418891.5, 9 August 2019.
138. Singh, D.; Rani, R.; Sharma, A.K.; Gupta, R.P.; Ramakumar, S.S.V.; Mathur, A.S. Extraction, separation and purification of fatty acid ethyl esters for biodiesel and DHA from Thraustochytrid biomass. *Biotechnol. J.* **2024**, *19*, 2300350. [CrossRef]
139. Sanchez-Machado, D.I.; Lopez-Cervantes, J.; Nunez-Gastelum, J.A.; Servin de la Mora-Lopez, G.; Lopez-Hernandez, J.; Paseiro-Losada, P. Effect of the refining process on *Moringa oleifera* seed oil quality. *Food Chem.* **2015**, *187*, 53–57. [CrossRef]
140. Mjøs, S.A.; Solvang, M. Geometrical isomerisation of eicosapentaenoic and docosahexaenoic acid at high temperatures. *Eur. J. Lipid Sci. Technol.* **2006**, *108*, 589–597. [CrossRef]
141. O'brien, R.D. *Fats and Oils: Formulating and Processing for Applications*; CRC Press: Boca Raton, FL, USA, 2008.
142. Zuo, W.; Hu, X.; Yang, Y.; Jiang, L.; Ren, L.; Huang, H. Development of an Improved Method to Determine Saturated Aliphatic Aldehydes in Docosahexaenoic Acid-Rich Oil: A Supplement to *p*-Anisidine Value. *Eur. J. Lipid Sci. Technol.* **2017**, *119*, 1700243. [CrossRef]
143. Yu, D.; Dong, T.; Zhang, L.; Zhou, X.; Wang, L.; Yang, F.; Liu, T. Effects of different deodorization methods on the oxidation of sterol components in rice bran oil. *Food Chem.* **2023**, *404*, 134568. [CrossRef]
144. Jomova, K.; Alomar, S.Y.; Alwasel, S.H.; Nepovimova, E.; Kuca, K.; Valko, M. Several lines of antioxidant defense against oxidative stress: Antioxidant enzymes, nanomaterials with multiple enzyme-mimicking activities, and low-molecular-weight antioxidants. *Arch. Toxicol.* **2024**, *98*, 1323–1367. [CrossRef] [PubMed]
145. Medina-Meza, I.G.; Barnaba, C. Kinetics of Cholesterol Oxidation in Model Systems and Foods: Current Status. *Food Eng. Rev.* **2013**, *5*, 171–184. [CrossRef]
146. Johnson, D.R.; Decker, E.A. The Role of Oxygen in Lipid Oxidation Reactions: A Review. *Annu. Rev. Food Sci. Technol.* **2015**, *6*, 171–190. [CrossRef] [PubMed]
147. Shahidi, F.; Zhong, Y. Lipid oxidation and improving the oxidative stability. *Chem. Soc. Rev.* **2010**, *39*, 4067–4079. [CrossRef]

148. Amarachukwu Uzombah, T. The Implications of Replacing Synthetic Antioxidants with Natural Ones in the Food Systems. In *Natural Food Additives*; Prieto, M.Á., Otero, P., Carpena Rodriguez, M., Eds.; IntechOpen: Rijeka, Croatia, 2022.
149. Sahari, M.A.; Moghimi, H.R.; Hadian, Z.; Barzegar, M.; Mohammadi, A. Physicochemical properties and antioxidant activity of α-tocopherol loaded nanoliposome's containing DHA and EPA. *Food Chem.* **2017**, *215*, 157–164. [CrossRef]
150. Jacobsen, C.; Hartvigsen, K.; Lund, P.; Thomsen, M.K.; Skibsted, L.H.; Hølmer, G.; Adler-Nissen, J.; Meyer, A.S. Oxidation in fish oil-enriched mayonnaise: 4. Effect of tocopherol concentration on oxidative deterioration. *Eur. Food Res. Technol.* **2001**, *212*, 308–318. [CrossRef]
151. Yang, J.; Goksen, G.; Zhang, W. Rosemary essential oil: Chemical and biological properties, with emphasis on its delivery systems for food preservation. *Food Control* **2023**, *154*, 110003. [CrossRef]
152. Qiu, K.; Wang, S.; Duan, F.; Sang, Z.; Wei, S.; Liu, H.; Tan, H. Rosemary: Unrevealing an old aromatic crop as a new source of promising functional food additive—A review. *Compr. Rev. Food Sci. Food Saf.* **2024**, *23*, e13273. [CrossRef]
153. Guo, M.; Yang, L.; Li, X.; Tang, H.; Li, X.; Xue, Y.; Duan, Z. Antioxidant Efficacy of Rosemary Extract in Improving the Oxidative Stability of Rapeseed Oil during Storage. *Foods* **2023**, *12*, 3583. [CrossRef]
154. Ozogul, Y.; Ayas, D.; Yazgan, H.; Ozogul, F.; Boga, E.K.; Ozyurt, G. The capability of rosemary extract in preventing oxidation of fish lipid. *Int. J. Food Sci. Technol.* **2010**, *45*, 1717–1723. [CrossRef]
155. Ren, L.-J.; Sun, X.-M.; Ji, X.-J.; Chen, S.-L.; Guo, D.-S.; Huang, H. Enhancement of docosahexaenoic acid synthesis by manipulation of antioxidant capacity and prevention of oxidative damage in *Schizochytrium* sp. *Bioresour. Technol.* **2017**, *223*, 141–148. [CrossRef] [PubMed]
156. Gao, H.-X.; Yu, J.; Chen, N.; Zeng, W.-C. Effects and mechanism of tea polyphenols on the quality of oil during frying process. *J. Food Sci.* **2020**, *85*, 3786–3796. [CrossRef] [PubMed]
157. Yin, F.; Sun, X.; Zheng, W.; Luo, X.; Zhang, Y.; Yin, L.; Jia, Q.; Fu, Y. Screening of highly effective mixed natural antioxidants to improve the oxidative stability of microalgal DHA-rich oil. *RSC Adv.* **2021**, *11*, 4991–4999. [CrossRef] [PubMed]
158. Pei, X.-C.; Yin, F.-W.; Zhong, X.; Liu, H.-L.; Song, L.; Zhao, G.-H.; Wang, Y.-F.; Zhou, D.-Y. Effects of different antioxidants and their combinations on the oxidative stability of DHA algae oil and walnut oil. *Food Sci. Nutr.* **2022**, *10*, 2804–2812. [CrossRef]
159. Huang, P.-W.; Yan, C.-X.; Sun, X.-M.; Huang, H. Economical downstream processing of microbial polyunsaturated fatty acids. *Trends Biotechnol.* **2023**, *41*, 857–859. [CrossRef]
160. Dai, F.; Chen, W.-F.; Zhou, B. Antioxidant synergism of green tea polyphenols with α-tocopherol and l-ascorbic acid in SDS micelles. *Biochimie* **2008**, *90*, 1499–1505. [CrossRef]
161. Pedrielli, P.; Skibsted, L.H. Antioxidant Synergy and Regeneration Effect of Quercetin, (−)-Epicatechin, and (+)-Catechin on α-Tocopherol in Homogeneous Solutions of Peroxidating Methyl Linoleate. *J. Agric. Food Chem.* **2002**, *50*, 7138–7144. [CrossRef]
162. Chen, X.; Li, H.; Zhang, B.; Deng, Z. The synergistic and antagonistic antioxidant interactions of dietary phytochemical combinations. *Crit. Rev. Food Sci. Nutr.* **2022**, *62*, 5658–5677. [CrossRef]
163. Enko, J.; Gliszczynska-Swiglo, A. Influence of the interactions between tea (*Camellia sinensis*) extracts and ascorbic acid on their antioxidant activity: Analysis with interaction indexes and isobolograms. *Food Addit. Contam. Part A Chem. Anal. Control Expo. Risk Assess.* **2015**, *32*, 1234–1242. [CrossRef]
164. Palafox-Carlos, H.; Gil-Chavez, J.; Sotelo-Mundo, R.R.; Namiesnik, J.; Gorinstein, S.; Gonzalez-Aguilar, G.A. Antioxidant Interactions between Major Phenolic Compounds Found in 'Ataulfo' Mango Pulp: Chlorogenic, Gallic, Protocatechuic and Vanillic Acids. *Molecules* **2012**, *17*, 12657–12664. [CrossRef]
165. Murugesan, N.; Damodaran, C.; Krishnamoorthy, S.; Raja, M. In-vitro evaluation of synergism in antioxidant efficiency of Quercetin and Resveratrol. *Chem. Biol. Lett.* **2023**, *10*, 534.
166. Chang, M.; Zhang, T.; Guo, X.; Liu, Y.; Liu, R.; Jin, Q.; Wang, X. Optimization of cultivation conditions for efficient production of carotenoid-rich DHA oil by *Schizochytrium* sp. S31. *Process Biochem.* **2020**, *94*, 190–197. [CrossRef]
167. Liu, R.; Xu, Y.; Zhang, T.; Gong, M.; Liu, R.; Chang, M.; Wang, X. Interactions between liposoluble antioxidants: A critical review. *Food Res. Int.* **2022**, *155*, 111104. [CrossRef]
168. Jomova, K.; Valko, M. Health protective effects of carotenoids and their interactions with other biological antioxidants. *Eur. J. Med. Chem.* **2013**, *70*, 102–110. [CrossRef]
169. Capitani, C.D.; Carvalho, A.C.L.; Botelho, P.B.; Carrapeiro, M.M.; Castro, I.A. Synergism on antioxidant activity between natural compounds optimized by response surface methodology. *Eur. J. Lipid Sci. Technol.* **2009**, *111*, 1100–1110. [CrossRef]
170. Shen, Y.; Guo, C.; Lu, T.; Ding, X.-Y.; Zhao, M.-T.; Zhang, M.; Liu, H.-L.; Song, L.; Zhou, D.-Y. Effects of gallic acid alkyl esters and their combinations with other antioxidants on oxidative stability of DHA algae oil. *Food Res. Int.* **2021**, *143*, 110280. [CrossRef]

Disclaimer/Publisher's Note: The statements, opinions and data contained in all publications are solely those of the individual author(s) and contributor(s) and not of MDPI and/or the editor(s). MDPI and/or the editor(s) disclaim responsibility for any injury to people or property resulting from any ideas, methods, instructions or products referred to in the content.

marine drugs

MDPI

Review

Fatty Acids in Cnidaria: Distribution and Specific Functions

Vasily I. Svetashev

A.V. Zhirmunsky National Scientific Center of Marine Biology, Far Eastern Branch, Russian Academy of Sciences, ul. Palchevskogo 17, Vladivostok 690041, Russia; vsvetashev@mail.ru

Abstract: The phylum Cnidaria comprises five main classes—Hydrozoa, Scyphozoa, Hexacorallia, Octocorallia and Cubozoa—that include such widely distributed and well-known animals as hard and soft corals, sea anemones, sea pens, gorgonians, hydroids, and jellyfish. Cnidarians play a very important role in marine ecosystems. The composition of their fatty acids (FAs) depends on food (plankton and particulate organic matter), symbiotic photosynthetic dinoflagellates and bacteria, and de novo biosynthesis in host tissues. In cnidarian lipids, besides the common FA characteristics of marine organisms, numerous new and rare FAs are also found. All Octocorallia species and some Scyphozoa jellyfish contain polyunsaturated FAs (PUFAs) with 24 and 26 carbon atoms. The coral families can be distinguished by specific FA profiles: the presence of uncommon FAs or high/low levels of common fatty acids. Many of the families have characteristic FAs: Acroporidae are characterized by 18:3n6, eicosapentaenoic acid (EPA) 20:5n3, 22:4n6, and 22:5n3; Pocilloporidae by 20:3n6, 20:4n3, and docosahexaenoic acid 22:6n3 (DHA); and Poritidae by arachidonic acid (AA) and DHA. The species of Faviidae show elevated concentrations of 18:3n6 and 22:5n3 acids. Dendrophylliidae, being azooxanthellate corals, have such dominant acids as EPA and 22:5n3 and a low content of DHA, which is the major PUFA in hermatypic corals. The major and characteristic PUFAs for Milleporidae (class Hydrozoa) are DHA and 22:5n6, though in scleractinian corals, the latter acid is found only in trace amounts.

Keywords: Cnidaria; Octocorallia; Hexacorallia; corals; fatty acids; uncommon fatty acids; PUFA; symbiosis; zooxanthellae; food chain markers

1. Introduction

The phylum Cnidaria is an ancient member of Metazoa, with a simple body organization. It comprises a diverse group of relatively primitive diploblastic animals characterized by having an extremely complex cellular organelle referred to as cnida (which in ancient Greek means "stinging nettle") or nematocysts. Cnidarians have a characteristic radial symmetry and the planula and polyp stages in their ontogeny [1].

The phylum Cnidaria consists of nearly 12,500 extant species from the classes Anthozoa, Medusazoa, Hydrozoa, and Scyphozoa [2]. Cnidarians play a very important role in marine ecosystems, of which tropical coral reefs are exceptionally productive [3]. The high levels of production in reef-building corals are achieved primarily due to their symbiotic algae. Zooxanthellae (unicellular algae) in corals account for 50–70% of the total primary production in most reefs. The number of coral species at depths greater than 50 m is usually higher than that on shallow, tropical coral reefs [4]. Jellyfish are attracting increasingly more attention because of the frequently documented cases of their mass proliferation, their potential capability of shaping the food web structure [5], and their importance in global biogeochemical cycles [6]. Cnidarian jellyfish are considered a sustainable source of high-value compounds that can find biotechnological applications, e.g., as antioxidants and nutraceuticals [7,8].

The lipid biochemistry of Cnidarians deserves special attention. The use of thin-layer chromatography (TLC) has shown that their total lipids contain phospholipids, sterols, free fatty acids (FAs), triacylglycerols (TGs), monoalkyl-diacylglycerols (MADAGs), and wax esters (WEs). The content of free FAs in total lipids is usually low, approximately 1.0–1.5%. The concentrations of storage neutral lipids such as TG, WE, and, less commonly, MADAGs vary between 20 and 30, 30 and 50, and 5 and 10%, respectively. The concentrations of structural lipids, sterols, and phospholipids (PLs) range between 5 and 10, and 10 and 20%, respectively [9]. Lipids of the sea anemone *Anthopleura elegantissima* have been found to include an unusual phosphosphingolipid with a 2-aminoethylphosphonic group, ceramide aminoethylphosphonate (CAEP) [10]. The latter, in turn, has been found as a major phospholipid (up to 20%) in gorgonarians and alcyonarians [11,12]. In the cold-water alcyonarian *Gersemia rubiformis*, the content of CAEP and N-methyl-CAEP reaches 29% of total phospholipids [13]. A possible role of CAEP is the stabilization of its own membranes against toxic actinoporins, which constitute the main family of pore-forming proteins from sea anemones [14]. In jellyfish, ceramide 2-aminoethylphosphonate is concentrated in the membranes of the tentacles (oral arms) bearing stinging cells, where it may resist hydrolysis due to the endogenous phospholipase A2 [15]. In this review, we compiled data on the FAs of the major Cnidaria taxa, with a special focus on characteristic components, marker FAs, the presence of symbionts, and new and uncommon fatty acids. It is relevant to note here that the commonly applied term 'coral' has no valid taxonomic definition. "Initially, the Mediterranean coral, *Corallium rubrum* (Anthozoa: Octocorallia), having a calcitic skeleton was referred to as 'coral'. Afterwards, the term was used not only for other octocorallians, such as the blue coral (Heliopora), but also for hexacorallians, such as black corals with horny skeleton, *Antipathes*, or reef-building corals (scleractinians or Madreporaria) with aragonitic skeleton, or even hydrozoans such as fire corals (Millepora). However, 'coral' is also used to designate Octocorallia without hard skeleton, alcyonarians (soft corals), which adds even more confusion. Scleractinian corals are known for their skeleton and are, in this case, called 'reef-building corals' or hermatypic corals ('herm' in ancient Greek means 'reef'). Hermatypic corals host unicellular symbiotic dinoflagellates (*Symbiodiniactaea*), commonly referred to as zooxanthellae, in their tissues. However, all scleractinians are not reef builders (ahermatypic corals), although they are characterized by their aragonitic skeleton" [16].

In the literature, different abbreviations are commonly used for the structures of FAs, e.g., 18:1ω9, 18:1n9, and 18:1n-9 for oleic acid, which are equivalent. Eicosapentaenoic acid (EPA) 20:5(n-3) in texts can be abbreviated as 20:5(n-3), 20:5n3, and 20:5ω3. Sometimes, abbreviations are used to indicate the position of the double bond counted from the carboxyl end, e.g., Δ7-18:1, Δ9,12-18:2, and Δ5,8,11,14,17-20:5. In this review, I will use the most simple abbreviations with one letter, like 20:5n3.

2. Some Methodological Notes

Comparing various data on FA composition that are scattered in numerous articles and obtained by different analytical protocols poses a major challenge. In some cases, it may lead to erroneous or not fully reliable results. The most common mistakes in marine FA analyses are the too short time of the GC run and the use of saponification or the basic transesterification of microalgae lipids.

Total Lipid Extraction and FA Analysis

Since some Cnidaria species (especially jellyfish) contain only ca. 0.2% total lipids (TLs), it is better to use the cost-effective method proposed by [17] for lipid extraction. For the isolation of TLs from stony corals, the extraction of crashed corals, instead of the preliminary isolation of soft tissues by air spraying, is a much more efficient approach [18]. For the preparation of FA methyl esters (FAMEs), the use of acid reagents (5% HCl or 1–2% sulfuric acid in methanol), 50 °C overnight or 80 °C for 90 min, is more preferable [19] because the preliminary saponification or transesterification with $NaOCH_3$ in methanol induces the decomposition and isomerization of octadecapentaenoic acid 18:5n3 into four isomers [20,21]. The methyl ester of octadecapentaenoic acid 18:5n3 on GC, separated on a Supelcowax 10 polar column, has an equivalent chain length (ECL) of 20.22, while isomers run much more latter, with ECL of 20.48, 20.88 (main peak), and 21.19. Accordingly, instead of the peak of 18:5n3 with an ECL of 20.22, which is located close to (even overlaps) the peak of 20:1n9, one can see artifact peaks with an ECL of 20.88. Identifying 18:5n3 on a non-polar column with 5% phenyl silicon such as SLB-5 or one similar, where the ECL is equal to 17.55, is much easier. For the identification of marine-derived FAs, FAME standard mixtures such as Supelco 37 are often used, but these do not contain the usual marine FAMEs such as 18:1n7, 18:4n3, 20:4n3, 22:5n3, and 22:5n6. Moreover, marine organisms may have PUFAs with C24 and C26 and even longer carbon chains. There are reports about the presence of elaidic acid (18:1n9 trans), which is not characteristic of marine samples. Besides the use of sophisticated equipment like GC–MS, there is a simple and convenient method of FAME identification carried out by the calculation of ECL values. It helps identify the most common and unusual FAs merely by comparing its own data with published tables of ECL values, e.g., in [22]. Moreover, it allows for the calculation or prediction of ECL for new FAs [23]. More details on ECL values for FAMEs on columns with different phases and their MS spectra are available at Chrombox.org [24]. For very-long-chain PUFAs with the carbon chains C24–C26, which are characteristic of the class Octocorallia, the analysis is preferably conducted under special conditions because these FAs have very long retention times (RTs). For example, on the common polar column Supelcowax 10 at 220 °C, the RT for the acids 22:6n3, 24:6n3, and 26:7n3 are 28.4, 48.1, and 86.7 min, respectively. On the non-polar column SLB-5ms with a temperature ramp of 170–280 °C (2°/min), the RT is quite moderate, <55 min, and the separation is good. An example of a chromatogram for total lipids in the jellyfish *Rhopilema esculentum* is shown in Figure 1.

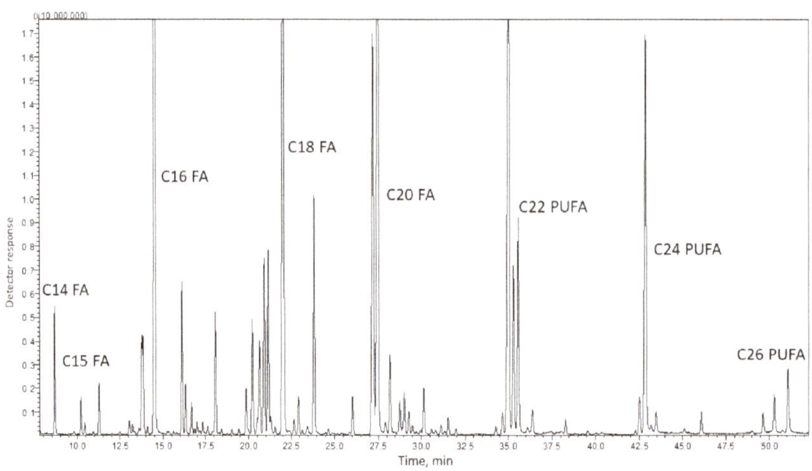

Figure 1. GC-MS chromatogram of medusa *Rhopilema esculentum* total lipid FAMEs on SLB-5ms column.

3. Fatty Acids of Hexacorallia

The class Hexacorallia comprises scleractinian, black corals, tube anemones, and sea anemones (with a total of nearly 4300 species) [1]. The major orders are Actiniaria (ca. 1200 species), Antipatharia, Ceriantharia, Corallimorpharia, Scleractinia (ca. 1300 species), and Zoantharia.

3.1. Fatty Acids of Actiniaria

Actiniaria are soft-bodied, solitary polyps with tentacles. They are predatory animals. In Actiniaria species, the major PUFAs are DHA, 22:5n3 (DPA3), EPA and 22:4n6, while n3 FAs prevail over n6 acids (with an n6/n3 ratio of 0.1–0.80) (Table 1). The content of arachidonic acid is 2.6–7.7% and is always lower than that of EPA. The total amount of C22 PUFA is greater than that of C20 PUFA. Abyssal and cold-water species accumulate monoenoic C20:1 and C22:1 acids. Together with marked concentrations of 18:1n9 and DHA, these indicate feeding on zooplankton. Of particular interest is the high concentrations of n3 docosapentaenoic acid and 22:4n6, which are homologs of arachidonic and eicosapentaenoic acids.

Table 1. Main FA composition of total lipids of Actiniaria species (% of total FA).

Fatty Acids	*Bathyphelli australais*	*Metridium dianthus*	*Metridium senile*	*Aiptasia pallida*	*Actinia equina*	*Anemonia viridis*	*Actinostola callosa*
16:0	9.7	4.6	12.6	21.2	28.5	13.7	7.9
16:1n7	1.5	0.6	3.3	3.0	3.5	0.5	2.5
18:0	3.2	5.9	9.9	7.7	11.3	9.1	6.1
18:1n9	9.3	2.7	9.2	5.3	8.5	1.4	5.9
18:1n7	5.5			1.1	7.1	1.1	
18:2n6	0.4		1.1	3.2	2.1	0.3	0.6
18:3n6			2.8	7.8		0.8	
18:4n3	1.1					1.4	
20:1n7	7.7	7.3				3.6	2.3
20:1n9	2.0				2.2	6.4	10.8
AA	3.1	2.6	7.7	4.80	2.8	5.8	
EPA	14.0	27.1	12.5	4.1	9.1	28.7	14.1
22:1	3.4	4.9	0.8				14.6
22:4n6	2.0	4.2	8.6	8.0	0.3	8.1	5.9
22:5n3	6.0	12.0	1.2	3.6		12.3	9.3
DHA	11.5	19.0	17.9	22.9	5.2	0.3	9.9
18:1n9/n7	1.7	NA	NA	5.0		1.3	NA
n6/n3	0.11			0.79	0.58	0.4	0.27
20:1+22:1	13.5	4.9	0.8		2.4	10.0	27.5
References	[25] abyssal	[26] abyssal	[27] littoral	[28] tropical	[29] temperate	[30] aquarium	[31] batial

In [32], the authors compared the effects of symbiotic dinoflagellates (*Symbiodinium muscatinei*) and chlorophytes (*Elliptochloris marina*) on lipids of the common sea anemone, *Anthopleura elegantissima*. In general, the authors noted an increased total concentration of FAs in *A. elegantissima* with symbionts, but the results were not complete (because a standard MEFA Supelco 37 mixture lacks some common marine acids such as 18:4n3, 18:5n3, and 18:1n7). In [33,34], a set of new FAs, including non-methylene interrupted (NMI) dienoic and trienoic acids C18–C23, brominated FA, etc., were identified in phospholipids. In abyssal Actiniaria species, the uncommon FA 21:4n7 was detected at 2.3% and tetracosahexaenoic acid (TPA) 24:6n3 at 0.8% [35]. In a discussion of the first communication on tetracosapolyenoic acids in Cnidaria [36], the authors mentioned that "among all other coelenterates studied (class Hydrozoa: order Leptolida: class Hexacorallia: orders Actiniaria, Zoantharia, Ceriantharia, insignificant amounts of TPA were found (up to 1%)".

In [37], the FA compositions of three species of bathyal sea anemones were studied, and high proportions of mono- and polyunsaturated fatty acids (MUFAs and PUFAs), as well as n3 FAs, and unusually low proportions of arachidonic acid were found. High values of 20:5n3, 20:1n9, and 22:1n11 suggested feeding on zooplankton. In [38], low contents of DHA in neutral lipids (NLs) and PLs (0.4–2.6%) and AA (2.7–4.7%) were reported for the red and green forms of the Black Sea *Actinia equina*. In the paper, the structure and marked concentrations of the less common fatty acids 22:4n6 and 22:5n3, previously found in another sea anemone species, were well described.

3.2. Fatty Acids of Antipatharia and Zoantharia

The major PUFAs in black thorny coral (*Stauropathes arctica*), which is a cold-water antipatharian, are fatty acids of the n3 family, 22:5n3 and EPA, constituting 17.1 and 10.6%, respectively [39], (Table 2). Arachidonic acid (AA) and DHA, which are common for Cnidaria, have only minor proportions, 1.5 and 1.6%, with a low n6/n3 ratio (0.11). The proportion of saturated acids is also low, 10%, while that of the monoenoic acids C20:1 and 22:1 is 34.5%. This suggests a mainly zooplankton-based diet. In *Palythoa* [40], octadecapentaenoic acid has been detected, which, together with 18:3n6 and 18:4n3, clearly indicates the presence of zooxanthellae.

Table 2. Major fatty acids of Antipatharia and Zoantharia corals (as % of total FAMEs).

Fatty Acids	Antipatharia *Stauropathes arctica*	Zoantharia *Palythoa* 5 Species	Zoantharia *Palythoa caesia*	Zoantharia *P. caribaeorum*
16:0	6.42	18.1–26.4	29.5	43
16:1n7	1.85	3.4–9.9	4.3	3.3
18:0	1.35	14.9–4.5	9.9	6.2
18:1n9	6.01	4.0–8.9	3.9	1.4
18:1n7	6.95	6.5–1.5	1.4	1.2
18:2n6	0.47	0.8–2.9	1.1	
18:3n6	0.04	1.7–3.6	2.3	
18:4n3	0.79	0.9–3.8	3.2	4.9
18:5n3		+	0.42	
20:1n9	14.47	0.5–1.5	0.7	0.8
20:1n7	1.85		0.2	
20:4n6	1.48	1.2–15.0	10.3	5.2
20:5n3	10.61	–	3.4	
22:6n3	1.61	0.0–1.9	2.3	
22:5n3	17.08	3.8–10.1	6.5	3.0
22:4n6	1.39	1.2–15.0	4.6	4.8
22:1n9	7.48		0.3	0.5
22:1n11(13)	8.65			
C20:1+C22:1	34.6		1.6	
Odd + Br	1.1	5.3–8.9		5.8
16:1n7+18:1 n7	8.8	5.1–17.4	5.7	5.5
References	[39]	[40]	[41]	[34]

In zoantharian species of the genus *Palythoa*, the major PUFAs are AA, DPA3, 22:4n6, and AA. The contents of EPA and DHA are low, as well as those of C20:1 and 22:1 acids. In [42], a statistical analysis of the distribution of 10 major PUFAs was carried out for 66 hexacoral and octocoral species. For nine species of *Palythoa* and *Zoanthus*, the Zoanthidae, the data were similar to those presented in Table 3. In [40,41], the uncommon acid 18:5n3, characteristic of symbiotic dinoflagellates, was found. Thus, the detection of the acids 18:3n6, 18:4n3, and 18:5n3 indicated the presence of symbiotic dinoflagellates [43]. In [41], two azooxanthellate species of *Palythoa* showed AA as a major fatty acid (8–10%), much lower levels of EPA and DHA (ca. 2%), and noticeable amounts of 22:4n6 (ca. 3.5%) and

22:5n3 (ca. 5%). The authors of [42] considered concentrations of only five major n3 and five n6 acids.

3.3. Order Scleractinia

According to [44], this order comprises 1363 extant species from the major families Acroporidae (270 species), Agariciidae (47), Caryophylliidae (307), Dendrophylliidae (187), Fungiidae (54), Lobophylliidae (60), Merulinidae (152), Pocilloporidae (55), Poritidae (100), and Turbinoliidae (70). Initially, the data on the FA composition of scleractinian corals were somewhat contradictory and incomplete. The first screening of FAs in Caribbean Scleractinia was conducted by [45]. However, since there were some drawbacks in the lipid isolation method and FAME analysis, the data presented have only a limited value. Later on, the authors of the study [46] compared the PUFA contents and compositions between eight species of corals from the depths of 2–5 and 25–30 m and suggested that the relatively high levels of PUFAs, such as 22:5n3 and 22:6n3 in the species from the depths of 25–30 m, compared to those in shallow-water species, might be derived from food (copepods), while for the species from the depths of 2–5 m, the main source of FAs might be symbiotic zooxanthellae. In [47], a more advanced FA analysis method was applied for Scleractinian corals from the families Acroporidae, Pocilloporidae, Poritidae, and Dendrophylliidae from Vietnam and the Seychelles (Table 3).

Table 3. FA composition (% of total FAMEs) of scleractinian corals from the families Acroporidae, Seriatoporidae, Poritidae, and azooxanthellate Dendrophylliidae from Vietnam (V) and the Seychelles (S) [47].

Fatty Acids	*Acropora nasuta*	*Acropora millepora*	*Acropora florida*	*Seriatopora caliendrum*	*Stylophora pistillata*	*Stylophora pistillata*	*Pocillopora damicornis*	*Pocillopora verrucosa*	*Porites lutea*	*Goniopora* sp.	*Tubastrea coccinea* [a]
	V	V	V	S	S	V	V	V	V	V	S
16:0	38.6	24.5	33.1	24.0	21.8	41.0	44.5	41.7	49.1	17.2	7.2
16:1n7	1.0	1.0	1.2	2.1	2.9	3.0	3.1	2.3	1.9	3.5	5.9
18:0	7.3	9.3	9.0	6.5	4.6	10.6	10.2	8.0	7.2	5.7	4.2
18:1n9	8.0	2.2	3.2	13.3	14.4	5.5	4.9	7.0	3.8	11.7	23.3
18:3n6	5.8	9.5	8.2	2.7	3.1	5.4	4.1	2.6	9.7	4.5	0.4
18:4n3	2.6	6.6	5.1	1.7	1.3	1.4	0.8	3.3	2.9	2.3	0.7
20:1	0.7	1.1	1.0	0.9	0.7	1.4	1.1	2.6	0.9	5.9	3.0
20:3n6	1.9	2.4	0.3	11.3	12.3	7.2	7.6	3.2	1.6	3.6	0.8
20:4n6	7.1	7.2	11.0	4.8	4.3	1.7	2.0	1.8	2.3	13.3	7.8
20:5n3	0.8	10.4	6.9	2.6	2.0	1.4	1.4	3.2	3.3	4.1	14.9
22:4n6	4.3	6.0	6.3	1.5	1.8	1.0	0.9	1.3	1.4	3.3	4.7
22:5n3	0.9	3.0	1.2	1.2	1.3	4.5	0.4	0.7	0.8	1.0	16.4
22:6n3	10.8	12.6	6.7	16.9	16.4	8.8	9.5	10.4	5.3	15.7	1.4
PUFA	40.4	59.6	49.4	48.5	50.4	31.5	30.5	30.0	29.3	53.8	52.0
(n3/n6)	0.6	1.2	0.7	0.9	1.0	0.7	1.2	1.7	0.8	0.8	2.1

[a] Azooxanthellate species.

There the major FAs were 16:0, 18:0, 18:1n9, 20:4n6, 20:5n3, 22:4n6, and 22:6n3. Some of the coral families had significant levels of characteristic FAs: 20:3n6 for Pocilloporidae, 18:3n6, 18:4n3, and 22:4n6 for Acroporidae, and 18:3n6 for Poritidae. The noticeable amounts of 18:3n6 and 18:4n3 indicated the input of symbiotic microalgae. Two asymbiotic species of *Tubastraea*, from the family Dendrophylliidae, showed the major FAs 18:1n-9 (23.3–26.4%), EPA, AA, and 22:5n3. The concentration of DHA, 18:4n3, and 18:3n6 in these species was low, nearly 2%. Of particular note was that the azooxanthellate species of *Tubastraea* had high concentrations of AA and EPA and their C22 homologs, 22:4n6 and 22:5n3. The contents of some common acids such as 18:1n7, 18:2n6, 20:1n7, 20:2n6, and 22:2n6 were no higher than 2%. The content of branched- and odd-chain acids of bacterial origin was also insignificant. In addition, the authors found differences in the FA composition between the same species from different geographic places and depths of sampling. An even more extensive FA study of 16 species of reef-building corals was published later [48]. There, in addition to the above-mentioned families, corals of Faviidae, Pectiniidae, and Fungiidae were also analyzed (Table 4).

Table 4. Total FA composition (as % of total lipids FAME) in scleractinian corals from the families Agariciidae (mean of 2 species), Dendrophylliidae (mean of 5 species), Euphylliidae (mean of 2 species), Faviidae (mean of 5 species), Fungiidae (mean of 3 species), and Pectiniidae (mean of 2 species) [9,48].

Fatty Acids	Agariciidae *Pavonaria*	Dendrophylliidae *Turbinaria* [a]	*Balanophillia* sp. [a]	Euphylliidae *Euphyllia*	Faviidae *Favia*	Fungiidae *Fungia*	Pectiniidae *Echinophyllia*
16:0	37.3	32.6	9.5	38.3	40.4	46.7	35.1
16:17	2.2	3.0	3.2	3.7	3.7	1.8	3.9
18:0	7.8	6.4	6.1	5.8	4.5	8.6	5.6
18:19	6.0	4.6	20.5	6.0	7.6	4.2	8.3
18:1n7	1.0	1.1	3.4	1.0	1.1	0.7	1.2
18:2n6	1.4	2.1	1.5	2.2	1.6	1.5	3.1
18:3n6	6.5	11.2	0.9	6.3	10.8	6.4	8.3
18:4n3	1.6	3.1	0.6	1.6	1.2	1.1	1.9
20:1n9	2.4	0.2	1.9	0.3	0.4	0.4	0.7
20:3n6	2.7	1.7	1.1	3.9	1.9	2.2	2.7
20:4n6	6.7	8.4	10.9	11.2	4.6	5.5	6.4
20:5n3	1.9	2.0	7.3	2.5	0.8	1.8	1.5
22:4n6	2.9	3.3	6.7	2.6	2.1	2.9	1.9
22:5n3	2.3	4.6	13.0	0.4	6.7	2.6	5.5
22:6n3	7.9	7.2	1.7	8.6	3.6	6.9	5.5
n6/n3	1.6	1.6	1.0	1.0	1.7	1.5	1.5

[a] Azooxanthellate species.

In general, the data reported by [9,47,48] were similar. In two *Favia* species, the major PUFAs 18:3n6, AA and 22:5n3 and a low level of EPA were found. In *Balanophillia* sp., in the family Dendrophylliidae, 18:1n9 was a major acid (20.3%). The polyunsaturated FAs were DHA (9.2%), AA (3.4%), DPA3 (2.3%), and 18:3n6 (2.4%). The levels of the rest of C18–C22 PUFAs were no higher than 2% each.

The most noteworthy results were obtained through the statistical processing of FA analysis data. In [47,48], principal component analysis (PCA) was performed, using 10 variables (the square root of the selected unsaturated fatty acid content), of the FAs 18:1n-9, 18:1n-7, 20:1n9, 20:1n7, 20:4n3, 20:5n3, 22:4n6, 22:5n3, and 22:6n3. Figure 2 shows the relationships between the FA compositions and taxonomy of 35 reef-building coral specimens. The four families of corals can be recognized in the score plot.

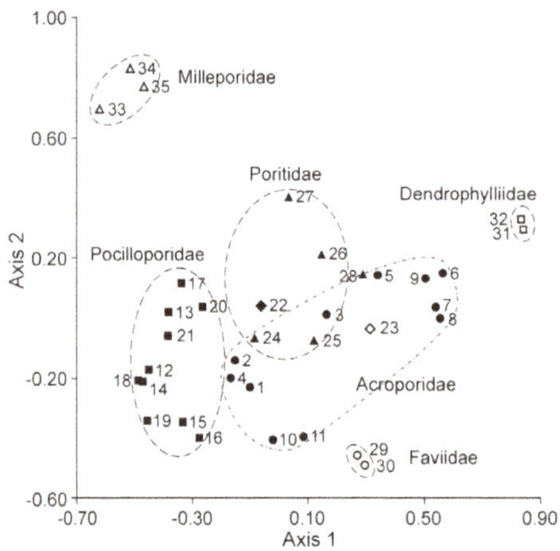

Figure 2. Multidimensional scale analysis performed using ten FAs selected as marker variables. Filled circles are for Acroporidae; filled squares for Pocilloporidae; filled triangles for Poritidae; open circles for Faviidae; open squares for Dendrophylliidae; filled rhombi for Pectiniidae; and open rhombi for Fungiidae. The plot is redrawn from [48].

4. Octocorallia

4.1. Alcyonacea

The FA composition of the families Alcyoniidae and Nephtheidae was analyzed in [48] and are here presented in Table 5. The polyunsaturated FAs 24:5n6 and 24:6n3 (TPA) were the most characteristic of all Alcyonacea, comprising 3.1–10.03% in them. The highest level of TPA was found in the azooxanthellate species of *Dendronephthya*. The major FAs were 16:0 (up to 35%), arachidonic acid (9.4–19.7%), and 18:3n6 (up to 10.6%). In all species, the EPA content was only 2.1–5.9%, and the DHA content was 2.1–6.9%. The n6/n3 ratio was always >1 (Table 6). Most species had high concentrations of the uncommon fatty acid 16:2n7 (0.3–9.4%). The contents of the C20:1 and C22:1 fatty acids, which indicate feeding on zooplankton, were low. In most species, there were marked concentrations of the C18 PUFAs 18:3n6 and 18:4n3, which are typical of symbiotic zooxanthellae [43]. Two (zooxanthellate and azooxanthellate) species of Nephtheidae had a difference in the contents of 18:3n6 (7.9–0.40%), EPA (5.9–1.4%), and TPA (6.4–10.3%). It was reported that the main sources of the fatty acids 16:2n-7, 20:4n6, and 20:5n3 were, respectively, protists, zooplankton, and phytoplankton that corals fed on [49]. Obviously, zooxanthellae was the main food source for Alcyonacea. Only *Lobophytum* and *Sarcophytum* had noticeable levels of bacterial FAs (8.3 and 3.9%). These data generally supported the results of previous studies where 10 *Dendronephthya* species had been analyzed (the major FAs were AA (15.3%) and TPA (17%)) [48].

Table 5. Main FA composition of total lipids in Alcyonacea (% of total FA) [50] (sp—means number of investigated species).

FA	Alcyoniidae					Nephtheidae	
	Klyxum molle	*Cladiella* 3 sp.	*Lobophytum* 5 sp.	*Sarcophytum* 6 sp.	*Sinularia* 12 sp.	*Nephthea*	*Dendroneph thya* 3 sp. [a]
16:0	13.60	34.60	29.24	29.29	27.43	20.30	28.67
16:1n7	5.20	4.00	2.64	2.87	3.15	0.90	2.17
16:2n7	0.40	0.30	5.92	9.43	4.74		4.43
18:0	6.00	8.30	6.06	5.29	6.22	15.80	8.60
18:1n9	5.00	4.90	2.46	3.16	2.38	4.10	3.10
18:1n7	0.30	0.40	0.70	0.57	0.43	0.10	0.87
18:2n6	1.70	2.60	0.45	0.20	1.02	2.40	1.00
18:3n6	10.60	9.40	0.17	0.37	7.40	7.90	0.40
18:3n3	0.20	0.10	0.24	0.77	0.31	0.20	0.33
18:4n3	8.70	3.70	2.86	4.30	2.79	2.60	1.97
20:1n9	0.20	0.20	0.16	1.05	0.15	0.20	0.30
20:4n6	19.70	9.40	19.50	18.27	15.78	18.00	15.33
20:4n3	0.30	0.10	0.58	0.47	0.90	0.70	0.87
20:5n3	4.80	2.10	1.96	1.53	2.78	5.90	1.43
22:5n6	0.10	0.10	0.30	0.27	0.23	0.20	1.10
22:6n3	6.90	5.20	2.10	2.50	4.82	3.20	2.47
24:5n6	5.50	2.20	4.82	5.63	4.98	4.80	8.40
24:6n3	1.20	0.90	1.18	0.83	1.94	1.60	1.63
C24 PUFA	6.70	3.10	6.00	6.46	6.93	6.40	10.03
n6/n3	1.73	1.99	2.88	2.39	2.25	2.49	2.94
Odd+Br	1.10	0.60	8.28	3.90	2.52	0.57	5.60

[a]—azoxanthellate species.

4.2. Gorgonacea

For Gorgonacea species, the absolute dominance of arachidonic acid (up to 47% in *Echinogorgia* sp.) and high levels of C24 PUFA were typical (Table 6). The azooxanthellate species *Leptogorgia piccola* had the highest content of TPA, 21.1%. Other major FAs were 16:1n7, 20:5n3, 22:5n6, 22:6n3, 24:5n6, and 24:6n3. The content of common C18 and C20 acids such as 18:2n6, 18:4n3, 20:1n9, and 20:3n6 was no higher than 2%. Bacterial odd- and branched-chain FAs were detected in quite noticeable amounts (up to 5.2% of total FAs). The average ratio of oleic to *cis*-vaccenic acids, 18:1n9/18:1n7, was 17.9 and 1.5 for zooxanthellate and azooxanthellate species, respectively. Thus, the zooxanthellate species of *Rumphella* showed marked levels of 18:3n6 and 18:4n3, characteristic of endophotosymbionts. The FAs of the n6 series were dominant PUFAs in the species under study; the n6/n3 ratio ranged from 2.7 to 17.1 with an average of 7.0. I did not find any significant difference between zooxanthellate and azooxanthellate coral species in the n6/n3 ratio. Large amounts (up to 10%) of uncommon furanoic FAs (F-acids, containing a furan ring in the hydrocarbon chain) were detected in azooxanthellate Gorgonacea [51].

Table 6. Main FA composition (% of total FAs) of zooxanthellate and azooxanthellate Gorgonacea species from the families Melithaeidae, Acanthogorgiidae, Nidaliidae, Plexauridae, Ellisellidae, Paramuriseidae, and Gorgoniidae [51] and data for *Leptogorgia piccolo* [52].

Fatty Acids	*Acabaria erythraea*	*Acanthogorgia isoxia*	*Chironephthya variabilis*	*Echinogorgia* sp.	*Ellisella plexauroides*	*Menella praelonga*	*Leptogorgia piccola*	*Rumphella aggregate* [Z]
16:1n7	1.10	2.10	1.50	1.60	1.07	1.87	4.8	2.65
16:0	7.50	11.87	11.07	14.00	8.90	8.83	9.7	32.95
18:3n6		0.10				0.10	0.6	1.10
18:4n3	0.40	0.17	0.10		0.13	0.27		2.15
18:2n6	0.90	1.03	1.37	0.70	0.90	0.83	1.0	0.65
18:1n9	2.40	3.27	4.03	3.10	2.67	2.13	2.5	3.85
18:1n7	1.90	2.13	1.63	2.20	2.53	1.93	3.2	0.35
18:0	6.50	6.17	5.80	5.30	7.40	5.30	7.0	9.60
20:4n6	37.20	38.77	40.43	47.60	39.30	39.70	20.5	13.15
20:5n3	1.70	3.27	1.90	2.20	1.97	3.67	8.0	2.05
22:5n6	5.70	1.13	0.70	0.20	0.90	1.27	1.5	0.10
22:6n3	3.90	2.53	1.37	0.80	2.90	2.60	3.8	1.60
22:4n6	0.60	3.83	0.70	0.40	8.97	3.93	2.8	0.65
24:5n6	14.50	7.50	12.33	8.90	3.10	9.13	15.8	3.40
24:6n3	2.30	2.40	1.30	2.60	1.40	2.93	5.3	0.20
Sum TPA	16.8	9.90	13.63	11.50	4.50	12.06	21.1	3.60
Odd+Br	2.50	4.53	5.67	5.20	5.17	4.20	2.3	3.0
20:1+22:1	0.60	1.60	1.05		1.30	1.85		0.60

[Z] Zooxanthellate species.

Table 7 shows the FAs of azooxanthellate, the zooxanthellate species of Gorgonacea, and gorgonarian *Bebryce* sp. with a sponge symbiont [51]. The arachidonic (20:4n6) and palmitic (16:0) acids were dominant in all the species. However, the azooxanthellate species had a 2–3-fold lower content of 16:0. The arachidonic acid content was 40.5%, and the sum of n6 acids reached 54.5% of total FAs in the azooxanthellate species. The contents of EPA and DHA were nearly 2.5%. Bacterial odd- and branched-chain FAs were detected in quite a noticeable amount, up to 6.1% of total FAs, in *Bebryce* sp. The total content of tetracosapolyenoic acids was substantial in all species and reached 11.4% in the azooxanthellate species. *Bebryce* sp. with a symbiotic sponge had 17.2% of demospongic acids with C25, C26, and C28 carbon atoms, which are major FAs in marine and fresh-water sponges of the class Demospongiae [53].

Table 7. Main FA (% of total FAMEs) composition of azooxanthellate, zooxanthellate species of Gorgonacea and *Bebryce* sp. with a sponge symbiont [51].

Fatty Acids	Azooxanthellate Gorgonacea Mean of 6 Species	Zooxanthellate *Rumphella aggregata*	*Bebryce* sp. with Sponge Symbiont
16:1n7	1.54	2.65	1.9
16:0	10.36	32.95	8.9
18:3n6	0.1	1.1	-
18:4n3	0.21	2.15	0.3
18:2n6	0.96	0.65	0.7
18:1n9	2.93	3.85	1.8
18:1n7	2.05	0.35	2.1
18:0	6.08	9.6	5.4
20:4n6	40.5	13.15	21.7
20:5n3	2.45	2.05	2.0
22:5n6	1.65	0.1	4.2
22:6n3	2.35	1.6	3.6
22:4n6	3.07	0.65	1.1
24:5n6	9.24	3.4	7.2
24:6n3	2.16	0.2	0.5
n6/n3	7.5	2.9	5.3
Odd & Br	4.55	3.0	6.1
20:1+22:1	1.07	0.6	1.2
Demospongic acids *			17.2%

* Demospongic acids: Δ5,9-25:2, Δ5,9-26:2, Δ5,9,19-26:3, and Δ5,9,19-28:3.

4.3. *Helioporacea and Stolonifera*

Helioporacea comprises only two monogeneric families that are unique among octocorals in producing calcified skeletons of crystalline aragonite. The well-known blue coral, *Heliopora coerulea*, is distributed widely across the Indo-Pacific where it is a common member of shallow coral reef communities. The enigmatic genus *Epiphaxum* is known from only a few localities at depths of 50–400 m. The phylogenetic relationships of these two families to one another and to other Octocorallia remain uncertain [1]. The FAs in *H. coerulea* have been analyzed only in two studies [48,54] (Table 8), whose data look contradictory: one sample has 18:3n6 as the major PUFA. Furthermore, both samples demonstrated the lowest level (0.2–0.9%) of arachidonic acid found in Octocorallia. Another feature of the FAs from *H. coerulea* is the lowest content of tetracosapolyenoic acids, with approximately 2% of 24:6n3.

Table 8. Major fatty acids (% of total FAMEs) in Helioporacea (*Heliopora coerulea*) and Stolonifera (*Clavularia* sp. and *Carijoa riisei*).

Fatty Acids	*H. coerulea*	*H. coerulea*	*Clavularia* sp.	*Carijoa riisei*	*Carijoa riisei*
16:0	40.9	45.5	14.7	7.4	20.4
16:1n7	3.1	2.3	1.6	0.7	0.6
18:0	7.3	6.6	2.7	4.7	24.6
18:1n9	3.1	2.4	4.0	4.9	8.2
18:1n7	0.2	0.8	0.3	-	0.2
18:2n6	2.1	3.2	6.3	1.8	1.6
18:3n6	15.1	1.0	0.5	9.4	4.8
18:4n3	3.5	2.7	8.3	10.6	4.6
20:3n6	0.9	0.3		0.9	0.9
20:4n6	0.2	0.6	21.6	17.7	7.4
20:5n3	5.4	10.2	7.5	11.4	5.8
22:4n6	0.4	-		0.7	0.2
22:5n3	0.5	1.8	2.6	-	0.1
22:6n3	4.7	9.9	10.5	7.3	3.2
24:5n6	-	-	2.4	8.9	4.0
24:6n3	1.7	2	0.4	2.1	0.0
References	[54]	[48]	[48]	[48]	[48]

The major PUFAs in *Carijoa riisei* (azooxanthellate) and *Clavularia* sp. are AA, 18:4n3, EPA, DHA, and 24:5n6. Of particular interest is the high concentrations of FAs 18:3n6 and 18:4n3, which, together with 18:5n3, are typical of symbiotic zooxanthellae. The study [42] showed AA as the major PUFA (24.2%) and EPA, 22:4n6, and DHA, with levels of nearly 3% each in *C. riisei*. In the cited article, data on only the five major n3 and five n6 PUFAs were presented.

4.4. *Pennatulacea*

Sea pens, or pennatulaceans, are a highly specialized group of Cnidaria. They are benthic sessile animals adapted to living partially buried in sediment. Sea pens are encountered all over the world's oceans and seas and at almost all depths (from the intertidal zone to depths greater than 6100 m). Many deep-sea species have nearly cosmopolitan distributions in such habitats [55]. Sea pens are colonial animals with multiple polyps possessing eight tentacles. A single polyp develops into a rigid, erect stalk (rachis) and loses its tentacles, forming a bulbous 'root' or peduncle at its base.

There are plenty of data on the FA composition of Pennatulacea species, but most publications do not show the presence of the tetracosapolyenoic acids 24:5n6 and 24:6n3, which are major components (constituting up to 20% in total) (Table 9). In deep-sea species, the prominent components were EPA, C20:1 and 22:1 acids, with n3 > n6 in most samples, but in the shallow-water *Rennilla koellikeri*, a high content of AA was found. In all species, the DHA content was low (within the range of 0.5–3.4%), and only in the NLs of the deep-sea *Pavonaria finmarchica* did it reach 9.4%. Information on the FAs of Pennatularia species is generally insufficient and somewhat contradictory. One of the issues is the use of different analytical protocols. In [41], the FAs of pennatulaceans had higher contents of 20:5n3, 22:5n3, 22:4n6, and 24:6n3 and lower contents of 18:4n3 and 24:5n6 compared to those of alcyonaceans. Also, the authors found a similarity of the FA profile with that of the azooxanthellate *Pseudopterogorgia* sp. [39] and carried out an extensive comparative analysis of FAs (without TPAs, as they used a too short analysis time) in cold-water Cnidaria, including 25 samples of sea pens, 16 samples of Gorgonians, and 32 samples of soft corals. Soft corals and gorgonians (the order Alcyonacea) were close in composition and likely fed on phytodetritus resulting from algae, macrophytes and/or foraminifera, while sea pens (the order Pennatulacea) seemed to consume more diatoms and/or zooplankton. The study [56] described the occurrence of TPAs in the stenophagous nudibranch mollusk *Armina maculata* that feeds exclusively on the sea pen *Veretillum cynomorium*. There is also experimental evidence that TPAs are retained in tissues of *A. maculata,* even after 30 days of starvation [57]. In abyssal Pennatulacea species, the uncommon acid 21:4(n-7) was found [35].

Table 9. Pennatulacea FA composition (% of total FAMEs) of genera *Pennatula*, *Pavonaria*, *Malacobelemnon*, and *Veretillum*.

FA	*Pennatula aculeate* (3 Specimens)	*Pavonaria finmarchica* PL	*P. finmarchica* NL	*Rennilla koellikeri* (2 Specimens)	*Veretillum cynomorium* (6 Specimens)	Sea Pens (25 Species)
16:0	7.3–8.9	4.9	8.7	16.1–17.7	8.3–10.9	12.6
16:1n-7	3.2–4.7	0.9	5.0	1.6–2.3	1.2–1.8	4.3
17:0	0.6–0.7	0.4	2.2	2.2–3.0	1.2–1.4	
18:0	2.1–2.6	1.3	0.6	8.0–9.5	5.9–6.8	2.6
18:1n-9	5.6–9.7	2.3	11.3	1.8–2.2	1.2–2.7	10.2
18:1n-7	5.1–7.2	–	–	3.3–4.2	2.0–2.7	3.7
20:1n-9	12.8–14.0	12.2	10.3	0.9–1.8	0.4–0.6	8.1
20:1n-7	3.0–3.4				0.7–1.1	
20:4n6	9.0–15.7	4.3	1.2	31.4–41.9	6.2–10.0	4.6
20:5n3	8.5–15.7	38.3	19.8	4.4–18.3	9.0–12.4	17.1
22:1n-11	10.9–11.4	0.3	2.1			8.2
22:1n-9	4.3–4.7				0.2	4.8
22:4n6	5.5–8.9	0.6	0.3		2.5–3.9	1.2
22:5n3	2.2–2.7	0.3	0.4		0.9–1.3	1.2
22:6n3	1.0–2.2	2.5	9.4	0.7–1.3	0.8–2.2	3.4
24:5n6	ND	6.0	3.6	ND	0.8–1.0	ND
24:6n3	ND	18.0	16.4	ND	7.7–11.5	ND
n3/n6		4.8	6.7	0.21–0.48	2.02–2.88	3.04
C20:1+C22:1	18.2–19.5	12.5	12.4			24.48
References	[31] deep sea	[36] deep sea	[36] deep sea	[58] shallow water	[59] shallow water	[39] deep sea

5. Medusozoa

5.1. Cubozoa, Order Cubomedusae

Cubozoa is a class of the phylum Cnidaria currently comprised of only 36 valid species [1]. Cubozoans are also known as box jellyfish, usually small in size. Their tentacles are located at the corners of the square umbrella margin. The venom of cubozoans is lethal to humans. The data on cubozoans' lipids and FAs are limited to a relatively old communication [60]. Four analyzed species showed similar FA compositions. The major PUFA were n3 family: 22:5n3, EPA, and DHA (Table 10). The content of AA was low; the level of n6 acid was increased due to the presence of the less common FA 22:4n6 only in *Maeotias inexpectata*. According to the published database of FA microalgae [61], significant amounts of the fatty acid were found in microalgae from Prymnesiophyceae, Cryptophycea, and Bacillariophyceae. These microalgae are likely used as food or/and symbionts. The contents of C18 PUFA and C20:1 and C22:1 acids were generally low, 2–5 and 1.1–2.0%, respectively.

Table 10. Major FA (% of total FAMEs) in the cubozoan families Olindiidae, Bougainvilliidae, Chirodropidae, and Tamoyidae [60].

FA	*Maeotias inexpectata*	*Nemopsis bachei*	*Chiropsalmus* sp.	*Tamoya haplonema*
16:0	9.5	8.5	17.6	8.5
16:1	2.6	1.8	4.9	2.0
18:0	10.1	9.1	10.4	9.3
18:1	4.7	3.7	6.6	6.6
18:2n6	1.0	1.2	1.2	1.2
18:3n3	0.4	2.0	0.8	0.5
18:4n3	0.5	2.3	0.23	0.7
20:0	0.3	1.7	0.3	0.2
20:1	0.8	2.0	1.3	1.1
22:1	0.5			0.5
AA	1.3	3.3	3.3	5.0
EPA	2.6	19.1	17.6	18.4
22:4n6	13.4	2.5	3.1	3.3
22:5n6	1.6	0.7		0.7
22:5n3	21.2	14.7	17.6	18.4
DHA	16.3	20.0	6.7	15.3
(n3)/n6	2.4	8.8	7.6	5.4

5.2. *Hydrozoa*

Hydrozoa is a speciose class of Cnidaria comprised of more than 3500 species. They form colonies of polyps and free-swimming medusa [1]. Most hydrozoans are predators or filter feeders. A few species have symbiotic zooxanthellae. Colonial polyps can secret chitinous or calcareous coatings similar to that in Scleractinia.

Family Milleporidae

Corals of the genus *Millepora* are known as fire coral for their painful stings to humans. They look like Scleractinia corals but are related to Hydrozoa. The species are characterized by the dominance of C22 PUFA 22:4n6, 22:5n6, and 22:6n3 and low levels of AA and EPA (Table 11). The n3/n6 ratio varies between 2.0 and 5.1. The fatty acids of *Millepora* species and scleractinian reef-building corals differ in many features (Table 4). The samples from Okinawa [62] had unexpectedly very low concentrations of PUFA, only 21.4%. This can be explained by the accumulation of neutral lipids with a high concentration of saturated and monoenoic acids.

Table 11. Major FA composition (% of total FAs) of hydrozoan corals of the genus *Millepora* from Vietnam (V) and the Seychelles (S) [47,62].

Fatty Acid	*Millepora* sp. V	*M. platyphylla* V	*M. dichotoma* V	*M. platyphylla* S	*M. dichotoma* S	*M. murrayi* Okinawa
16:0	6.3	23.6	19.8	17.6	18.9	29.0
16:1n7	-	0.1	-	0.2	-	1.0
18:0	7.1	15.4	15.3	21.3	19.4	17.4
18:1n9	1.4	6.1	3.9	2.4	3.2	2.8
18:2n6	0.6		0.1′	0.4	0.7	2.1
18:3n6		0.2		0.2	0.4	2.4
18:4n3	6.9	1.5	1.9	5.2	4.8	10.8
20:0		3.3	5.5	-	-	-
20:1n9	0.3	0.2	0.4	6.3	6.1	-
20:4n6	0.4	0.7	-	0.3	1.0	0.3
20:5n3	1.1	0.4	0.8	0.4	0.6	-
22:4n6	3.8	2.6	3.5	4.6	3.7	-
22:5n6	8.3	6.8	7.3	10.0	8.5	-
22:5n3	-	0.4	1.1	0.7	0.9	0.3
22:6n3	61.5	32.0	33.3	27.5	28.0	1.7
Saturated	14.0			39.0	40.0	62,2
PUFA	83.9			51.0	49.6	21.4
n3/n6	5.1	2.3	3.3	2.0	2.3	

5.3. *Families Physaliidae and Velellidae*

The Portuguese man-of-war *Physalia physalis* (the family Physaliidae) and the wind sailor *Velella velella* (the family Velellidae) are two pleustonic hydrozoan species. *Physalia physalis* inhabits tropical and subtropical waters, while *V. velella* occurs in warm and temperate waters. In both species, DHA, EPA and 18:4n3 as major PUFAs and a very low content of n6 acids were found, while the n3/n6 ratios were 10.8 and 12.4 [63] (Table 12). The presence of C18 PUFA suggested the existence of zooxanthellae. The difference between the two species was not significant. However, previous data on *Physalia* FAs [64] and unpublished results showed arachidonic acid as a major component, and the n3/n6 ratios were only 1.0 and 1.7. In all species, the contents of C20:1 and C22:1 and odd- and branched-chain acids were low.

Table 12. FA composition (% of total FAs) of the hydrozoan families Physaliidae and Velellidae.

Fatty Acids	*Physalia physalis*	*Physalia physalis*	*Physalia physalis*	*Velella velella*
16:0	22.6	17.5	22.4	16.0
16:1n7	3.9	1.2	0.8	1.2
7M7-16:1	3.2	6.1		
17:0	1.0	1.1	1.2	0.4
18:0	9.1	15.4	9.9	5.0
18:1n9	5.1	3.4	4.9	7.3
18:1n7	1.6	0.9	1.2	0.5
18:2n6	1.4	0.6		1.2
18:3n6	0.5	0.8		0.9
18:3n3	0.9	0.4		1.0
18:4n3	1.2	0.2	2.1	3.8
20:1n9	0.6	1.2	0.5	4.2
AA	9.1	13.2	0	0.3
20:3n3		0.1	2.3	0.1
EPA	7.7	5.4	6.5	7.8
22:4n6	0.7	2.2	0.3	0.5
22:5n6	4.3	6.7	1.6	0.5
22:5n3	1.1	1.4	1.9	1.4
DHA	15.5	16.2	22.9	27.6
n3/n6	1.7	1.0	12.5	10.9
Reference	[64]	-	[63]	[63]

5.4. *Hydromedusae*

Hydromedusae are a diverse group (>800 species worldwide), in which most representatives are <10 mm in size. All species are carnivorous, capturing prey with specialized stinging cells referred to as nematocysts. Most hydrozoans show an alternation between the polyp and medusa phases. The difference between most hydrozoans and scyphozoans is that the polyp stage in the former usually predominates, with medusae small in size or sometimes absent. The major FAs in all species were DHA and EPA (Table 13). The contents of arachidonic and other n6 acids were generally low, 0–1.4%, and the n6/n3 ratio was no greater than 0.1%. Only *Aequorea victoria* was distinguished by marked quantities of both 20:4n6 and 22:5n6. In most species, the presence of high levels of such acids as 20:1, 22:1, and 18:1n9 was associated with carnivorous feeding.

Table 13. Main FA composition (% of total FAs) of the Hydromedusae families [a] Thecata, [b] Rhopalonematidae, [c] Bythotiaridae, and [d] Physaliidae.

FA	*Aequorea victoria* [a]	*Arctapodema ampla* [b]	*Calycopsis borchgrevinki* [c]		*Dimophyes arctica* [d]	*Diphyes antarctica* [d]	
16:0	7.7	14.9	16.6	11.6	18.9	18	20.1
18:0	8.9	5.1	4.8	7.8	9.3	8.8	7.6
16:1n7	2.6	2.8	9.2	5.2	15.3	4.0	3.8
18:1n9	5.4	5.7	26.4	14.3	17.2	8.7	7.3
18:1n7	1.1	2	3.7	1.9	4.1	2.0	4.1
20:1n9	2.1	1.8	2.5	4.4	1.1	1.2	1.7
20:1n7	3.1	2	2.8	6.9	1.0	3.6	0.6
22:1n9	6.2	2.7	0	0.3	1.3	0.1	0
18:2n6	0.4	0.9	1.5	0.9	1.0	2.2	1.9
18:4n3	0.4	0.2	0.1	0.1	0.2	2.1	0.6
20:4n6	9.6		1.4		0	1.4	1.2
20:5n3	9.6	16.4	9.6	9.9	8.2	16.5	19.1
22:4n6	0.3			2.1			
22:5n6	4.4						
22:5n3	1.5	0.2	2.1	5.6	0.1	0.2	0.3
22:6n3	18.8	26.4	7.0	16.1	10.4	16.9	17.6
Odd+Br	1.7	1.7	0.3	1.2	0.8	1.1	1.0
20:1+22:1	11.4	6.5	5.3	12.5	3.4	5.1	2.3
n6/n3	0.17	0.02	0.08	0.09	0.05	0.06	0.05
References	[65]	[66]	[67]	[66]	[67]	[67]	[67]

6. Scyphozoa

The class Scyphozoa, or 'true medusae', includes the orders Coronatae (mainly deep-sea species), Rhizostomeae, and Semaeostomeae. The total number of species is greater than 220 [1]. The most commonly known scyphozoans of the genera *Aurelia*, *Cyanea*, and *Chrysaora* are usually found near beaches. Jellyfish range in size from a dozen of millimeters to more than two meters in diameter. Jellyfish often get caught in fishing gear and nets along with fish and then are discarded. Nevertheless, the jellyfish catch accounts for about 3% of total fish landing [68]. Jellyfish can be used as a sustainable source of high-value compounds with biotechnological applications [7]. The largest (in size) species is *Cyanea arctica*, whose tentacles may reach over 40 m in length [1]. All medusae are carnivorous, but one of the *Cassiopea* species, Rhizostomeae, is known to possess symbiotic zooxanthellae. The major PUFAs are best known and studied: EPA, DHA and AA. The proportion of C20:1+C22:1 was high, up to 17.8% (Table 14).

Table 14. Major FA compositions (% of total FAs) of the Scyphozoa order Semaeostomeae.

FA	*Aurelia aurita*	*Aurelia aurita*	*Cyanea lamarckii*	*Cyanea capillata*	*Chrysaora isosceles*
14:0	2.3–3.3	3.3	3.5	1.4–2.2	3.3
16:0	21.7–28.8	16.0	19.0	10.1–14.4	9.5
17:0	1.3–2.8		1.0	1.8–2.1	0.6
18:0	12.5–17.8	6.4	12.0	4.5–7.1	7.1
16:1n7	3.1–6.0	4.6	4.7	2.3–5.2	3.7
18:1n9	1.4–2.6	8.9	5.5	4.6–7.2	4.4
18:1n7	2.2–2.4	2.6	3.1		1.5
20:1	0.3–0.6	20.4.8	3.4	6.8–12.5	6.6
22:1	3.0–10.0	3.2	2.5	1.9–2.4	6.1
18:3n3	0.6–0.8	4.5	0.4	0.3–0.8	0.6
18:4n3	0.4	0.1	0.6	0.5–0.6	1.3
20:4n6	2.8–8.6	6.7	8.7	6.2–9.1	5.4
20:5n3	10.2–15.6	8.5	13.8	9.8–19.4	20.0
22:4n6		0.6		0.9–2.0	
22:5n6				0.2–1.0	
22:5n3	1.3–2.5	0.3	3.2	2.0–4.8	5.4
22:6n3	3.0–6.1	7.0	12.1	10.4–20.4	19.7
C20:1+C22:1	3.3–10.6	17.8	5.9	9.2–14.4	12.7
Reference	[69]	[70]	[71]	[72]	[71]

For *Atolla* species, Coronatae, the major FAs were EPA, DHA, and DPA3 (Table 15). They showed marked concentrations of the monoenoic acids 16:1n7, 18:1n7 and n9, and 20:1n9. These fatty acids are typical of carnivorous species. Three Rhizostomeae species had 18:0,18:1n9, EPA and DHA as major FAs and much lower contents of 16:1n7, 18:1n7 and C20:1 and C22:1 than those recorded from *Atolla*. The FA composition of the zooxanthellate *Cassiopea* looked somewhat different [73]. The presence of 18:3n6 and 18:4n3 indicated symbiotic dinoflagellates, but the PUFA content was low, only 18.4%. Also, there were very low concentrations of EPA and DHA.

Tetracosapolyenoic acids have long been known as characteristic of Octocorallia. They have been found as major acids in Alcyonacea, Gorgonacea, Pennatulacea, and Stolonifera. However, Helioporacea shows some distinguishing features. Specimens of this taxon collected from Vietnam had calcified skeletons of crystalline aragonite and contained only small amounts of 24:6n3 (almost 2.0%), which were the lowest for Cnidaria. I found four papers where TPAs were detected: *Catostylus tagi* contained only one TPA, 24:5n6 and *Pelagia noctiluca* had the rare acid 24:4n6, while *Aurelia aurita* and *Rhopilema asamushi* had mainly 24:6n3 (Table 16).

Table 15. Major FAs (% of total FAs) of the Scyphozoa orders Coronatae and Rhizostomeae.

FA	Coronatae	Rhizostomeae			
	Atolla wyvillei	*Cotylorhiza tuberculata*	*Rhizostoma octopus*	*Phyllorhiza punctata*	*Cassiopea* sp. [Z]
14:0	2.0–6.1	2.9	5.1	7.7	3.1
16:0	16.0–20.8	26.1	27.3	26.0	26.7
18:0	2.8–11.9	24.2	21.7	17.7	5.4
16:1n7	5.3–2.6	1.2	3.8	3.8	1.9
18:1n9	11.8–9.0	12.8	6.8	9.3	7.25
18:1n7	7.2–4.1	1.2	3.5		
20:1n9	4.6–7.3		0.8		
22:1n9	0.4–0.5		-		1.5
18:2n6	2.5–1.4	8.3	1.6	2.4	1.65
18:3n3	1.0–0.5		1.6	1.2	
18:3n6					0.9
18:4n3	0.9–5.0		2.8		4.1
20:3n6				5.1	
20:4n6	0–4.2	5.3	2.8	6.2	4.5
20:5n3	16.6–15.0	5.1	9.7	6.3	1.0
22:5n3	2.4–9.7		1.3		1.0
22:6n3	5.6–3.9	7.2	5.3	7.4	2.6
C20:1+C22:1	4.8–7.8				
Reference	[67]	[72]	[70]	[7]	[73]

[Z] Zooxanthellate species.

Table 16. Major FA composition (% of total FAs) of Scyphozoa medusae with C24 and C26 PUFAs.

Fatty Acids	*Aurelia aurita*	*Aurelia aurita*	*Rhopilema asamushi*	*Catostylus tagi*	*Pelagia noctiluca*
20:4n6	9.9 ± 2.3	1.84 ± 0.06	8.39 ± 3.85	7.5	18.5–9.0
20:5n3	14.1 ± 1.9	33.28 ± 0.80	13.05 ± 3.90	13.5	14.6–10.1
22:6n3	9.8 ± 1.6	11.20 ± 1.25	12.29 ± 1.97	11.2	9.7–5.6
22:4n6	0.6 ± 0.0	0.26 ± 0.02	2.35 ± 1.96		4.0–2.4
22:5n3	1.1 ± 0.2	5.03 ± 0.41	5.07 ± 1.54	2.1	
24:5n6	1.1 ± 0.2	0.00 ± 0.00	0.44 ± 0.24	3.9	
24:6n3	9.3 ± 1.8	9.48 ± 1.00	6.00 ± 1.96		
24:4n6			0.69 ± 0.62		2.4–1.1
24:5n3		0.10 ± 0.01	0.31 ± 0.06		5.3–2.0
24:4n3			0.47 ± 0.17		
26:7n3			0.55 ± 0.27		
26:6n3			0.44 ± 0.16		
26:5n3			0.63 ± 0.39		
(n3)/n6	2.61	21.06	3.09	3.7	
∑C24 PUFA	10.4	9.58	7.90	3.9	7.7–3.1
∑C26 PUFA			1.62		
References	[74]	[75]	[75]	[76]	[77]

Recent data on the presence of tetracosapolyenoic acids in Scyphozoa medusae suggest a wider distribution of tetracosapolyenoic acids. As experiments have shown, TPA causes a more pronounced inhibition of the synthesis of triglycerides and sterol esters than EPA and DHA [78]. The hexacosapolyenoic acids (HPAs) 26:5n3, 26:6n3, and 26:7n3 were found in *Rhopilema asamushi*. Their chemical structure was confirmed by GC–MS, and their total content in *Rh. asamushi* was 1.6% [75]. All members of this group had high proportions of PUFAs (48.3–66.0%); the n3/n6 ratio was 2.6–21.1. The total proportion of tetracosapolyenoic acids in Scyphomedusae was 3.1–10.4%.

7. Summary

This review provides an overview of the available data on fatty acid (FA) compositions for various cnidarian taxa. Obviously, for reliable results, standard methods are necessary that allow us to analyze a very extensive set of C12–C28 fatty acids and those with 1–7 double bonds. Otherwise, some important FAs can be omitted because of too long retention times, like C24–C26 PUFAs or demospongic acids C24–C28 with 2–3 double bonds, or can be destroyed, like 18:5n3 in basic transesterification.

7.1. Class Hexacorallia

In Actiniaria species, the major PUFAs were DHA, 22:5n3 (DPA3), EPA, and 22:4n6, with n3 FAs prevailing over n6 FAs. The content of arachidonic acid was relatively low, 2.6–7.7%, and was always lower than EPA. The total amount of C22 PUFA was higher than that of C20 PUFA. Abyssal and cold-water species accumulated monoenoic C20:1 and C22:1 acids. Together with noticeable concentrations of 18:1n9 and DHA, they indicated feeding on zooplankton. Of particular interest are the markedly high values of 22:5n3 and 22:4n6.

The order Antipatharia: *Stauropathes arctica* showed a somewhat simplified FA composition, the highest levels of C20 and C22 monoenoic acids (34.6%), EPA and DPA3 as major PUFAs (with a total of 27.7%), and surprisingly low contents of AA and DHA. The zoantharian genus *Palythoa* showed the dominance of the n6 acids AA and 22:4n6 and only traces of C20:1 and C22:1. The presence of 18:3n6, 18:4n3, and even 18:5n3 indicated the contribution of symbiotic zooxanthellae. The input of bacterial FAs was also noticeable.

The order Scleractinia: To date, these corals have been quite comprehensively investigated. The major unsaturated FAs in them were 18:1(n-9), 20:4n6, 20:5n3, 22:4n6, and 22:6n3. Some of the coral families had significant levels of characteristic FAs: 20:3n6 for Pocilloporidae, 18:3n6, 18:4n3, and 22:4n6 for Acroporidae, and 18:3n6 for Poritidae. The noticeable amounts of 18:3n6 and 18:4n3 indicated the input of symbiotic microalgae. The asymbiotic specimens of *Tubastrea* showed a significant difference from the symbiotic specimens. The contents of major FAs such as 18:1n9 (23.3–26.4%), 20:5n3 (10.9–14.9%), and 22:5n6 (16.4–17.3%) were multifold higher than in the symbiotic species. Also, the content of 22:6n3 was surprisingly low (only 1.3–1.4%), while in the symbiotic species, the level of DHAs ranged between 5.3 and 16.9%.

7.2. Class Octocorallia

The fundamental difference between Octocorallia and Hexacorallia consists of the presence of the tetracosapolyenoic acids 24:6(n3) and 24:5n6 in the former. These FAs are common and major components of Alcyonacea, Gorgonacea, Helioporacea, Pennatulacea, and Stolonifera.

The order Alcyonacea: This order was characterized by the dominance of n6 acids (with the n6/n3 ratio being 2–3). In all species, the major PUFAs were AA, 18:3n6, DHA, EPA, 24:5n6, and 18:4n3. The high concentrations of the FAs 18:3n6 and 18:4n3 indicated the major role of zooxanthellae in the food balance in Alcyonacea. The levels of bacterial odd- and branched chain acids in Lobophytum, Sarcophytum, and Dendronephthya were unusually high. The asymbiotic species had the highest concentration of tetracosapolyenoic acids, 10.0%. The absence of zooxanthellae and lack of TPA in any available food suggest the biosynthesis of TPA in Octocorallia.

The order Gorgonacea: Most Gorgonacea do not have zooxanthellae. They were characterized by marked contents of major PUFAs such as arachidonic acid (up to 47.6% in *Echinogorgia* sp.) and TPA (with a total of 4.5–21.1%, mainly 24:5n6). The low levels of EPA, C20:1 and 22:1 and a noticeable concentration of bacterial FA suggest the suspension feeding mode. Zooxanthellate *Rumphella* species were distinguished by the presence of FAs typical of symbiont microalgae, 18:3n6 and 18:4n3, and a lower proportion of TPA. The FA composition of *Bebryce* sp. with an unusual sponge symbiont included 17.2% of the demospongic acids Δ5,9-25:2, Δ5,9-26:2, Δ5,9,19-26:3, and Δ5,9,19-28:3, which are major FAs in marine and freshwater sponges of the class Demospongiae [53]. Moreover, this species had 8.8% of odd- and branched-chain bacterial FAs.

The order Helioporacea is unique among octocorals in producing calcified skeletons of crystalline aragonite. Lipids of *Heliopora* had high concentrations of the saturated acids 16:0 and 18:0 and an unexpectedly low content of AA, less than 1%. Another feature of the FAs from *H. coerulea* was the lowest content of tetracosapolyenoic acids, with approximately 2% of 24:6n3. The major PUFAs in *Carijoa riisei* (azooxanthellate) and *Clavularia* sp. were AA, 18:4n-3, EPA, DHA, and 24:5n-6. The noticeable proportion of 18:3n6 and 18:4n3 suggests the essential role of zooxanthellae.

The order Pennatulacea: There are plenty of data on the FA composition of Pennatulacea species, but most of the publications do not report the presence of the tetracosapolyenoic acids 24:5n6 and 24:6n3, which are major components (with a total amount of up to 20%). In the species with detected TPA, the rest of the PUFAs were EPA, AA, 22:4n6, a low level of DHA, and noticeable levels of C20:1 and C22:1. A comparative analysis of 25 sea pen samples showed that these animals apparently consumed more diatoms and/or zooplankton.

7.3. Class Medusozoa

The order Cubomedusae is a non-speciose axon of the phylum Cnidaria, also known as box jellyfish, usually small in size. The data on FAs in cubozoans are limited to a relatively old communication. The major PUFAs were the n3 family: 22:5n3, EPA, and DHA. They had noticeable levels of the symbiotic acids 18:3n6 and 18:4n3, a high n3/n6 ratio (5.4–8.8), and low contents of C20:1 and C22:1 acids, which may be explained by the contribution of symbiotic algae and feeding on diatoms. *Maeotias inexpectata* differed by low levels of AA and EPA and an unexpectedly high content of AA and EPA homologs, the acids 22:4n6 (13.4%) and 22:5(n3) (21.2%).

7.4. Class Hydrozoa

The hydrozoan family Milleporiidae is commonly known as 'stinging coral' or 'fire coral'. Millepores are voracious zooplankton feeders that can also obtain part of their nutrition from autotrophic sources, i.e., photosynthetic production by symbiotic zooxanthellae. The dominance of C22 PUFAs, especially DHA, 22:5n6 and 22:4n6, is characteristic of the *Millepora* species. In some species, C20:1 acid is present in noticeable amounts. In addition, they contain 18:3n6 and 18:4n3, characteristic of symbiotic microalgae. These data and the exceptionally high level of DHA suggest polytrophic feeding on zooplankton, microalgae, and symbiotic zooxanthellae.

Physalia (from the family Siphonophora) and *Velella* (Anthoathecata) are pleustonic hydrozoan genera. In both, DHA, EPA, and 18:4n3 as major PUFAs and a very low content of n6 acids were found. The n3/n6 ratios were high, 10.8 and 12.4, respectively. The presence of C18 PUFA suggests the existence of the zooxanthellae symbiont. However, two previous studies on *Physalis* showed AA as a major acid, and the n3/n6 ratios were 1.0 and 1.7. The low values of C20:1 and C22:1 and the high concentrations of DHA indicated feeding on microalgae.

In the Hydromedusae families Thecata, Rhopalonematidae, Bythotiaridae, and Physalidae, the major FAs were DHA and EPA. The contents of arachidonic and other n6 acids were generally low, 0–1.4%, and the n6/n3 ratio was no greater than 0.1%. Only *Aequorea victoria* differed by marked contents of both 20:4n6 and 22:5n6. In most species, the presence of high levels of such acids as 20:1, 22:1 and 18:1n9 could be associated with carnivorous feeding.

7.5. Class Scyphozoa

The class Scyphozoa, or 'true medusae', consists of the orders Coronatae (mainly deep-sea species), Rhizostomeae, and Semaeostomeae. All medusae are carnivorous, but one of the *Cassiopea* species (Rhizostomeae) is known to possess symbiotic zooxanthellae. The FA compositions of common medusae from the order Semaeostomeae were characterized by such major components as EPA, DHA and AA. Low contents of C18 PUFA and significant levels of C20:1 andC22:1 and 18:1n9 are typical of carnivorous feeding. *Atolla wyvillei* (Coronatae) showed an FA composition similar to that of the Semaeostomeae species. The rest of the Rhizostomeae species demonstrated the lack of C20:1 and C22:1 acids, which may be explained by feeding on microalgae and unspecific suspended particulate matter in the water column.

Tetracosapolyenoic acids have long been known as characteristic of Octocorallia. Nevertheless, I found four papers where TPAs were reported for medusae: *Catostylus tagi* contained only one TPA, 24:5n6 and *Pelagia noctiluca* had the rare acid 24:4n6, while *Aurelia aurita* and *Rhopilema asamushi* had mainly 24:6n3. These data on the presence of tetracosapolyenoic acids in Scyphozoa medusae suggest a wider distribution of tetracosapolyenoic acids. The total proportion of tetracosapolyenoic acids in Scyphomedusae ranged between 3.1 and 10.4%.

In this review, I presented contemporary data on fatty acids in the main taxa of the ancient marine phylum Cnidaria. Numerous cnidarian species inhabit very different biotopes, from tropical and arctic seas to abyssal depth. This investigation on FA distribution in cnidarians gives information on the influence of many factors such as the taxonomy, depth, temperature, feed and presence of symbionts on the lipid and FA biochemistry of Cnidaria.

Funding: This research received no external funding.

Institutional Review Board Statement: Not applicable.

Informed Consent Statement: Not applicable.

Data Availability Statement: The data presented in this study are available on request from the corresponding author.

Acknowledgments: Special thanks are due to E. Shvetsov (from the A.V. Zhirmunsky National Scientific Center of Marine Biology, Far Eastern Branch, Russian Academy of Sciences) for proofreading the English of the manuscript.

Conflicts of Interest: The author declare no conflict of interest.

References

1. Daly, M.; Brugler, M.R.; Cartwright, P. The phylum Cnidaria: A review of phylogenetic patterns and diversity 300 years after Linnaeus. *Zootaxa* **2007**, *1668*, 127–182. [CrossRef]
2. WorMS. Available online: http://www.marinespecies.org (accessed on 10 November 2024).
3. Crossland, C.J.; Hatcher, B.G.; Smith, S.V. Role of coral reefs in global ocean production. *Coral Reefs* **1991**, *10*, 55–64. [CrossRef]
4. Cairns, S. Deep-water corals: An overview with special reference to diversity and distribution of deep-water Scleractinia. *Bull. Mar. Sci.* **2007**, *81*, 311–322.
5. West, E.J.; Pitt, K.A.; Welsh, D.T.; Koop, K.; Rissik, D. Top-down and bottom-up influences of jellyfish on primary productivity and planktonic assemblages. *Limnol. Oceanogr.* **2009**, *54*, 2058–2071. [CrossRef]
6. Lebrato, M.; Mendes, P.D.J.; Steinberg, D.K.; Cartes, J.E.; Jones, B.M.; Birsa, L.M.; Benavides, R.; Oschlies, A. Jelly biomass sinking speed reveals a fast carbon export mechanism. *Limnol. Oceanogr.* **2013**, *58*, 1113–1122. [CrossRef]
7. D'Ambra, I.; Merquiol, L. Jellyfish from Fisheries By-Catches as a Sustainable Source of High-Value Compounds with Biotechnological Applications. *Mar. Drugs* **2022**, *20*, 266. [CrossRef]
8. Prieto, L.; Enrique-Navarro, A.; Volsi, R.L.; Ortega, M.J. The Large Jellyfish *Rhizostoma luteum* as Sustainable a Resource for Antioxidant Properties, Nutraceutical Value and Biomedical Applications. *Mar. Drugs* **2018**, *16*, 396. [CrossRef]
9. Imbs, A.B.; Latyshev, N.A.; Dautova, T.N.; Latypov, Y.Y. Distribution of lipids and fatty acids in corals by their taxonomic position and presence of zooxanthellae. *Mar. Ecol. Progr. Ser.* **2010**, *409*, 65–75. [CrossRef]
10. Simon, G.; Rouser, G. Phospholipids of the Sea Anemone: Quantitative Distribution; Absence of Carbon-Phosphorus Linkages in Glycerol Phospholipids; Structural Elucidation of Ceramide Aminoethylphosphonate. *Lipids* **1967**, *2*, 55–59. [CrossRef] [PubMed]
11. Lam, N.C.; Nguyen, K.H.; Stekhov, V.B.; Svetashev, V.I. Phospholipids and fatty acids of gorgonian corals. *Sov. J. Mar. Biol.* **1981**, *7*, 366–369.
12. Latyshev, N.A.; Nguen, H.K.; Do, N.T.; Svetashev, V.I. Composition and Seasonal Variations of Phospholipids Content of Alcyonarians. *Russ. J. Mar. Biol.* **1986**, *3*, 52–56.

13. Imbs, A.B.; Demina, O.A.; Demidkova, D.A. Lipid Class and Fatty Acid Composition of the Boreal Soft Coral *Gersemia rubiformis*. *Lipids* **2006**, *41*, 721–725. [CrossRef] [PubMed]
14. Honma, T.; Shiomi, K. Peptide toxins in sea anemones: Structural and functional aspects. *Mar. Biotechnol.* **2006**, *8*, 1–10. [CrossRef]
15. Christie, W.W. Phosphonolipids. The Lipid Map. Available online: https://www.lipidmaps.org/ (accessed on 20 June 2024).
16. Allemand, D.; Ferrier-Pagès, C.; Furla, P.; Houlbrèque, F.; Puverel, S.; Reynaud, S.; Tambutté, E.; Tambutté, S.; Zoccola, D. Biomineralisation in reef-building corals: From molecular mechanisms to environmental control. *Comptes Rendus Palevol.* **2004**, *3*, 453–467. [CrossRef]
17. Bligh, E.G.; Dyer, W.J. A rapid method of total lipid extraction and purification. *Can. J. Biochem. Physiol.* **1959**, *37*, 911–917. [CrossRef]
18. Conlan, J.A.; Melissa, M.R.; Francis, D.S. A comparison of two common sample preparation techniques for lipid and fatty acid analysis in three different coral morphotypes reveals quantitative and qualitative differences. *PeerJ* **2017**, *5*, e3645. [CrossRef] [PubMed]
19. Christie, W.W. (Ed.) *Advances in Lipid Methodology—Two*; Oily Press: Dundee, UK, 1993; pp. 69–111.
20. Ghioni, C.; Porter, A.E.A.; Sadler, I.H.; Tocher, D.R.; Sargent, J.R. Cultured fish cells metabolize octadecapentaenoic acid (all-*cis* D3,6,9,12,15–18:5) to octadecatetraenoic acid (all-*cis* D6,9,12,15–18:4) via its 2-*trans* intermediate (*trans* D2, all-*cis* D6,9,12,15–18:5). *Lipids* **2001**, *36*, 145–152. [CrossRef] [PubMed]
21. Svetashev, V.I.; Imbs, A.B. Isomerization of octadecapentaenoic acid (18:5n3) in algal lipid samples under derivatization for GC and GC-MS analysis. *J. Phycol.* **2014**, *50*, 322–327. [CrossRef] [PubMed]
22. Stránský, K.; Jursík, T.; Vítek, A.; Skořepa, J. An improved method of characterizing fatty acids by equivalent chain length values. *J. High Resol. Chromatogr.* **1992**, *15*, 730–740. [CrossRef]
23. Wolff, R.L.; Christie, W.W. Structures, practical sources (gymnosperm seeds), gas-liquid chromatographic data (equivalent chain lengths), and mass spectrometric characteristics of all-cis Δ5-olefinic acids. *Eur. J. Lipid Sci. Technol.* **2002**, *104*, 234–244. [CrossRef]
24. Chrombox. Available online: https://www.chrombox.org/ (accessed on 20 August 2024).
25. Drazen, J.C.; Phleger, C.F.; Guest, M.A.; Nichols, P.D. Lipid, sterols and fatty acids of abyssal polychaetes, crustaceans, and a cnidarian from the northeast Pacific Ocean: Food web implications. *Mar. Ecol. Prog. Ser.* **2008**, *372*, 157–167. [CrossRef]
26. Hooper, S.N.; Ackman, R.G. Trans-6-hexadecenoic acid and the corresponding alcohol in lipids in the sea anemone *Metridium dianthus*. *Lipids* **1971**, *6*, 341–346. [CrossRef]
27. Mason, W.T. Isolation and characterization of the lipids of the sea anemone, *Metridium senile*. *Biochim. Biophys. Acta* **1972**, *280*, 538–544. [CrossRef] [PubMed]
28. Leal, M.C.; Nunes, C.; Kempf, S.; Reis, A.; da Silva, T.L.; Serôdio, J.; Cleary, D.F.R.; Calado, R. Effect of light, temperature and diet on the fatty acid profile of the tropical sea anemone *Aiptasia pallida*. *Aquac. Nutr.* **2013**, *19*, 818–826. [CrossRef]
29. Lanza, M.; Casili, G.; Torre, G.L.; Daniele, G.; Rotondo, A.; Esposito, E.; Ardizzone, A.; Rando, R.; Bartolomeo, G.; Albergamo, A.; et al. Antioxidants Properties of a New Food Supplement Containing Actinia equina Extract. *Antioxidants* **2020**, *9*, 945. [CrossRef] [PubMed]
30. Revel, J.; Massi, L.; Mehiri, M.; Boutoute, M.; Mayzaud, P.; Capron, L.; Sabourault, C. Differential distribution of lipids in epidermis, gastrodermis and hosted *Symbiodinium* in the sea anemone *Anemonia viridis*. *Comp. Biochem. Physiol. Part A* **2016**, *191*, 140–151. [CrossRef]
31. Parzanini, C.; Parrish, C.C.; Hamel, J.-F.; Mercier, A. Functional diversity and nutritional content in a deep-sea faunal assemblage through total lipid, lipid class, and fatty acid analyses. *PLoS ONE* **2018**, *13*, e0207395. [CrossRef] [PubMed]
32. Quesada, A.; Schoo, K.; Bingham, B. Effect of symbiotic state on the fatty acid composition of *Anthopleura elegantissima*. *Mar. Ecol. Prog. Ser.* **2016**, *545*, 175–187. [CrossRef]
33. Carballeira, N.M.; Medina, J.R. New D5,9 fatty acids in the phospholipids of the sea anemone *Stoichactic helianthus*. *J. Nat. Prod.* **1994**, *57*, 1688–1695. [CrossRef]
34. Carballeira, N.M.; Reys, M. Identification of a New 6-Bromo-5,9-eicosadienoic Acid from the Anemone *Condylactis gigantea* and the Zoanthid Palythoa caribaeorum. *J. Nat. Prod.* **1995**, *58*, 1689–1694. [CrossRef] [PubMed]
35. Kharlamenko, V.I.; Odintsova, N.A. Unusual methylene-interrupted polyunsaturated fatty acids of abyssal and hadal invertebrates. *Progr. Oceanogr.* **2019**, *178*, 102132. [CrossRef]
36. Vysotskii, M.V.; Svetashev, V.I. Identification, isolation and characterization of tetracosapolyenoic acids in lipids of marine coelenterates. *Biochim. Biophys. Acta* **1991**, *1083*, 161–165. [CrossRef] [PubMed]
37. Sun, J.; Hamel, J.F.; Shikon, V.; Cossignani, E.F.; Mercier, A. Trophic ecology, diet and feeding behaviour of three bathyal sea anemones (Actiniaria: Cnidaria) in the Northwest Atlantic. *Deep Sea Res. Part I Oceanogr. Res. Pap.* **2022**, *179*, 103678. [CrossRef]
38. Stefanov, K.L.; Christie, W.W.; Brechany, E.Y.; Popov, S.S.; Andreev, S.N. Lipid composition of the red and green forms of *Actinia equinia* from the Black sea. *Comp. Biochem. Physiol.* **1992**, *103B*, 687–690. [CrossRef]
39. Salvo, F.; Hamoutene, D.; Hayes, V.E.W.; Edinger, E.N.; Parrish, C.C. Investigation of trophic ecology in Newfoundland cold-water deep-sea corals using lipid class and fatty acid analyses. *Coral Reefs* **2018**, *37*, 157–171. [CrossRef]

40. Miralles, J.; Diop, M.; Ferrer, A.; Kornprobst, J.M. Fatty acid composition of five Zoantharia from the Senegalese coast: *Palythoa dartevellei* Pax, *P. monodi* Pax, *P. variabilis* Duerden, *P. senegalensis* Pax and their associated zooxanthellae, and *P. senegambiensis* Carter and its commensal, the Decapoda *Diogenes ovatus* Miers. *Comp. Biochem. Physiol.* **1989**, *94B*, 91–97.
41. Imbs, A.B. Lipid class and fatty acid compositions of the zoanthid *Palythoa caesia* (Anthozoa: Hexacorallia: Zoanthidea) and its chemotaxonomic relations with corals. *Biochem. Syst. Ecol.* **2014**, *54*, 213–218. [CrossRef]
42. Figueiredo, C.; Baptista, M.; Rosa, I.C.; Lopes, A.R.; Dionísio, G.; Rocha, R.J.M.; Cruz, I.C.S.; Kikuchi, R.K.P.; Simoes, N.; Leal, M.C. 3D chemoecology and chemotaxonomy of corals using fatty acid biomarkers: Latitude, longitude and depth. *Biochem. Syst. Ecol.* **2017**, *70*, 35–42. [CrossRef]
43. Bishop, D.G.; Kenrick, J.R. Fatty acid composition of symbiotic zooxanthellae in relation to their hosts. *Lipids* **1980**, *15*, 799–804. [CrossRef]
44. Hoeksema, B.W.; Cairns, S. World List of Scleractinia. Available online: https://www.marinespecies.org/scleractinia/ (accessed on 20 June 2024).
45. Meyers, P.A. Fatty acids and hydrocarbons of Caribbean corals. In *Proceedings: Third International Coral Reef Symposium*; Taylor, D.L., Ed.; University of Miami: Coral Gables, FL, USA, 1977; Volume 1, pp. 529–536.
46. Meyers, P.A. Polyunsaturated fatty acids in coral: Indicators of nutritional sources. *Mar. Biol. Lett.* **1979**, *1*, 69–75.
47. Latyshev, N.A.; Naumenko, N.V.; Svetashev, V.I.; Latypov, Y.Y. Fatty acids of reef-building corals. *Mar. Ecol. Progr. Ser.* **1991**, *76*, 295–301. [CrossRef]
48. Imbs, A.B.; Demidkova, D.A.; Latypov, Y.Y.; Pham, L.Q. Application of Fatty Acids for Chemotaxonomy of Reef-Building Corals. *Lipids* **2007**, *42*, 1035–1046. [CrossRef] [PubMed]
49. Dalsgaard, J.; St John, M.; Kattner, G.; Müller-Navarra, D.; Hagen, W. Fatty acid trophic markers in the pelagic marine environment. *Adv. Mar. Biol.* **2003**, *46*, 225–340. [CrossRef] [PubMed]
50. Imbs, A.B.; Latyshev, N.A.; Zhukova, N.V.; Dautova, T.N. Comparison of fatty acid compositions of azooxanthellate *Dendronephthya* and zooxanthellate soft coral species. *Comp. Biochem. Physiol. Part B* **2007**, *148*, 314–321. [CrossRef]
51. Imbs, A.B.; Demidkova, D.A.; Dautova, T.N.; Latyshev, N.A. Fatty acid biomarkers of symbionts and unusual inhibition of tetracosapolyenoic acid biosynthesis in corals (Octocorallia). *Lipids* **2009**, *44*, 325–335. [CrossRef] [PubMed]
52. Mirallès, J.; Barnathan, G.; Renée, G.; Sall, T.; Samb, A.; Gaydou, E.M.; Kornprobst, J.M. New branched fatty acids from the Senegalese Gorgonian *Leptogorgia piccola* (white and yellow morphs). *Lipids* **1995**, *30*, 459–466. [CrossRef] [PubMed]
53. Kornprobst, J.-M.; Barnathan, G. Demospongic Acids Revisited. *Mar. Drugs* **2010**, *8*, 2569–2577. [CrossRef]
54. Svetashev, V.I.; Vysotskii, M.V. Fatty acids of *Heliopora coerulea* and chemotaxonomic significance of tetracosapolyenoic acids in Coelenterata. *Comp. Biochem. Physiol.* **1998**, *119B*, 73–75. [CrossRef]
55. Williams, G.C. Living genera of sea pens (Coelenterata: Octocorallia: Pennatulacea): Illustrated key and synopses. *Zool. J. Linn. Soc.* **1995**, *113*, 93–140. [CrossRef]
56. Rosa, R.; Santos, C.; Lopes, A.; Baptista, M.; Sampaio, E.; Rosa, I.C.; Repolho, T. Stenophagy as a Pathway for the Occurrence of Unique Coral Fatty Acid Biomarkers in Higher Branches of the Marine Tree of Life: The Nudibranch Armina maculata Case. *Ocean Sci. J.* **2019**, *54*, 297–305. [CrossRef]
57. Pernet, V.P.; Gavino, V.G.; Gavino, G.G.; Anctil, M.A. Variations of lipid and fatty acid contents during the reproductive cycle of the anthozoan Renilla koellikeri. *J. Comp. Physiol. B.* **2002**, *172*, 455–465. [PubMed]
58. García-Matucheski, S.; Muniain, C.; Cutignano, A.; Cimino, G.; Faimali, M.; Piazza, V.; Aristizábal, E.; Fontana, A. Renillenoic acids: Feeding deterrence and antifouling properties of conjugated fatty acids in Patagonian sea pen. *J. Exp. Mar. Biol. Ecol.* **2012**, *416–417*, 208–214. [CrossRef]
59. Baptista, M.; Lopes, V.M.; Pimentel, M.S.; Bandarra, N.; Narciso, L.; Marques, A.; Rosa, R. Temporal fatty acid dynamics of the octocoral sea pen *Veretillum cynomorium*. *Comp. Biochem. Physiol. B Biochem. Mol. Biol.* **2012**, *161*, 178–187. [CrossRef]
60. Joseph, J.D. Lipid composition of marine and estuarine invertebrates: Porifera and Cnidaria. *Prog. Lipid Res.* **1979**, *18*, 1–30. [CrossRef] [PubMed]
61. Lang, I.; Hodac, L.; Friedl, T.; Feussner, I. Fatty acid profiles and their distribution patterns in microalgae: A comprehensive analysis of more than 2000 strains from the SAG culture collection. *BMC Plant Biol.* **2011**, *11*, 124. [CrossRef] [PubMed]
62. Yamashiro, H.; Oku, H.; Onaga, K. Effect of bleaching on lipid content and composition of Okinawan corals. *Fish. Sci.* **2005**, *71*, 448–453. [CrossRef]
63. Lopes, A.R.; Baptista, M.; Rosa, I.C.; Dionísio, G.; Gomes-Pereira, J.; Paula, H.R.; Figueiredo, C.; Bandarra, N.; Calado, R.; Rosa, R. "Gone with the wind": Fatty acid biomarkers and chemotaxonomy of stranded pleustonic hydrozoans (*Velella velella* and *Physalia physalis*). *Biochem. System. Ecol.* **2016**, *66*, 297–306. [CrossRef]
64. Stillway, L.W. Fatty acids of the Portuguese man-of-war *Physalia physalis*. *Comp. Biochem. Physiol.* **1976**, *53B*, 535–537. [CrossRef] [PubMed]
65. Aral, M.N.; Ford, J.A.; Whyte, J.N.C. Biochemical composition of fed and starved *Aequorea victoria* (Murbach et Shearer, 1902). *J. Exp. Mar. Biol. Ecol.* **1989**, *127*, 289–299. [CrossRef]

66. Phleger, C.F.; Nichols, P.D.; Virtue, P. Lipids and trophodynamics of Antarctic zooplankton. Comp. *Biochem. Physiol.-B* **1998**, *120*, 311–323. [CrossRef]
67. Nelson, M.M.; Phleger, C.F.; Mooney, B.D.; Nichols, P.D. Lipids of Gelatinous Antarctic Zooplankton: Cnidaria and Ctenophora. *Lipids* **2000**, *35*, 551–559. [CrossRef] [PubMed]
68. Brotz, L.; Schiariti, A.; Lopez-Martínez, J.; Alvarez-Tello, J.; Hsieh, Y.-H.P.; Jones, R.P.; Quiñones, J.; Dong, Z.; Morandini, A.C.; Preciado, M. Jellyfish fisheries in the Americas: Origin, state of the art, and perspectives on new fishing grounds. *Rev. Fish Biol. Fish.* **2017**, *27*, 1–29. [CrossRef]
69. Fukuda, Y.; Naganuma, T. Potential dietary effects on the fatty acid composition of the common jellyfish *Aurelia aurita*. *Mar. Biol.* **2001**, *138*, 1029–1035. [CrossRef]
70. Hooper, S.N.; Ackman, R.G. Presence of trans-6-hexadecenoic acid in the white jellyfish Aurelia aurita Lamarck and in a caribbean gorgonian. *Lipids* **1972**, *7*, 624–626. [CrossRef]
71. Holland, D.L.; Davenport, J.; East, J. The fatty acid composition of the leatherback turtle *Dermochelys coriacea* and its jellyfish prey. *J. Mar. Biol. Assoc. UK* **1990**, *70*, 761–770. [CrossRef]
72. Sipos, J.C.; Ackman, R.G. Jellyfish (*Cyanea capillata*) Lipids: Fatty acid composition. *J. Fish. Res. Board Canada* **1968**, *25*, 1561–1569. [CrossRef]
73. Mortillaro, J.M.; Pitt, K.A.; Lee, S.Y.; Meziane, T. Light intensity influences the production and translocation of fatty acids by zooxanthellae in the jellyfish *Cassiopea* sp. *J. Exp. Mar. Biol. Ecol.* **2009**, *278*, 22–30. [CrossRef]
74. Nichols, P.D.; Danaher, K.T.; Koslow, J.A. Occurrence of high levels of tetracosahexaenoic acid in the jellyfish *Aurelia* sp. *Lipids* **2003**, *38*, 1207–1210. [CrossRef] [PubMed]
75. Svetashev, V.I. Fatty acids of the medusa *Aurelia aurita* (Linnaeus, 1758) and *Rhopilema esculentum* (Kishinouye, 1891): The presence of families of polyenoic acids with 24 and 26 carbon atoms. *Rus. J. Mar. Biol.* **2019**, *45*, 113–117. [CrossRef]
76. Morais, Z.B.; Pintão, A.M.; Costa, I.M.; Calejo, M.T.; Bandarra, N.M.; Abreu, P. Composition and in vitro antioxidant effects of jellyfish *Catostylus tagi* from Sado Estuary (SW Portugal). *J. Aquat. Food Prod. Technol.* **2009**, *18*, 90–107. [CrossRef]
77. Milisenda, G.; Rossi, S.; Vizzini, S.; Fuentes, V.L.; Purcell, J.E.; Tilves, U.; Piraino, S. Seasonal variability of diet and trophic level of the gelatinous predator *Pelagia noctiluca* (Scyphozoa). *Sci. Rep.* **2018**, *8*, 12140. [CrossRef]
78. Nagao, K.; Nakamitsu, K.; Ishida, H.; Yoshinaga, K.; Nagai, T.; Mizobe, H.; Kojima, K.; Yanagita, T.; Beppu, F.; Gotoh, N. Comparison of the lipid-lowering effects of four different n–3 highly unsaturated fatty acids in HepG2 cells. *J. Oleo Sci.* **2014**, *63*, 979–985. [CrossRef]

Disclaimer/Publisher's Note: The statements, opinions and data contained in all publications are solely those of the individual author(s) and contributor(s) and not of MDPI and/or the editor(s). MDPI and/or the editor(s) disclaim responsibility for any injury to people or property resulting from any ideas, methods, instructions or products referred to in the content.

marine drugs

MDPI

Review

Marine Microalgal Products with Activities against Age-Related Cardiovascular Diseases

Nova Yurika [1,2,†], Eleonora Montuori [2,3,†] and Chiara Lauritano [2,*]

1 Marine Biology Research Group, Ghent University, Krijgslaan 281, B-9000 Gent, Belgium; nova.yurika@imbrsea.eu
2 Ecosustainable Marine Biotechnology, Stazione Zoologica Anton Dohrn, Via Acton 55, 80133 Napoli, Italy; eleonora.montuori@studenti.unime.it
3 Department of Chemical, Biological, Pharmaceutical and Environmental Sciences, University of Messina, Viale F. Stagno d'Alcontres 31, 98166 Messina, Italy
* Correspondence: chiara.lauritano@szn.it; Tel.: +39-0815833221
† These authors contributed equally to this work.

Abstract: Heart disease is one of the leading causes of death worldwide, and it is estimated that 17.9 million people die of it each year. The risk factors for cardiovascular diseases are attributable to an unhealthy and sedentary lifestyle, poor nutrition, stress, genetic predisposition, diabetes, obesity, and aging. Marine microalgae have been the subject of numerous studies for their potential activity against several human diseases. They produce a plethora of primary and secondary metabolites such as essential nutrients, vitamins, pigments, and omega-3 fatty acid. Many of these molecules have antioxidant properties and have been shown to play a role in the prevention of heart diseases. The aim of this review is to summarize recent studies on the discovery of marine microalgal compounds and bioactivities for cardiovascular diseases, including in vitro and in vivo studies, showing and discussing recent discoveries and trends. The most promising results were found for microalgal polysaccharides, peptides and carotenoids. In conclusion, the overall data summarized here show that microalgae-based supplementation has the potential to improve age-related cardiovascular diseases and we expect more clinical studies in the future.

Keywords: cardiovascular diseases; marine microalgae; antioxidants; age-related diseases; bioactive compounds; marine natural products

1. Introduction

Marine microalgae have been the subject of numerous studies for their potential activities against several human diseases [1,2]. Microalgae have attracted a lot of attention in recent years owing to their biodiversity in terms of species, adapted to live in different environments, and in terms of chemical diversity, ranging from lipids and carbohydrates to complex polyketides. In addition, their use has been considered eco-sustainable and eco-friendly owing to their high growth rates and the possibility of culturing these both indoors and outdoors at an industrial scale. Various studies have also shown that these microorganisms are a promising source of beneficial nutrients for heart health. They produce a plethora of metabolites such as essential nutrients, vitamins, pigments, omega-3 fatty acid, and several antioxidant molecules which may play a role in the prevention of heart disease [3,4]. In particular, the n-3 long-chain polyunsaturated fatty acids (n-3 LC-PUFAs), such as eicosapentaenoic (EPA) and docosahexaenoic (DHA) acids, are known for their beneficial effects on the cardiovascular system [5,6] and to have protective effects against atherosclerotic, arrhythmic and thrombotic diseases [7,8]. These fatty acids have been reported to reduce cholesterol levels in the blood, lower blood pressure, and reduce inflammation [5,6]. The European Food Safety Authority (EFSA) recommends an intake of 250 mg for EPA plus DHA for adults, 100 mg DHA for infants (>6 months) and young children <24 months, and to increase the dose during pregnancy and lactation [9]. Furthermore, marine microalgae also contain other antioxidant molecules, such as pigments [10] and vitamins, like vitamin E, vitamin A and vitamin of complex B. These antioxidants have been reported to play a significant role in the prevention of cardiovascular diseases (CVD) [11,12]. For instance, it has been shown that pre-treatments with antioxidants, such as vitamins C and E, can mitigate endothelial dysfunction due to high-fat meals [13].

According to the World Health Organization (WHO), heart disease is one of the leading causes of death worldwide (https://www.who.int/health-topics/cardiovascular-diseases/#tab=tab_1 accessed on 29 January 2024), with about 17.9 million deaths globally each year. Cardiovascular diseases are the most prevalent age-related diseases [14,15]. As the pace of population aging around the world is increasing dramatically, old population presents one of the greatest challenges for the social and health care systems worldwide, especially in low-income and middle-income countries [16]. For older patients, hypertension, hyperlipidemia and diabetes are also frequent negatively influencing cardiovascular events [17]. Overall, cardiovascular disease prevention in older adults should be established based on the individuals, based on their estimated life expectancy, time to benefit, comorbidities, and preferences (e.g., more plant-based and low-fat diet, exercise when possible, quitting smoking, etc.).

The risk factors for cardiovascular diseases are mainly attributable to an unhealthy lifestyle, poor nutrition, sedentary lifestyle, stress, genetic predisposition, diabetes, obesity and aging (Figure 1) [18]. As reported in Izzo et al. [19], both age and gender are risk factors. Older females are more susceptible to cardiovascular disease compared to men of the same age. In both cases, in both men and women, these diseases are related to a decrease in sex hormones [19,20]. Possible women-specific risk factors that have been considered include gestational diabetes mellitus, pregnancy-induced hypertension, and preeclampsia, as well as reproductive endocrine disorders, including polycystic ovary syndrome and menopause [20]. CVD risk factors are highly prevalent in some countries and vary according to socioeconomic, gender, and educational levels [18]. In Pakistan, smoking (46%), family history (43%), hypertension (37%), dyslipidemia (33%), diabetes mellitus (18%) and overweight (63.3%) are the most common risk factors found in CVD patients under 45 years of age [21]. In the UK, the rate of hypertension has been reported as the highest risk, approximately 65%, followed by smoking (44.2%), high cholesterol (38.7%), diabetes (12%), overweight (5.13%), male gender (4.6%), and female gender (5.6%). In adults, metabolic risk factors tend to increase with age [22]. Additional factors, including frailty, obesity, and diabetes could complicate and enhance CVD risk factors amongst the elderly [23,24]. With advancing age, the heart undergoes structural and functional changes that make it more susceptible to pathologies such as heart failure, arrhythmia and atherosclerosis [25]. The onset of various health issues is related to the subsequent contribution of damages to the blood vessels and the heart itself. Obesity can cause an increase in blood cholesterol, which leads to a greater predisposition to the development of atherosclerotic diseases.

In this review, we reported marine microalgal compounds with beneficial and preventive activities against heart diseases related to aging. When available, we also discussed doses and mechanisms of action for both in vitro and in vivo studies. We showed that the most bioactive molecules from microalgae reported for CVDs were polysaccharides, peptides, carotenoids and lipids.

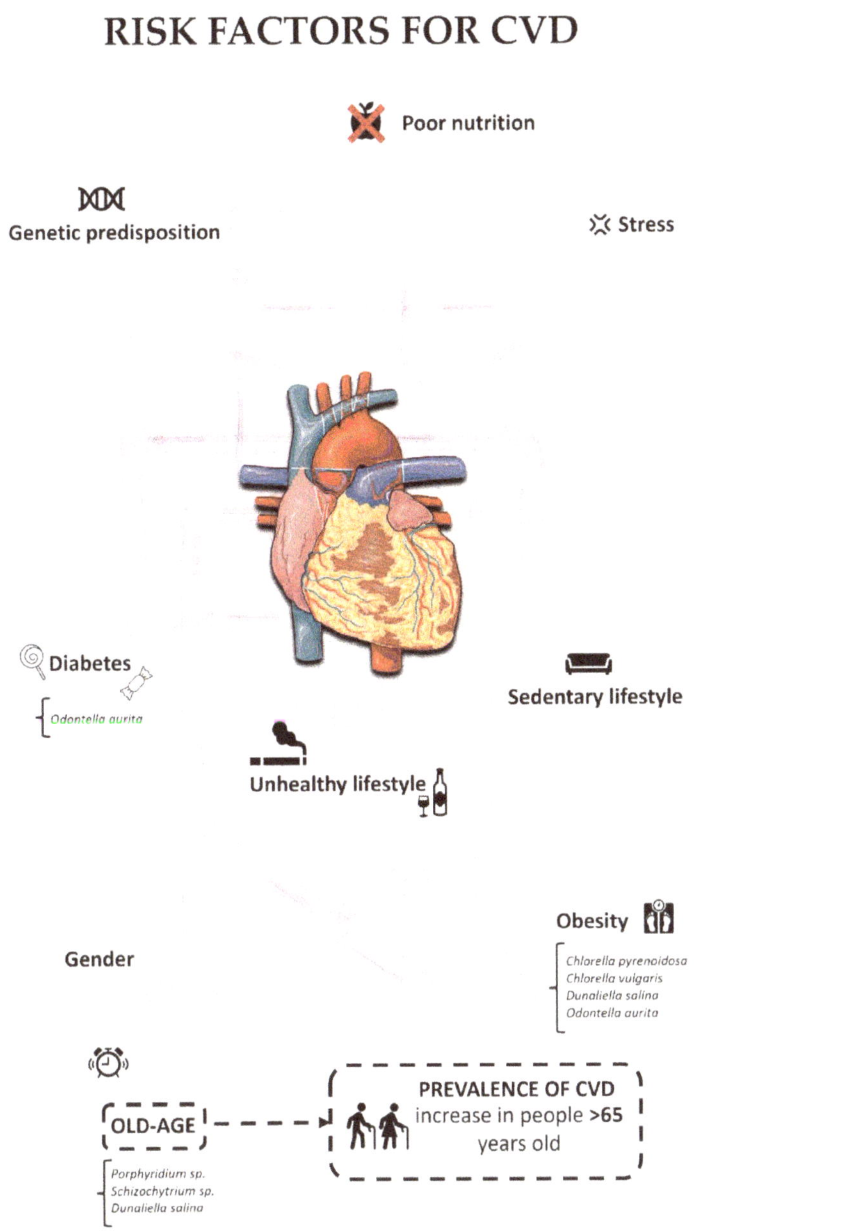

Figure 1. A schematic representation of principal risk factors of cardiovascular diseases and main microalgae which have shown potential beneficial activities.

2. Polysaccharides

Several studies have reported the potential of polysaccharides (PSs) to improve endothelial dysfunction, defined as functional, structural, and communication changes between the vascular endothelium and muscle cells [26]. For example, Levy-Ontman et al. in 2017 [27] evaluated the anti-inflammatory and vasodilation properties of polysaccharides produced by *Porphyridium* sp. using human coronary artery endothelial cells (HCAECs).

The authors showed that polysaccharides were able to attenuate inflammatory processes by interfering with tumor necrosis factor-alpha (TNF-α)-induced inflammation. In cells pre-treated with polysaccharides, there was an up-regulation of adhesion molecule 1 (ICAM-1) and vascular cell adhesion molecule 1 (VCAM-1), nuclear factor kappa-B (NF-kB) translocation, and attenuated inhibitor of nuclear factor kappa B (IκB) degradation. Polysaccharides improved endothelial function as measured by increased nitric oxide NO formation and decreased endothelin 1 (ET-1) protein expression [27]. Hamias et al. in 2018 [28] studied the ability of polysaccharides (PSs) from *Porphyridium* sp. to improve endothelial state and found that PSs attenuated inflammatory atherosclerotic pathways up-regulated by Angiotensin II (Ang II). When HCAECs were pre-treated with PSs (500 μg/mL) under Ang II induction, PSs were able to down-regulate the NF-kB activation and suppress adhesion molecule ICAM-1 and VCAM-1 up-regulation in a dose-dependent manner. Furthermore, polysaccharides enhanced nitric oxide (NO) and endothelial nitric oxide synthase (eNOS) production, and reduced ET-1 expression levels [28].

3. Peptides

3.1. *In Vitro*

In addition to polysaccharides, peptides can counteract the pathological processes by mimicking the function of mediators or modulating the activities and expression of mediators involved in hypertension, hypercholesterolemia, diabetes, inflammation and oxidative stress [29]. Lin et al. [30] studied *Isochrysis zhanjiangensis*, which was suggested to inhibit vascular injury and angiogenesis, and to have a protective effect on CVDs. They characterized the production and the activity of an octapeptide (ICE) isolated from *I. zhanjiangensis*, demonstrating that ICE was able to decrease ROS production in lipopolysaccharide (LPS)-induced HUVECs (concentrations of ICE 1, 10, 20, and 50 μM). The peptide could reduce cell damage by increasing the expression of antioxidant enzymes, such as the antioxidant enzymes superoxide dismutase (SOD), glutathione peroxidase-1 (GPX), and haem oxygenase 1 (HO-1). In addition, it also inhibited pro-inflammatory mediators tumor necrosis factor (TNF)-α, cytokine interleukin-6 (IL-6), and ICAM-1 [30] (Table 1). Vo et al. in 2013 [31] isolated two peptides with aminoacidic sequences of LDAVNR for peptide 1 and MMLDF for peptide 2 from the peptidic hydrolysates of *Spirulina maxima*. These peptides showed anti-inflammatory properties in histamine-induced EA.hy926 endothelial cells (used for cardiovascular disease research) with a decrease in interleukin (IL)-8 expression, measured by the ELISA assay. It is known that endothelial inflammation is a risk factor for atherosclerosis and the authors suggested these two peptides for possible anti-atherosclerotic activity [31].

Jiang et al. in 2021 [32] suggested that microalgal compounds may have great potential as a healthier anti-hypertensive treatment substitution to conventional anti-hypertensive drugs causing side effects. Hypertension, a risk factor for the development of CVDs, consists of a sustained increase in arterial pressure above 140/90 mm Hg [33]. In particular, the authors showed that peptides from microalgae are promising angiotensin-converting enzyme (ACE) inhibitors (Table 1). Renin–angiotensin–aldosterone system (RAAS) hyperactivity is involved in the progression of vascular disease. The key effector peptide of the RAAS, angiotensin II (Ang II), is generated by angiotensin I through endothelial angiotensin-converting enzyme (ACE). Inhibition of RAAS is recommended for managing most cardiovascular diseases, particularly hypertension, heart failure, acute myocardial infarction, and stroke [34]. ACE inhibitors and angiotensin receptor blockers (ARBs) are commonly prescribed medication for primary hypertension [35] and other chronic conditions, including heart failure, by reducing systolic function. Chen et al. in 2020 [36] purified and identified a peptide (PIZ protein hydrolysate) produced by *Isochrysis zhanjiangensis* that was able to inhibit ACE. The ACE activity calculated from the amount of hippuric acid liberated from hippuryl-His-Leu (HHL) showed that PIZ acts as a mixed non-competitive inhibitor of ACE at an IC_{50} value of 61.38 μM. Pretreatment with PIZ 10 μM for 24 h on human umbilical vein endothelial cells (HUVECs) inhibited the NF-κB pathway by protecting inhibitor IκBα degradation and down-regulating NF-κB expression. In addition, they showed that PIZ had modest ACE inhibitory effects due to its ability to reduce inflammatory cytokine expression (NO, COX-2, and ICAM-1) and block the production of ET-1. ICAM-1 and MCP-1 levels were significantly suppressed by PIZ in a dose-dependent manner. Cell treatment with PIZ 10 μM decreased the expression levels of inflammatory cytokines COX-2 and slightly inhibited iNOS and ET-1 production, thereby improving endothelial dysfunction, reducing oxidative stress, and decreasing the risk of hypertension [36]. Samarakoon et al. [37] showed that pepsin hydrolysate from *Nannochloropsis oculate* exhibited ACE inhibitory activity. They demonstrated that the IC_{50} values of purified ACE inhibitory peptides were 123 μM and 173 μM and identified Gly-Met-Asn-Asn-Leu-Thr-Pro (GMNNLTP; MW, 728 Da) and Leu-Glu-Gln (LEQ; MW, 369 Da) as novel peptides, respectively [37]. Wu et al. in 2015 [38] reported that a purified peptide (Tyr-Met-Gly-Leu-Asp-Leu-Lys) from *Isochrysis galbana* showed potent ACE inhibitory activity with an IC_{50} of 36.1 μM. In 2017, Heo et al. [39] conducted a study to produce an ACE inhibitory peptide from marine *Spirulina* sp. The ACE inhibitory peptide (Thr-Met-Glu-Pro-Gly-Lys-Pro) showed the strongest ACE activity at an IC_{50} value of 0.3 mg/mL. In addition, the human umbilical vein endothelial cells (HUVECs) were treated for 1 h with aliquots of purified peptide (62.5, 125 and 250 μM) and subsequently incubated for 24 h with Ang II (1 μM). They showed that ACE inhibitory peptide inhibited NO and ROS generation, and suppressed the expression of inducible nitric oxide synthase (iNOS) and ET-1 [39].

Cunha et al. in 2022 [40] also showed that water-soluble hydrolysates rich in proteins/peptides from the microalgae *Chlorella vulgaris* had anti-hypertensive potential by measuring the percentage inhibition of the ACE enzyme (IC_{50}: 286 μg protein/mL) [40]. Recently, Pei et al. [41] showed that nonapeptide ETT (Glu-Met-Phe-Gly-Thr-Ser-Ser-Glu-Thr) from *Isochrysis zhanjiangensis* showed excellent effects in regulating hypertension by inhibiting ROS up-regulation of oxidized low-density lipoprotein receptor-1 (LOX-1) and ROS levels in Ang II-induced human umbilical vein endothelial cells (HUVECs). In addition, ETT inhibited the expression of various inflammatory mediators and the expression of related cytokines (IL-1β, IL-8, TNF-α, iNOS, COX-2, ET-1, AT-1) as well as cell adhesion molecules (ICAM-1 and VCAM-1) in a dose-dependent manner (10, 50, and 100 μM) [41]. Alzahrani et al. in 2018 [42] screened the anti-hypertension activities of *Nitzschia laevis* in vitro. The author showed that trypsin hydrolysates from this species had antagonist effects toward the ACE enzyme (IC_{50} 1.63 ± 0.01 mg/mL), higher than that of *Spirulina* and *Chlorella* [42]. Verspreet et al. [43] screened five microalgae (i.e., *Chlamydomonas nivalis*, *Porphyridium purpureum*, *Chlorella vulgaris*, *Nannochloropsis gaditana*, and *Scenedesmus* sp.) with respect to their ability to inhibit ACE by measuring the activity owing to an ACE-1 inhibition kit. The ACE inhibition bioassay showed that all microalgae tested inhibited ACE by 73.4–87.1% when tested at a concentration of 1 mg/mL [43].

3.2. *In Vivo*

Regarding in vivo experiments, the activities found were mainly related to anti-hypertension. Ko et al. [44] found that a purified peptide (Val–Glu–Gly–Tyr) from marine *Chlorella ellipsoidea* acted as a competitive inhibitor against ACE with an IC_{50} value of 128.4 μM. Furthermore, they tested the anti-hypertensive effects of the purified peptide by measuring the change in systolic blood pressure at 2, 4, 6 and 8 h after oral administration of the peptide (10 mg/kg of body weight) and showed that purified peptide was able to significantly decrease systolic blood pressure in rats [44].

Barkia et al. in 2019 [45] screened six strains of marine diatoms and found that papain hydrolysates had ACE inhibitory activity in vitro (2 mg/mL), with the highest activity obtained from *Bellerochea malleus*. Furthermore, in vivo assays showed that *Bellerochea malleus* hydrolysates reduced systolic and diastolic blood pressure in male spontaneously hypertensive rats after 5 days of hydrolysate treatment at doses of 75 and 100 mg/kg body weight [45]. Hayes et al. in 2023 [46] studied hydrolysate and bioactive peptides from the red microalga *Porphyridium* sp., namely, GVDYVRFF, AIPAAPAAPAGPKLY, and LIHADPPGVGL, and assessed the anti-hypertensive activity using spontaneously hypertensive rats. The *Porphyridium* sp. hydrolysate was also included in a food carrier (jelly candies; 0.5 g of the hydrolysate). Hydrolysate and hydrolysate–jelly candies reduced systolic blood pressure by −1.54 mm Hg and −6.17 mm Hg, respectively, while Captopril® reduced systolic blood pressure by −18.21 mm Hg after 24 h [46].

4. Carotenoids

4.1. In Vitro

Microalgae are known to produce a variety of pigments with various color shades and biological activities, including carotenoids [3,12]. Zuluaga et al. in 2018 [47] reported astaxanthin protective actions against ischemia and reperfusion (I/R) injury. Astaxanthin also ameliorates myocardial cell oxidative stress injury [47]. Astaxanthin from the freshwater microalga *Haematococcus pluvialis* has also been shown to prevent oxidative stress on human endothelial cells (HUVECs) without toxicity up to a dose of 10 µg/mL [48].

4.2. In Vivo

El-baz et al. in 2018 [49] demonstrated that β-carotene rich *Dunaliella salina* carotenoid fraction (250 g/kg) as well as the whole biomass (250 mg/kg) had protective potentials against cardiac disfunction in a group of rats injected with D-galactose (200 mg/kg). *Dunaliella salina* β-carotene and biomass exhibited potent antioxidant activity and significant reducing capacity of homocysteine, IL-6 and iNOS. In another study, El-Baz et al. [50] examined the effects of zeaxanthin heneicosylate (ZH) isolated from *Dunaliella salina* on cardiac dysfunction. The study was performed in vivo in rats, by injecting D-galactose in rats for 8 weeks following orally treated with ZH (250 µg/kg) for a period of 28 days. ZH improved cardiac aging manifestation, including irregular heartbeat and increased NF-κB. ZH injected rats ameliorated NF-κB and restored superoxide dismutase (SOD). SOD is an antioxidant enzyme that has been shown to protect the heart against oxidative stress, and ischemic damage, and hypertrophy after myocardial infarction [51]. Oral administration of ZH up-regulated retinoic acid receptor alpha (RAR-α) gene expression in cardiac tissue. RAR-α plays important roles in cardiac regeneration after myocardial infarction. Depletion of RA pathway leads to cardiomyocyte apoptosis after myocardial infarction [52]. El-Baz et al. [53] conducted a study on the carotenoid rich fraction of the microalgae *Dunaliella salina* activity against inflammation-associated cardiac dysfunction in cardiac-obese rats induced by high fat diet, demonstrating that the carotenoid rich fraction increased adiponectin and glucagon serum level. The histopathological examination of rat treated with the carotenoid rich fraction showed the absence of fibrosis and severe congestion in the myocardial blood vessels [53].

5. Lipids and Other Bioactive Extracts and Molecules

In addition to polysaccharides, peptides and pigments, other molecules from microalgae have shown promising results. In particular, Dahli et al. [54] demonstrated that lyso-diacylglyceryltrimethylhomoserine (lyso-DGTS) isolated from *Nannochloropsis* sp. ethanolic extract might be useful for the prevention of atherosclerotic risk factors by showing increased activities of recombinant paraoxonase 1 (rePON1) lactonase [54].

A mixture of omega-3 polyunsaturated fatty acids (35%) from *Schizochytrium* sp., extra virgin olive oil (75%) and algae oil (25%) was reported to activate the phosphoinositide 3 kinase (PI3K/Akt) pathway that is known to repair vascular endothelium. Aortic rings from old rats treated with the oil mixture (2.5 mL/kg) showed a decreased response to the vasoconstrictor Ang II [55]. Haimeur et al. [56] assessed the effects of two n-3 PUFA from freeze-dried *Odontella aurita* on risk factors for CVDs. A rat group fed with the high-fat diet supplemented with *Odontella aurita* displayed a significantly lower body weight and reduced insulinemia, as well as a reduced serum lipid level, reduced platelet aggregation and oxidative status induced by high fat intake. The authors reported that *Odontella aurita* was more effective than the fish oil in reducing the hepatic triacyglycerol levels and in preventing high-fat diet-induced steatosis [56].

Dudek et al. [57] summarized, in a review, the beneficial role of dietary silicon in the prevention of age-related diseases. Vide et al. [58] reported the effects of *Spirulina* and dietary silicon-enriched *Spirulina* (SES) on atherosclerosis. Hamsters on a high-fat diet were treated with *Spirulina* or SES at a dose 57 mg/kg body weight daily, corresponding to 0.57 mg of silicon/kg body weight. The results showed that in the SES group, there was a reduction in inflammation by lowering the levels of TNF-α, IL-6, as well as a reduction in the number of polymorphonuclear cells and prevention of the activity of NF-κB. Both SES and *Spirulina* itself similarly protected against oxidative stress by reducing the activity of nicotinamide adenine dinucleotide phosphate oxidase (NOX) and maintaining the activity of the antioxidant SOD and glutathione peroxidase [58].

Quagliariello et al. in 2022 [59] reported that *Spirulina platensis*, *Ganoderma lucidum* and *Moringa oleifera* were able to improve cardiac function by reducing inflammation and cardiotoxicity induced by anthracyclines, adjuvant therapies for cancers. Female mice were treated with doxorubicin (DOXO) or a combination of *Spirulina*, *Ganoderma lucidum*, and *Moringa oleifera* (Singo). Following that, they analyzed the myocardial expressions of nucleotide-binding domain, leucine-rich–containing family, pyrin domain-containing-3 (NLRP3), galectin-3 and calgranulin S100, and 13 cytokines through ELISA methods. The authors also assessed myocardial fibrosis, necrosis, and hypertrophy through immunohistochemistry. In addition, they performed tests on human cardiomyocytes by exposing them to DOXO (200 nM) alone or in combination with Singo (at 10, 25 and 50 µg/mL) for 24 and 48 h. The results showed that Singo reduced NLRP3 and p65/NF-kB levels in human cardiomyocytes exposed to Singo at 10, 15 and 50 µg/mL and reduced cytokine levels (the concentration of Singo was 25 µg/mL). Immunohistochemistry analysis indicated that Singo (at 12 mg/kg) reduced fibrosis and hypertrophy in the myocardial tissues of mice during exposure to DOXO [59].

Umei et al. in 2022 [60] demonstrated that oral administration of *Euglena gracilis* was beneficial to improve cardiac function in a mice model of isoproterenol-induced heart failure. A group of mice were injected with isoproterenol (ISO) (20 mg/kg/day) for 7 days. They showed that oral administration of *Euglena gracilis* (2%), in combination with an AIN93G diet, alleviated cardiac dysfunction [60]. Song et al. [51] tested *Dunaliella salina*'s protective effects on myocardial ischemia/reperfusion injury (MIRI) in the Langendorff perfused heart model in mice. The authors reported that *D. salina* (500 mg/kg) was able to improve left ventricle function, reduce the rate of malignant arrhythmia and infarct size, and increase the antioxidant superoxide dismutase. In a recent study published by Tsai et al. in 2023 [61], *D. salina* was reported to have cardioprotective effects against myocardial ischemia/reperfusion (I/R) injury. A group of rats was subjected to surgical procedures for inducing myocardial I/R injury. *D. salina* extract treatment (0.1 mg/kg) was able to decrease myocardial infarct size and attenuate the expressions of cyclooxygenase-2 (COX-2) and the activity of STAT1, janus kinase 2 (JAK2), inhibitor of IκB, NF-κB [61]. Yang et al. [62] showed that *Chlorella pyrenoidosa* was able to lower the blood pressure in rats fed a diet containing N ω-nitro-L-arginine methyl ester hydrochloride (L-NAME), which induced endothelial dysfunction (40 mg/kg). Rats consuming 4 and 8% *Chlorella* had significantly lower ACE activity in the aorta and reduced TNF-α concentrations in the aorta and heart. Histopathological results showed that *Chlorella* consumption reduced the injury scale of the coronary arteries, ventricles, and septum of the heart [62].

Clinical Studies

Recently, Sandgruber et al. [63] completed a clinical trial with 80 young and healthy participants who consumed a smoothie enriched with either 15 g of *Chlorella pyrenoidosa* dry weight (d.w.) or 15 g of *Microchloropsis salina* d.w. for 14 days. They demonstrated that regular consumption of *Chlorella pyrenoidosa* ameliorated CVD factors such as total cholesterol, LDL cholesterol, the LDL–cholesterol to HDL–cholesterol ratio, and non-HDL cholesterol, possibly due to its rich vitamin D2 source. *Microchloropsis salina* improved the fatty acid distribution in plasma lipids by increasing the LC n3 PUFA content and reducing the n6/n3 PUFA ratio [63]. Clinical studies with *Chlorella* were also conducted by Shimada et al. [64] with eighty subjects with systolic blood pressure of 130–159 mmHg or diastolic blood pressure of 85–99 mmHg. The subjects took γ-Aminobutyric Acid (GABA)-rich *Chlorella* (20 mg as γ-aminobutyric acid or placebo twice daily for 12 weeks) as a dietary supplement. Systolic blood pressure decreased significantly compared with placebo, with a higher reduction in the subjects with borderline hypertension than in the subjects with high–normal blood pressure [64]. A randomized triple-blind placebo-controlled clinical trial study conducted by Ghaem et al. [65] in 2021 involved 41 patients with hypertension consuming a salad dressing containing 2 g of *Spirulina platensis* powder for two months. The results showed that the *Spirulina* dressing significantly decreased systolic blood pressure, diastolic blood pressure, serum triglyceride, total cholesterol, and low-density lipoprotein (LDL) levels in comparison to placebo controls [65]. Bioactive compounds and extracts from microalgae for CVDs are summarized in Table 1.

Table 1. The table reports marine microalgal bioactive compounds with potential beneficial activities for cardiovascular diseases. Microalgae, activity observed, compound, concentration (Conc.) used, and model are reported. Abbreviations: CVDs for cardiovascular diseases, DHA for Docosahexaenoic acid, EPA for Eicosapentaenoic acid, EVOO for extra virgin olive oil, IC_{50} for inhibitory concentration values, NLRP3 for NOD-, LRR- and pyrin domain-containing protein 3.

Microalgae	Activity Observed	Compound	Conc.	Model	Reference
Polysaccharides					
Porphyridium sp. (Rhodophyta/Porphyridiophyceae)	Preserve endothelial function, anti-inflammatory	Polysaccharides	50 µg/mL	In Vitro: Human coronary artery endothelial cells (HCAECs)	[27]
Porphyridium sp. (Rhodophyta/Porphyridiophyceae)	Preserve endothelial function, anti-atherosclerosis	Polysaccharide	500 mg/mL	In Vitro: Human coronary artery endothelial cells (HCAEC)	[28]
Peptides					
Spirulina maxima (Cyanobacteria/Cyanophyceae)	Anti-atherosclerosis	Peptic hydrolysates of *Spirulina*	200 µM	In Vitro: EA.hy926 endothelial cell	[31]
Isochrysis zhanjiangensis (Haptophyta/Coccolithophyceae)	Inhibit vascular injury and angiogenesis	Octapeptide (IEC; Ile-Ile-Ala-Val-Glu-Ala-Gly-Cys)	1, 10, 20, and 50 µM	In Vitro: Human umbilical vein endothelial cells (HUVECs)	[30]
Isochrysis zhanjiangensis (Haptophyta/Coccolithophyceae)	Anti-hypertensive, angiotensin-converting enzyme (ACE) inhibitors	Peptide (PIZ; Phe-Glu-Ile-His-Cys-Cys)	IC_{50} = 61.38 µM	In Vitro: Hippuryl-His-Leu (HHL) HHL assay	[36]
Chlamydomonas nivalis (Chlorophyta/Chlorophyceae), *Porphyridium purpureum* (Rhodophyta/Porphyridiophyceae), *Chlorella vulgaris* (Chlorophyta/Trebouxiophyceae), *Nannochloropsis gaditana* (Heterokontophyta/Eustigmatophyceae), and *Scenedesmus* sp. (Chlorophyta/Chlorophyceae)	Angiotensin-converting enzyme (ACE) inhibitors	-	1 mg/mL	In Vitro: Hippuryl-His-Leu (HHL) HHL assay	[43]

Table 1. *Cont.*

Microalgae	Activity Observed	Compound	Conc.	Model	Reference
Chlorella vulgaris (Chlorophyta/Trebouxiophyceae)	Anti-hypertensive, angiotensin-converting enzyme (ACE) inhibitors	Water-soluble hydrolysates rich in proteins/peptides	IC_{50}: 286 µg protein/mL	In Vitro: Hippuryl-His-Leu (HHL) HHL assay	[40]
Nannochloropsis oculate (Heterokontophyta/ Eustigmatophyceae)	Angiotensin-converting enzyme (ACE) inhibitors	Peptides: Gly-Met-Asn-Asn-Leu-Thr-Pro (GMNNLTP; MW, 728 Da) and Leu-Glu-Gln (LEQ; MW, 369 Da),	IC_{50}: 123 IC_{50} = 173 µM, respectively	In Vitro: Hippuryl-His-Leu (HHL) HHL assay	[37]
Nitzschia laevis (Heterokontophyta/ Bacillariophyceae)	Angiotensin-converting enzyme (ACE) inhibitors	-	IC_{50} = 1.63 ± 0.01 mg/mL	In Vitro: Hippuryl-His-Leu (HHL) HHL assay	[42]
Isochrysis galbana (Haptophyta/Coccolithophyceae)	Angiotensin-converting enzyme (ACE) inhibitors	Peptide: (Tyr-Met-Gly-Leu-Asp-Leu-Lys)	IC_{50} = 36.1 µM	In Vitro: Hippuryl-His-Leu (HHL) HHL assay	[38]
Marine *Spirulina* sp. (Cyanobacteria/Cyanophyceae)	Anti-hypertensive, angiotensin-converting enzyme (ACE) inhibitors	Peptide (Thr-Met-Glu-Pro-Gly-Lys-Pro)	IC_{50} = 0.3 mg/mL	In Vitro: Hippuryl-His-Leu (HHL) HHL assay	[39]
Isochrysis zhanjiangensis (Haptophyta/Coccolithophyceae)	Anti-atherosclerosis, anti-apoptosis and anti-inflammation	Nonapeptide named ETT (Glu-Met-Phe-Gly-Thr-Ser-SerGlu-Thr)	IC_{50} = 15.08 µM	In Vitro: Hippuryl-His-Leu (HHL) HHL assay	[41]
Chlorella ellipsoidea (Chlorophyta/Trebouxiophyceae)	Anti-hypertensive, angiotensin-converting enzyme (ACE) inhibitors	Peptide (Val–Glu–Gly–Tyr)	In Vitro: IC_{50} = 128.4 µM In Vivo: 10 mg/kg of body weight	In Vitro: Hippuryl-His-Leu (HHL) HHL assay In Vivo: Rats	[44]
Bellerochea malleus (Heterokontophyta/Mediophyceae)	Anti-hypertensive, ACE-inhibitory activities,	Papain hydrolysates	In Vitro: 2 mg m/L; In Vivo: the dose of 400 mg/kg body weight	In Vitro: Hippuryl-His-Leu (HHL) HHL assay In Vivo: Rats	[45]
Porphyridium sp. (Rhodophyta/Porphyridiophyceae)	Anti-hypertensive	Peptide: GVDYVRFF, AIPAAPAAPAG-PKLY, and LIHADPPGVGL	-	In Vivo: Rats	[46]
Carotenoids					
Dunaliella salina (Chlorophyta/Chlorophyceae)	Ameliorate age-associated cardiac dysfunction	Zeaxanthin heneicosylate (ZH)	250 µg/kg	In Vivo: Rats	[49]
Dunaliella salina (Chlorophyta/Chlorophyceae)	Improve cardiac tissue fibrosis and congestion in the myocardial blood vessels	Carotenoid rich fraction	150 mg/kg body weight	In Vivo: Rats	[50]

Table 1. *Cont.*

Microalgae	Activity Observed	Compound	Conc.	Model	Reference
Haematococcus pluvialis	Antioxidant	Astaxanthin	10 µg/mL	In Vitro: Human endothelial cells (HUVECs)	[48]
Dunaliella salina	Protective potentials against cardiac dysfunction Antioxidant	β-carotene rich *Dunaliella salina* carotenoid fraction	250 mg/kg	In Vivo: Rats	[53]
Dunaliella salina (Chlorophyta/Chlorophyceae)	Improve Myocardial ischemia-reperfusion injury (MIRI), improve left ventricle function and reduce the rate of malignant arrhythmia	-	500 mg/kg	Langendorff perfused heart model in mice	[51]
Chlorella sp. (Chlorophyta/Trebouxiophyceae)	Anti-hypertensive	-	20 mg	Clinical trials	[64]
Spirulina platensis (Cyanobacteria/Cyanophyceae)	Anti-hypertensive	-	2 g	Clinical trials	[65]
Lipids and other bioactive extracts and molecules					
Nannochloropsis sp. (Heterokontophyta/ Eustigmatophyceae)	Anti-atherosclerosis	Lyso-diacylglyceryltr-imethylhomoserine (lyso-DGTS)	1.43 mg/mL	In Vivo: Mice	[54]
A Mixture of *Schizochytrium* sp. and Extra Virgin Olive Oils (not found in algaebase, but found in wikipedia)	Attenuate aging-induced endothelial dysfunction	2.5 mL/kg of a mixture of 75% of EVOO (*Cornicabra* variety; 80% oleic acid and 63.49 mg/g of secoiridoids) and 25% of Algae oil (*Schizochytrium* spp.: 35% DHA, 20% EPA and 5% Docosapentaenoic (DPA))	Omega-3 polyunsaturated fatty acids (ω-3 PUFA)	In Vivo: Male Wistar rats	[55]
Freeze-dried *Odontella aurita* (Heterokontophyta/Mediophyceae)	Anti-atherosclerosis, reduced insulinemia, serum lipid levels, platelet aggregation and oxidative status	Marine omega-3	12% (*w*/*w*) of freeze-dried *O. aurita*	In Vivo: Male Wistar rats	[56]
Spirulina sp. Cyanobacteria/Cyanophyceae)	Anti-atherosclerosis	Dietary silicon-enriched *Spirulina* (SES)	Hamster on a high-fat diet were treated with *Spirulina* or SES at a dose 57 mg/kg body weight daily,	In Vivo: Hamster	[57,58]

Table 1. *Cont.*

Microalgae	Activity Observed	Compound	Conc.	Model	Reference
Spirulina platensis, Ganoderma lucidum and *Moringa oleifera*	Reduction in NLRP3 and p65/NF-kB levels in human cardiomyocytes. Reduction in fibrosis and hypertrophy in the myocardial tissues of mice	Singo (*Spirulina platensis, Ganoderma lucidum* and *Moringa oleifera*)	In Vitro: 10, 15 and 50 μg/mL In Vivo: 12 mg/kg	In Vitro: Human cardiomyocyte. In Vivo: Mice	[59]
Dunaliella salina	Cardioprotective effects against myocardial ischemia/reperfusion (I/R) injury	*Dunaliella salina* extract	0.1 mg/kg	In Vivo: Rats	[61]
Euglena gracilis	Improvement in cardiac function	-	*Euglena gracilis* 2%	In Vivo: Mice	[66]
Chlorella pyrenoidosa	Ameliorative effects on CVDs factors	-	15 g for 14 days	Clinical trials	[63]
Microchloropsis salina	Improvement in fatty acid distribution in plasma lipids	-	15 g for 14 days	Clinical trials	[63]
Chlorella pyrenoidosa	Anti-hypertensive	-	40 mg/Kg	In Vivo: Rats	[62]

6. Conclusions

Overall, this review highlights that the most common compounds with bioactivities useful for cardiovascular diseases are omega-3, pigments, peptides, and carbohydrates. The most abundant phyla of microalgae that have shown beneficial activities for heart-related diseases were Chlorophyta (i.e., *Chlorella* sp., *Chlamydomonas nivalis, Chlorella vulgaris, Chlorella ellipsoidea, Scenedesmus* sp., *Dunaliella salina*), followed by Heterokontophyta and Rhodophyta. In general, the most common mechanisms of action involved in the protective role of microalgal extracts and compounds for cardiovascular diseases are antioxidant and anti-inflammatory activity by reducing free radicals and inhibiting the release of inflammatory mediators (Figure 2).

As regards patents, the WO2019026067A1 relates to extracts of the microalga *Nannochloropsis* and their uses. According to the patent, the nutraceutical composition of WO2019026067A1 (https://patents.google.com/patent/WO2019026067A1/en; accessed on 14 March 2024) may be used for ameliorating conditions associated with atherogenesis and preventing atherosclerotic cardiovascular diseases and associated conditions, such as heart attack, stroke, and high blood pressure. An example of a product is Spirulysat®, a product produced by AlgoSource (https://algosource.com/healthcare/preventive-cardiovascular-care/; accessed on 14 March 2024), based on *Spirulina* extracts, rich in phycocyanins. AlgoSource suggests this product for cardiovascular disease prevention. In particular, Spirulysat® was suggested to prevent the formation of atheroma plaques (https://algosource.com/healthcare/preventive-cardiovascular-care/; accessed on 14 March 2024). Owing to their rapid growth, the possibility of applying metabolic engineering, and multiple bioactive metabolites, marine microalgae represent a great sustainable source of molecules for an industry-scale production of ingredients for functional foods, cosmeceuticals and possible future drugs.

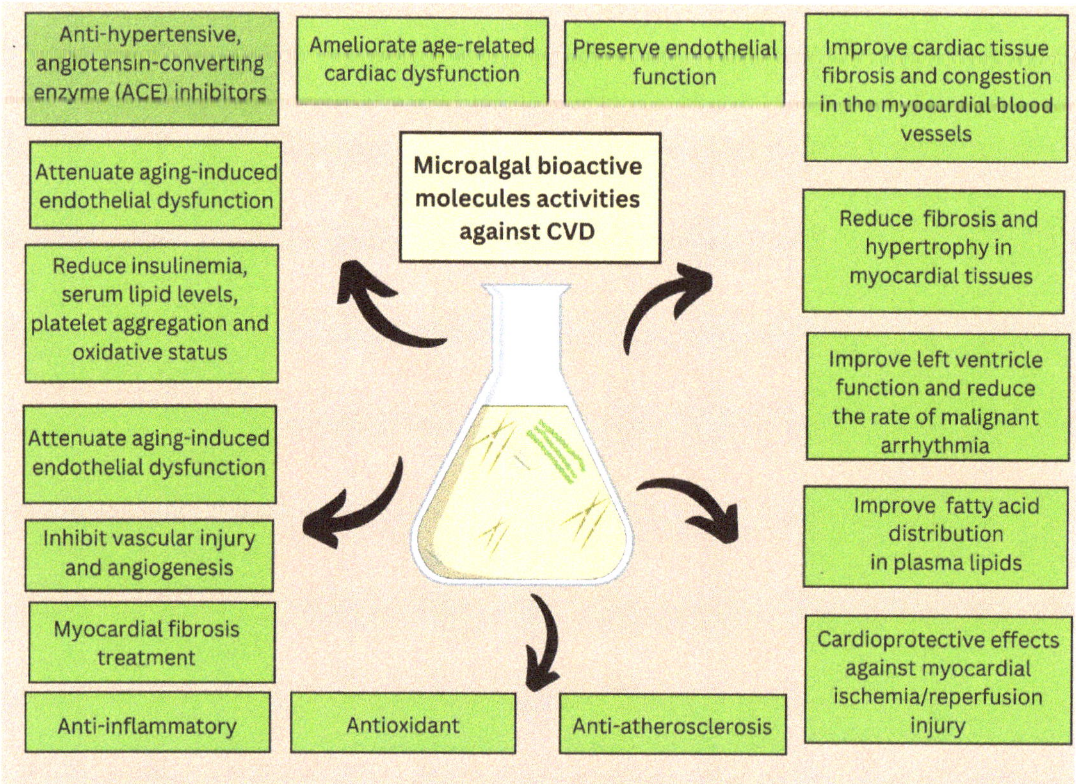

Figure 2. A schematic representation of microalgal bioactive molecules for different age-related cardiovascular disease applications. CVD abbreviation stands for cardiovascular disease.

Author Contributions: Conceptualization, C.L.; writing—original draft preparation, N.Y., E.M. and C.L.; writing—review and editing, N.Y., E.M. and C.L. All authors have read and agreed to the published version of the manuscript.

Funding: This research received no external funding.

Acknowledgments: The authors thank Servier Medical Art (SMART; https://smart.servier.com/ accessed on 5 March 2024) and the IAN/UMCES Symbol and Image Libraries (Integration and Application Network; ian.umces.edu/media-library accessed on 27 March 2024) for the element in Figures 1 and 2.

Conflicts of Interest: The authors declare no conflict of interest.

References

1. Montuori, E.; Capalbo, A.; Lauritano, C. Marine Compounds for Melanoma Treatment and Prevention. *Int. J. Mol. Sci.* **2022**, *23*, 10284. [CrossRef] [PubMed]
2. Montuori, E.; Hyde, C.A.C.; Crea, F.; Golding, J.; Lauritano, C. Marine Natural Products with Activities against Prostate Cancer: Recent Discoveries. *Int. J. Mol. Sci.* **2023**, *24*, 1435. [CrossRef]
3. Saide, A.; Martínez, K.A.; Ianora, A.; Lauritano, C. Unlocking the Health Potential of Microalgae as Sustainable Sources of Bioactive Compounds. *Int. J. Mol. Sci.* **2021**, *22*, 4383. [CrossRef] [PubMed]
4. Manson, J.E.; Cook, N.R.; Lee, I.-M.; Christen, W.; Bassuk, S.S.; Mora, S.; Gibson, H.; Albert, C.M.; Gordon, D.; Copeland, T.; et al. Marine N−3 Fatty Acids and Prevention of Cardiovascular Disease and Cancer. *N. Engl. J. Med.* **2019**, *380*, 23–32. [CrossRef] [PubMed]

5. Khan, S.U.; Lone, A.N.; Khan, M.S.; Virani, S.S.; Blumenthal, R.S.; Nasir, K.; Miller, M.; Michos, E.D.; Ballantyne, C.M.; Boden, W.E.; et al. Effect of Omega-3 Fatty Acids on Cardiovascular Outcomes: A Systematic Review and Meta-Analysis. *eClinicalMedicine* **2021**, *38*, 100997. [CrossRef] [PubMed]
6. Sherratt, S.C.R.; Libby, P.; Budoff, M.J.; Bhatt, D.L.; Mason, R.P. Role of Omega-3 Fatty Acids in Cardiovascular Disease: The Debate Continues. *Curr. Atheroscler. Rep.* **2023**, *25*, 1–17. [CrossRef] [PubMed]
7. Marhuenda, J.; Villaño, D.; Cerdá, B.; Pilar Zafrilla, M. Cardiovascular Disease and Nutrition. In *Nutrition in Health and Disease—Our Challenges Now and Forthcoming Time*; Mózsik, G., Figler, M., Eds.; IntechOpen: London, UK, 2019; ISBN 978-1-78984-007-0.
8. Albracht-Schulte, K.; Kalupahana, N.S.; Ramalingam, L.; Wang, S.; Rahman, S.M.; Robert-McComb, J.; Moustaid-Moussa, N. Omega-3 Fatty Acids in Obesity and Metabolic Syndrome: A Mechanistic Update. *J. Nutr. Biochem.* **2018**, *58*, 1–16. [CrossRef] [PubMed]
9. EFSA Panel on Dietetic Products, Nutrition, and Allergies (NDA). Scientific Opinion on Dietary Reference Values for Fats, Including Saturated Fatty Acids, Polyunsaturated Fatty Acids, Monounsaturated Fatty Acids, Trans Fatty Acids, and Cholesterol. *EFSA J.* **2010**, *8*, 1461. [CrossRef]
10. Lauritano, C.; Montuori, E.; De Falco, G.; Carrella, S. In Silico Methodologies to Improve Antioxidants' Characterization from Marine Organisms. *Antioxidants* **2023**, *12*, 710. [CrossRef] [PubMed]
11. Fisher, M.; Lees, K.; Spence, J.D. Nutrition and Stroke Prevention. *Stroke* **2006**, *37*, 2430–2435. [CrossRef]
12. Raposo, M.F.D.J.; De Morais, A.M.M.B. Microalgae for the Prevention of Cardiovascular Disease and Stroke. *Life Sci.* **2015**, *125*, 32–41. [CrossRef] [PubMed]
13. Plotnick, G.D.; Corretti, M.C.; Vogel, R.A. Effect of Antioxidant Vitamins on the Transient Impairment of Endothelium-Dependent Brachial Artery Vasoactivity Following a Single High-Fat Meal. *JAMA* **1997**, *278*, 1682–1686. [CrossRef]
14. Prince, M.J.; Wu, F.; Guo, Y.; Gutierrez Robledo, L.M.; O'Donnell, M.; Sullivan, R.; Yusuf, S. The Burden of Disease in Older People and Implications for Health Policy and Practice. *Lancet* **2015**, *385*, 549–562. [CrossRef] [PubMed]
15. Lakatta, E.G. Age-Associated Cardiovascular Changes in Health: Impact on Cardiovascular Disease in Older Persons. *Heart Fail. Rev.* **2002**, *7*, 29–49. [CrossRef]
16. Sleeman, K.E.; De Brito, M.; Etkind, S.; Nkhoma, K.; Guo, P.; Higginson, I.J.; Gomes, B.; Harding, R. The Escalating Global Burden of Serious Health-Related Suffering: Projections to 2060 by World Regions, Age Groups, and Health Conditions. *Lancet Glob. Health* **2019**, *7*, e883–e892. [CrossRef] [PubMed]
17. Lettino, M.; Mascherbauer, J.; Nordaby, M.; Ziegler, A.; Collet, J.P.; Derumeaux, G.; Hohnloser, S.H.; Leclercq, C.; O'Neill, D.E.; Visseren, F.; et al. Cardiovascular Disease in the Elderly: Proceedings of the European Society of Cardiology—Cardiovascular Round Table. *Eur. J. Prev. Cardiol.* **2022**, *29*, 1412–1424. [CrossRef] [PubMed]
18. Adhikary, D.; Barman, S.; Ranjan, R.; Stone, H. A Systematic Review of Major Cardiovascular Risk Factors: A Growing Global Health Concern. *Cureus* **2022**, *14*, e30119. [CrossRef] [PubMed]
19. Izzo, C.; Carrizzo, A.; Alfano, A.; Virtuoso, N.; Capunzo, M.; Calabrese, M.; De Simone, E.; Sciarretta, S.; Frati, G.; Oliveti, M.; et al. The Impact of Aging on Cardio and Cerebrovascular Diseases. *Int. J. Mol. Sci.* **2018**, *19*, 481. [CrossRef]
20. Gao, Z.; Chen, Z.; Sun, A.; Deng, X. Gender Differences in Cardiovascular Disease. *Med. Nov. Technol. Devices* **2019**, *4*, 100025. [CrossRef]
21. Nadeem, M.; Ahmed, S.S.; Mansoor, S.; Farooq, S. Risk Factors for Coronary Heart Disease in Patients below 45 Years of Age. *Pak. J. Med. Sci.* **2012**, *29*, 91–96. [CrossRef]
22. Tuomilehto, J. Impact of Age on Cardiovascular Risk: Implications for Cardiovascular Disease Management. *Atheroscler. Suppl.* **2004**, *5*, 9–17. [CrossRef] [PubMed]
23. Rodgers, J.L.; Jones, J.; Bolleddu, S.I.; Vanthenapalli, S.; Rodgers, L.E.; Shah, K.; Karia, K.; Panguluri, S.K. Cardiovascular Risks Associated with Gender and Aging. *J. Cardiovasc. Dev. Dis.* **2019**, *6*, 19. [CrossRef] [PubMed]
24. Ciumărnean, L.; Milaciu, M.V.; Negrean, V.; Orășan, O.H.; Vesa, S.C.; Sălăgean, O.; Iluţ, S.; Vlaicu, S.I. Cardiovascular Risk Factors and Physical Activity for the Prevention of Cardiovascular Diseases in the Elderly. *Int. J. Environ. Res. Public Health* **2021**, *19*, 207. [CrossRef] [PubMed]
25. Singam, N.S.V.; Fine, C.; Fleg, J.L. Cardiac Changes Associated with Vascular Aging. *Clin. Cardiol.* **2020**, *43*, 92–98. [CrossRef] [PubMed]
26. El Assar, M.; Angulo, J.; Vallejo, S.; Peiró, C.; Sánchez-Ferrer, C.F.; Rodríguez-Mañas, L. Mechanisms Involved in the Aging-Induced Vascular Dysfunction. *Front. Physiol.* **2012**, *3*, 132. [CrossRef] [PubMed]
27. Levy-Ontman, O.; Huleihel, M.; Hamias, R.; Wolak, T.; Paran, E. An Anti-Inflammatory Effect of Red Microalga Polysaccharides in Coronary Artery Endothelial Cells. *Atherosclerosis* **2017**, *264*, 11–18. [CrossRef] [PubMed]
28. Hamias, R.; Wolak, T.; Huleihel, M.; Paran, E.; Levy-Ontman, O. Red Alga Polysaccharides Attenuate Angiotensin II-Induced Inflammation in Coronary Endothelial Cells. *Biochem. Biophys. Res. Commun.* **2018**, *500*, 944–951. [CrossRef]
29. Li, Y.; Lammi, C.; Boschin, G.; Arnoldi, A.; Aiello, G. Recent Advances in Microalgae Peptides: Cardiovascular Health Benefits and Analysis. *J. Agric. Food Chem.* **2019**, *67*, 11825–11838. [CrossRef] [PubMed]
30. Lin, L.; He, Y.-L.; Tang, Y.; Hong, P.; Zhou, C.; Sun, S.; Qian, Z.-J. Mechanism Analysis of Octapeptide from Microalgae, *Isochrysis zhanjiangensis* for Suppressing Vascular Injury and Angiogenesis in Human Umbilical Vein Endothelial Cell. *Int. Immunopharmacol.* **2022**, *111*, 109149. [CrossRef]

31. Vo, T.-S.; Ryu, B.; Kim, S.-K. Purification of Novel Anti-Inflammatory Peptides from Enzymatic Hydrolysate of the Edible Microalgal *Spirulina maxima*. *J. Funct. Foods* **2013**, *5*, 1336–1346. [CrossRef]
32. Jiang, Q.; Chen, Q.; Zhang, T.; Liu, M.; Duan, S.; Sun, X. The Antihypertensive Effects and Potential Molecular Mechanism of Microalgal Angiotensin I-Converting Enzyme Inhibitor-Like Peptides: A Mini Review. *Int. J. Mol. Sci.* **2021**, *22*, 4068. [CrossRef] [PubMed]
33. Giles, T.D.; Materson, B.J.; Cohn, J.N.; Kostis, J.B. Definition and Classification of Hypertension: An Update. *J. Clin. Hypertens.* **2009**, *11*, 611–614. [CrossRef]
34. Van Thiel, B.S.; Van Der Pluijm, I.; Te Riet, L.; Essers, J.; Danser, A.H.J. The Renin–Angiotensin System and Its Involvement in Vascular Disease. *Eur. J. Pharmacol.* **2015**, *763*, 3–14. [CrossRef]
35. Li, E.C.; Heran, B.S.; Wright, J.M. Angiotensin Converting Enzyme (ACE) Inhibitors versus Angiotensin Receptor Blockers for Primary Hypertension. *Cochrane Database Syst. Rev.* **2014**, *2014*, CD009096. [CrossRef]
36. Chen, L.; Giesy, J.P.; Adamovsky, O.; Svirčev, Z.; Meriluoto, J.; Codd, G.A.; Mijovic, B.; Shi, T.; Tuo, X.; Li, S.-C.; et al. Challenges of Using Blooms of *Microcystis* spp. in Animal Feeds: A Comprehensive Review of Nutritional, Toxicological and Microbial Health Evaluation. *Sci. Total Environ.* **2021**, *764*, 142319. [CrossRef]
37. Samarakoon, K.W.; O-Nam, K.; Ko, J.-Y.; Lee, J.-H.; Kang, M.-C.; Kim, D.; Lee, J.B.; Lee, J.-S.; Jeon, Y.-J. Purification and Identification of Novel Angiotensin-I Converting Enzyme (ACE) Inhibitory Peptides from Cultured Marine Microalgae (*Nannochloropsis oculata*) Protein Hydrolysate. *J. Appl. Phycol.* **2013**, *25*, 1595–1606. [CrossRef]
38. Wu, H.; Xu, N.; Sun, X.; Yu, H.; Zhou, C. Hydrolysis and Purification of ACE Inhibitory Peptides from the Marine Microalga *Isochrysis galbana*. *J. Appl. Phycol.* **2015**, *27*, 351–361. [CrossRef]
39. Heo, S.-J.; Yoon, W.-J.; Kim, K.-N.; Oh, C.; Choi, Y.-U.; Yoon, K.-T.; Kang, D.-H.; Qian, Z.-J.; Choi, I.-W.; Jung, W.-K. Anti-Inflammatory Effect of Fucoxanthin Derivatives Isolated from *Sargassum siliquastrum* in Lipopolysaccharide-Stimulated RAW 264.7 Macrophage. *Food Chem. Toxicol.* **2012**, *50*, 3336–3342. [CrossRef]
40. Cunha, S.A.; Coscueta, E.R.; Nova, P.; Silva, J.L.; Pintado, M.M. Bioactive Hydrolysates from *Chlorella vulgaris*: Optimal Process and Bioactive Properties. *Molecules* **2022**, *27*, 2505. [CrossRef]
41. Pei, Y.; Lui, Y.; Cai, S.; Zhou, C.; Hong, P.; Qian, Z.-J. A Novel Peptide Isolated from Microalgae *Isochrysis zhanjiangensis* Exhibits Anti-Apoptosis and Anti-Inflammation in Ox-LDL Induced HUVEC to Improve Atherosclerosis. *Plant Foods Hum. Nutr.* **2022**, *77*, 181–189. [CrossRef]
42. Alzahrani, M.A.J. Proteins and Their Enzymatic Hydrolysates from the Marine Diatom *Nitzschia laevis* and Screening for Their In Vitro Antioxidant, Antihypertension, Anti-Inflammatory and Antimicrobial Activities. Ph.D. Thesis, University of Auckland, Auckland, New Zealand, 2018. [CrossRef]
43. Verspreet, J.; Soetemans, L.; Gargan, C.; Hayes, M.; Bastiaens, L. Nutritional Profiling and Preliminary Bioactivity Screening of Five Micro-Algae Strains Cultivated in Northwest Europe. *Foods* **2021**, *10*, 1516. [CrossRef]
44. Ko, S.-C.; Kang, N.; Kim, E.-A.; Kang, M.C.; Lee, S.-H.; Kang, S.-M.; Lee, J.-B.; Jeon, B.-T.; Kim, S.-K.; Park, S.-J.; et al. A Novel Angiotensin I-Converting Enzyme (ACE) Inhibitory Peptide from a Marine *Chlorella ellipsoidea* and Its Antihypertensive Effect in Spontaneously Hypertensive Rats. *Process Biochem.* **2012**, *47*, 2005–2011. [CrossRef]
45. Barkia, I.; Al-Haj, L.; Abdul Hamid, A.; Zakaria, M.; Saari, N.; Zadjali, F. Indigenous Marine Diatoms as Novel Sources of Bioactive Peptides with Antihypertensive and Antioxidant Properties. *Int. J. Food Sci. Technol.* **2019**, *54*, 1514–1522. [CrossRef]
46. Hayes, M.; Aluko, R.E.; Aurino, E.; Mora, L. Generation of Bioactive Peptides from *Porphyridium* sp. and Assessment of Their Potential for Use in the Prevention of Hypertension, Inflammation and Pain. *Mar. Drugs* **2023**, *21*, 422. [CrossRef] [PubMed]
47. Zuluaga, M.; Gueguen, V.; Letourneur, D.; Pavon-Djavid, G. Astaxanthin-Antioxidant Impact on Excessive Reactive Oxygen Species Generation Induced by Ischemia and Reperfusion Injury. *Chem.-Biol. Interact.* **2018**, *279*, 145–158. [CrossRef] [PubMed]
48. Cutrell, S.; Alhomoud, I.S.; Mehta, A.; Talasaz, A.H.; Van Tassell, B.; Dixon, D.L. ACE-Inhibitors in Hypertension: A Historical Perspective and Current Insights. *Curr. Hypertens. Rep.* **2023**, *25*, 243–250. [CrossRef] [PubMed]
49. El-Baz, F.; Abdel Jaleel, G.; Saleh, D.; Hussein, R. Protective and Therapeutic Potentials of *Dunaliella salina* on Aging-Associated Cardiac Dysfunction in Rats. *Asian Pac. J. Trop. Biomed.* **2018**, *8*, 403. [CrossRef]
50. El-Baz, F.K.; Hussein, R.A.; Saleh, D.O.; Abdel Jaleel, G.A.R. Zeaxanthin Isolated from *Dunaliella salina* Microalgae Ameliorates Age Associated Cardiac Dysfunction in Rats through Stimulation of Retinoid Receptors. *Mar. Drugs* **2019**, *17*, 290. [CrossRef] [PubMed]
51. Song, J.; Li, H.; Zhang, Y.; Wang, T.; Dong, Y.; Shui, H.; Du, J. Effect of *Dunaliella salina* on Myocardial Ischemia-Reperfusion Injury through KEAP1/NRF2 Pathway Activation and JAK2/STAT3 Pathway Inhibition. *Gene Protein Dis.* **2023**, *2*, 387. [CrossRef]
52. Da Silva, F.; Jian Motamedi, F.; Weerasinghe Arachchige, L.C.; Tison, A.; Bradford, S.T.; Lefebvre, J.; Dolle, P.; Ghyselinck, N.B.; Wagner, K.D.; Schedl, A. Retinoic Acid Signaling Is Directly Activated in Cardiomyocytes and Protects Mouse Hearts from Apoptosis after Myocardial Infarction. *eLife* **2021**, *10*, e68280. [CrossRef]
53. El-Baz, F.K.; Aly, H.F.; Abd-Alla, H.I. The Ameliorating Effect of Carotenoid Rich Fraction Extracted from *Dunaliella salina* Microalga against Inflammation- Associated Cardiac Dysfunction in Obese Rats. *Toxicol. Rep.* **2020**, *7*, 118–124. [CrossRef]
54. Dahli, L.; Atrahimovich, D.; Vaya, J.; Khatib, S. Lyso-DGTS Lipid Isolated from Microalgae Enhances PON1 Activities In Vitro and In Vivo, Increases PON1 Penetration into Macrophages and Decreases Cellular Lipid Accumulation. *BioFactors* **2018**, *44*, 299–310. [CrossRef] [PubMed]

55. González-Hedström, D.; Amor, S.; De La Fuente-Fernández, M.; Tejera-Muñoz, A.; Priego, T.; Martín, A.I.; López-Calderón, A.; Inarejos-García, A.M.; García-Villalón, Á.L.; Granado, M. A Mixture of Algae and Extra Virgin Olive Oils Attenuates the Cardiometabolic Alterations Associated with Aging in Male Wistar Rats. *Antioxidants* **2020**, *9*, 483. [CrossRef]
56. Haimeur, A.; Ulmann, L.; Mimouni, V.; Guéno, F.; Pineau-Vincent, F.; Meskini, N.; Tremblin, G. The Role of *Odontella aurita*, a Marine Diatom Rich in EPA, as a Dietary Supplement in Dyslipidemia, Platelet Function and Oxidative Stress in High-Fat Fed Rats. *Lipids Health Dis.* **2012**, *11*, 147. [CrossRef] [PubMed]
57. Dudek, Ł.; Kochman, W.; Dziedzic, E. Silicon in Prevention of Atherosclerosis and Other Age-Related Diseases. *Front. Cardiovasc. Med.* **2024**, *11*, 1370536. [CrossRef] [PubMed]
58. Vidé, J.; Virsolvy, A.; Romain, C.; Ramos, J.; Jouy, N.; Richard, S.; Cristol, J.-P.; Gaillet, S.; Rouanet, J.-M. Dietary Silicon-Enriched Spirulina Improves Early Atherosclerosis Markers in Hamsters on a High-Fat Diet. *Nutrition* **2015**, *31*, 1148–1154. [CrossRef]
59. Quagliariello, V.; Basilicata, M.G.; Pepe, G.; De Anseris, R.; Di Mauro, A.; Scognamiglio, G.; Palma, G.; Vestuto, V.; Buccolo, S.; Luciano, A.; et al. Combination of *Spirulina platensis*, *Ganoderma lucidum* and *Moringa oleifera* Improves Cardiac Functions and Reduces Pro-Inflammatory Biomarkers in Preclinical Models of Short-Term Doxorubicin-Mediated Cardiotoxicity: New Frontiers in Cardioncology? *J. Cardiovasc. Dev. Dis.* **2022**, *9*, 423. [CrossRef]
60. Umei, M.; Akazawa, H.; Saga-Kamo, A.; Yagi, H.; Liu, Q.; Matsuoka, R.; Kadowaki, H.; Shindo, A.; Nakashima, A.; Yasuda, K.; et al. Oral Administration of *Euglena gracilis* Z Alleviates Constipation and Cardiac Dysfunction in a Mouse Model of Isoproterenol-Induced Heart Failure. *Circ. Rep.* **2022**, *4*, 83–91. [CrossRef]
61. Tsai, C.-F.; Lin, H.-W.; Liao, J.-M.; Chen, K.-M.; Tsai, J.-W.; Chang, C.-S.; Chou, C.-Y.; Su, H.-H.; Liu, P.-H.; Chu, Y.-C.; et al. *Dunaliella salina* Alga Protects against Myocardial Ischemia/Reperfusion Injury by Attenuating TLR4 Signaling. *Int. J. Mol. Sci.* **2023**, *24*, 3871. [CrossRef]
62. Yang, S.-C.; Yang, H.-Y.; Yang, Y.-C.; Peng, H.-C.; Ho, P.-Y. *Chlorella pyrenoidosa* Ameliorated L-NAME-Induced Hypertension and Cardiorenal Remodeling in Rats. *Eur. J. Nutr.* **2013**, *52*, 601–608. [CrossRef]
63. Sandgruber, F.; Höger, A.-L.; Kunze, J.; Schenz, B.; Griehl, C.; Kiehntopf, M.; Kipp, K.; Kühn, J.; Stangl, G.I.; Lorkowski, S.; et al. Impact of Regular Intake of Microalgae on Nutrient Supply and Cardiovascular Risk Factors: Results from the NovAL Intervention Study. *Nutrients* **2023**, *15*, 1645. [CrossRef] [PubMed]
64. Shimada, M.; Hasegawa, T.; Nishimura, C.; Kan, H.; Kanno, T.; Nakamura, T.; Matsubayashi, T. Anti-Hypertensive Effect of γ-Aminobutyric Acid (GABA)-Rich *Chlorella* on High-Normal Blood Pressure and Borderline Hypertension in Placebo-Controlled Double Blind Study. *Clin. Exp. Hypertens.* **2009**, *31*, 342–354. [CrossRef] [PubMed]
65. Ghaem Far, Z.; Babajafari, S.; Kojuri, J.; Mohammadi, S.; Nouri, M.; Rostamizadeh, P.; Rahmani, M.H.; Azadian, M.; Ashrafi-Dehkordi, E.; Zareifard, A.; et al. Antihypertensive and Antihyperlipemic of Spirulina (*Arthrospira platensis*) Sauce on Patients with Hypertension: A Randomized Triple-blind Placebo-controlled Clinical Trial. *Phytother. Res.* **2021**, *35*, 6181–6190. [CrossRef]
66. AlKahtane, A.A.; Abushouk, A.I.; Mohammed, E.T.; ALNasser, M.; Alarifi, S.; Ali, D.; Alessia, M.S.; Almeer, R.S.; AlBasher, G.; Alkahtani, S.; et al. Fucoidan Alleviates Microcystin-LR-Induced Hepatic, Renal, and Cardiac Oxidative Stress and Inflammatory Injuries in Mice. *Environ. Sci. Pollut. Res.* **2020**, *27*, 2935–2944. [CrossRef] [PubMed]

Disclaimer/Publisher's Note: The statements, opinions and data contained in all publications are solely those of the individual author(s) and contributor(s) and not of MDPI and/or the editor(s). MDPI and/or the editor(s) disclaim responsibility for any injury to people or property resulting from any ideas, methods, instructions or products referred to in the content.

MDPI AG
Grosspeteranlage 5
4052 Basel
Switzerland
Tel.: +41 61 683 77 34

Marine Drugs Editorial Office
E-mail: marinedrugs@mdpi.com
www.mdpi.com/journal/marinedrugs

Disclaimer/Publisher's Note: The title and front matter of this reprint are at the discretion of the Guest Editors. The publisher is not responsible for their content or any associated concerns. The statements, opinions and data contained in all individual articles are solely those of the individual Editors and contributors and not of MDPI. MDPI disclaims responsibility for any injury to people or property resulting from any ideas, methods, instructions or products referred to in the content.

www.ingramcontent.com/pod-product-compliance
Lightning Source LLC
Chambersburg PA
CBHW041951260926
51850CB00019B/2334
* 9 7 8 3 7 2 5 8 5 5 0 7 0 *